MINERALOGY FOR STUDE

M. H. BATTEY

Department of Geology,
University of Newcastle upon Tyne

MINERALOGY FOR STUDENTS

SECOND EDITION

Longman Scientific & Technical
Longman Group UK Limited,
Longman House, Burnt Mill, Harlow
Essex CM20 2JE, England
and Associated Companies throughout the world

First published by Oliver & Boyd 1972
Reprinted (with corrections) by Longman Group Ltd.
   1975, 1979
Second edition 1981
Fourth impression 1986
Reprinted by Longman Scientific & Technical 1988, 1990

British Library Cataloguing in Publication Data

Battey, Maurice Hugh
   Mineralogy for students. — 2nd ed.
   1. Mineralogy
   I. Title
   549 QE363.2   80-41124

ISBN 0-582-44005-X

Printed in Malaysia
by Polygraphic Marketing Sdn. Bhd.,
Balakong, Selangor Darul Ehsan

PREFACE TO THE SECOND EDITION

The far-reaching changes of emphasis in geology in recent years have underlined the importance of a basic understanding of minerals, and of the messages they carry, in the interpretation of planetary evolution. Imprinted in the minerals and mineral assemblages is the record of the temperature, pressure and chemical conditions under which they were formed, whether it be in hydrothermal alteration of deep sea lavas, in fragments brought up in diamond pipes from depths of hundreds of kilometres within the Earth's mantle, or in samples from the Moon's surface. The recognition of minerals continues to be an essential part of the study of geological material, and a knowledge of their atomic structure, chemistry and occurrence vital to its interpretation.

The study of minerals as sources of chemical elements for technology and industry is of continuing importance, while technological advance has found many new uses for minerals as crystalline compounds whose intrinsic atomic structures confer upon them valuable electronic and other physical properties.

The second edition of this book, like the first, aims to provide a unified account of the principles and techniques of mineral study, together with descriptions of the commoner minerals of all chemical classes and geological environments. The atomic structures of the minerals are presented as the fundamental basis of their physical properties, as well as a powerful aid to remembering their chemical formulae and dynamic interrelationships.

The microscopic identification of rock-forming minerals forms an important part of most geology courses, as part of the basis for all rock classification. To assist the student working with the polarising microscope, this second edition has been expanded to include a new series of diagrams showing the optical orientations of the commoner rock-forming minerals. It is hoped that these diagrams will make for easier appreciation of the relationships between optical and physical directions of minerals seen in thin section, and thus supplement the data given under the articles on each mineral in the main text. For convenience of use at the bench, these diagrams are gathered together at the end of the book.

It remains the writer's hope that the university or college student will find here most of what he needs to know about mineralogy up to the level of a first degree in geology or mining, and that the material presented will furnish a firm basis for the study of petrology and the larger interpretations of planetary history.

ACKNOWLEDGEMENTS

The writer is indebted to the following individuals and institutions for permission to redraw diagrams or copy tables from the sources noted.

Akademische Verlagsgesellschaft, Geest u. Pontig K.-G.
Strunz, H., 1957. *Mineralogische Tabellen*, 3rd edn, Fig. 64, p. 52.
American Journal of Science.
Tuttle, O. F., 1952. *Bowen Volume*, Fig. 1, p. 557; Fig. 2, p. 559.
Cambridge University Press.
Murray, R. J., 1954. *Geological Magazine*, **91**, p. 22.
Cornell University Press.
Pauling, L., 1960. *The nature of the chemical bond*, 3rd edn, p. 98. (Distributed in the U.K. and Commonwealth (excluding Canada) by Oxford University Press.)
The Director, Geophysical Laboratory, Carnegie Institution of Washington.
Gilbert, M. C., Bell, P. M. and Richardson, S. W., 1969. *Carnegie Institution Yearbook 67*, Fig. 37, p. 136.
Joint Committee on Powder Diffraction Standards.
Card 9-427 of the Powder Diffraction File.
Longmans Green & Co.
Williams, D. E. G., 1966. *The magnetic properties of matter*, Fig. 9.1, p. 193.
McGraw-Hill Publishing Co.
Azaroff, L. V. and Buerger, M. J., 1958. *The powder method in X-ray crystallography*, Fig. 3, p. 47.
Jenkins, F. A. and White, H. E., 1957. *Fundamentals of optics*, Fig. 25G, p. 516; Fig. 25M, p. 521.

Macmillan & Co.
Dent, L. S. and Smith, J. V., 1958. *Nature*, **181**, Fig. 2, p. 1794.
Bovenkerk, H. P., Bundy, F. P., Hall, T. H., Strong, H. M. and Wentdorf, R. H., 1959. *Nature*, **184**, Fig. 7, p. 1094.
The Mineralogical Society of America.
Smith, J. R., 1958. *American Mineralogist*, **43**, 1188, Fig. 2.
The Council of the Mineralogical Society of Great Britain and Northern Ireland. Papers in the *Mineralogical Magazine* as under:
Leake, B. E., 1968, **36**, p. 746, Figs 1 and 2.
Muir, I. D., 1951, **29**, p. 713, Fig. 4.
        1955, **30**, p. 549, Fig. 2.
Smith, J. V. and Yoder, H. S., 1956, **31**, p. 213, Fig. 2.
Zussman, J., 1954, **30**, p. 498, Fig. 3(*b*).
Hey, M. H., 1954, **30**, p. 284, Fig. 4.
Oxford University Press.
Coulson, C. A., 1961. *Valence*, Fig. 2.4, p. 25; Fig. 2.9, p. 38; Fig. 5.3, p. 116; Fig. 8.3, p. 199.
Pergamon Press Ltd.
Ahrens, L. H., 1952. *Geochimica et Cosmochimica Acta*, **2**, p. 168.
The Physical Society and the Institute of Physics.
Tabor, D., 1954. *Proceedings of the Physical Society*, **B, 67**, p. 251, Fig. 1.
The Editor, *Zeitschrift für Kristallographie*.
Bragg, W. L., 1929, **70**, p. 488, Fig. 5b.

Thanks are also extended to colleagues in the Department of Geology at the University of Newcastle upon Tyne, and especially to John Lee for much helpful discussion. Miss Brenda Imrie ably typed some very difficult manuscript. My wife and Mrs. Jean McCallum kindly read the proofs. To these and all the other people who have helped in the preparation of this book, I am greatly indebted.

March, 1971.

M. H. BATTEY

SI units

SI units are used in this book. The following conversion factors may be useful.

1 nanometre (nm) $= 10^{-9}$m $= 1$ millimicron $= 10$Å

1 meganewton/m^2 (MN/m^2) $= 10$ bar

1 joule (J) $= 0.239$ calorie (thermochemical)

CONTENTS

INTRODUCTION

Mineralogy is the oldest organized branch of earth science. Indeed, it can rank with agriculture and animal husbandry amongst the oldest practical sciences of mankind. The recognition and use of the ores of copper probably dates back to 6000 B.C. From those times to this the subject has undergone a continuing expansion and diversification, and at the present time the exploitation of mineral resources and the discovery of new possibilities for the use of minerals in technology are proceeding at a greater rate than ever before. Study of the properties of crystals has become the new science of solid state physics, with which mineralogy has close ties.

Many minerals possess great aesthetic beauty. Besides this, there are two aspects of a mineral that govern its usefulness and its interest. One is its value as an ore. This value resides in its chemical composition, which determines what elements may be won from it by smelting or by breaking down the structure in some other way. Chalcocite, galena and sphalerite (the sulphides of copper, lead and zinc), cassiterite (the oxide of tin) and a host of other minerals possess this type of value.

The other value resides in the unique structural pattern, and hence the properties, of the mineral as a crystalline material. The value placed upon diamond (whether for adornment or as an abrasive) is an obvious example of this; graphite, the other form of carbon, whether as a lubricant or as a lead-pencil, provides another. This second value includes such properties as the generation of electric charges by stressing a crystal, which makes quartz useful in pressure gauges and electric oscillators, and allied to this is the whole field of transistor technology and of phosphors, involving the excita-

tion of electrons of impurity ions locked in crystal structures.

These two aspects, composition and crystal structure, are the basic concepts in the science of mineralogy. They are brought together in the *definition of a mineral as a solid body, formed by natural processes, that has a regular atomic arrangement which sets limits to its range of chemical composition and gives it a characteristic crystal shape.*

This definition excludes amorphous substances. Two amorphous substances, opal and lechatelierite, are in fact described in this book, simply because of their interest in relation to other, crystalline, forms of silica. Amber, coal, bitumens and petroleum oils are excluded, as having variable composition and no regular atomic structure. Although our definition of a mineral leaves out a few substances commonly thought of as minerals, it is probably as clear-cut and unequivocal as any definition in natural history, and this book is devoted to the description of minerals so defined.

Not the least fascinating aspect of minerals is the way they grow, by the addition of suitable atoms rigorously selected from a whole range of ions or molecules in solution. These are meticulously built into a framework of high perfection, which is always the same (within narrow limits) for the same substance, and is closely adjusted to the environment in which it grows. When we consider how this is achieved, the analogy to the growth of living molecules cannot fail to impress us, and the boundary between the world of living and non-living matter becomes as fine as a knife-edge.

In considering the growth of crystals, we pass naturally into the study of the association of minerals, the mutual relations of different species and hence to the study of mineral aggregates, or *rocks*, which is the science of petrology. The two subjects inevitably join hands, and it is rather difficult to draw the line separating a mineral description from an account of its petrological environment. Beyond this, again, lie more hypothetical ideas about the significance of the mineral in understanding the evolution of the earth's crust. One of the chief reasons for studying mineralogy is to go on to the study of rocks and crustal processes; but a textbook of mineralogy cannot become a textbook of petrology. This book does not give an extended account of the physical chemistry of mineral formation and the genetic relationships of minerals; but it is hoped that it will provide a basis for the further exploration of these fields by the student.

PART I PRINCIPLES AND METHODS

1

THE PRINCIPLES OF CRYSTAL STRUCTURE

INTRODUCTION

We have seen that one of the principal attributes of a mineral is the possession of a regular crystalline form, characteristic of each species. The crystalline form is the outward expression of a definite internal structural arrangement of the atoms of which the mineral is composed and this structural pattern, in both its symmetry and its dimensions, is one of the essential characters that defines a mineral.

The recognition of the importance of the structural arrangement of the atoms began in 1912, when von Laue first showed that a crystal could be used as a three-dimensional grating to diffract a beam of the newly discovered Roentgen rays, or X-rays. We shall return in a later chapter to the methods of X-ray study of crystals. It will suffice here to say that from the angles at which they are diffracted we can determine the distances between the planes of atoms that build a crystal, and that by the study of the intensities of the diffracted beams we can make deductions about the positions of the atoms of the different elements in these planes. Armed with an accurate knowledge of the bulk chemical composition of a crystalline compound, all the elements present may be assigned appropriate positions in the structure. This provides us with a picture of the geometrical framework of the crystal.

To this geometrical picture of the structure must be added the dynamic aspect of the forces of cohesion between the atoms which we derive from chemistry—that is, the nature of the chemical bond. These two aspects of the crystal structure constantly interact and can, together, lead to an understanding of the crystallographic form, the limits of variation in chemical composition, and the physical properties of each mineral species. Moreover, the laws govern-ing the building of crystal structures explain many features of the distribution and mutual association of the rarer elements of the earth's crust, by showing where these may fit into minerals composed principally of more common elements.

THE STRUCTURE OF ATOMS AND THE PERIODIC SYSTEM

An atom consists of a nucleus surrounded by an envelope of electrons. It is the electronic envelope of the atom, and its interaction with the electrons of other atoms, that determine its behaviour in chemical combination. With the vastly more energetic inter-actions of particles in the atomic nucleus we are not concerned. For us, the nucleus may be considered as an entity composed of positively charged particles, called protons, and neutral particles, called neutrons, which together make up practically the whole mass of the atom. Around the nucleus lies the envelope, or cloud, of electrons, of negligible mass, each carrying a unit of negative electrical charge and equalling in number the protons (units of positive charge) in the nucleus, so producing the neutral atom. When the number of neutrons in the atomic nucleus of an element is varied, an *isotope* of the element is produced. This is an atom with a different atomic weight from, but the same chemical be-haviour as, the original atom. The atomic weight is changed because the number of neutrons is changed but this involves no change in the electrostatic charge on the nucleus, and hence no change in the number of surrounding electrons; and it is upon the complement of electrons that the chemical properties depend.

Our knowledge of electronic structure rests largely on the study of atomic spectra, the radiation

3

emitted by the atom under the influence of excitation by heat, or other cause. In the early picture of the atom its structure was likened to that of the solar system, with the nucleus playing the part of the sun and the electrons that of the planets. In 1913 Nils Bohr, introducing the concepts of the quantum theory, put forward the idea that the electrons were restricted to a certain number of fixed orbital zones, each associated with a definite amount of energy. An electron might change its orbital zone, and in doing so it emitted, or absorbed, a fixed quantity of energy in the form of light or X-rays, emission being associated with infall of an electron to an inner zone and absorption with its promotion to a more distant orbital zone.

DESIGNATION OF AN ELECTRON ORBITAL

1. The permitted orbital zones, or *quantum shells*, of Bohr's model are lettered in sequence outwards K, L, M, N, O, P and Q, and have the Principal Quantum Numbers 1 to 7.

2. Careful study of atomic spectra showed that, within these quantum shells, electrons occupied subsidiary orbits which are classified as of *s, p, d* and *f* type.

An electron is thus designated by its principal quantum number and its subsidiary letter symbol. The number of subsidiary orbits in any shell is limited and we have the following possibilities.

Principal Quantum Shell	K (1)	L (2)	M (3)	N (4)	O (5)	P (6)	Q (7)
Electron Symbol	1s	2s	3s	4s	5s	6s	7s
		2p	3p	4p	5p	6p	
			3d	4d	5d	6d	
				4f	5f		

3. A further distinction must be made to characterize an electron, namely its spin, which may be positive or negative and which furnishes it with a *spin quantum number*, $+\frac{1}{2}$ or $-\frac{1}{2}$.

4. Finally, there is a subdivision of *p* and *d* and *f* orbits related to the plane in which the electron moves. This leads to a *magnetic quantum number*, so-called because, though the energies of electrons in different subdivisions here are normally the same, when the atom is placed in a strong magnetic field they display energy differences leading to a splitting of spectral lines.

PICTURING THE ELECTRONIC STRUCTURE

Although we have been visualizing the electronic envelope as a set of orbits around the nucleus of the atom, the modern treatment of electronic structure and chemical bonding does not use the idea of circular or elliptical paths of electronic particles. There is a principle of theoretical physics (the Heisenberg Uncertainty Principle) which shows that we cannot measure the position of an electron exactly at any given moment, nor can we measure its velocity. The more accurately we attempt to measure the one, the less accurately shall we know the other, so that 'we must abandon the hopeless task of trying to follow an electron in its orbit' (C. A. Coulson).

On the other hand, the discovery of electron diffraction by the planes of atoms in crystals showed that the electron has associated with it a wave motion, and these electron waves (de Broglie waves) may be analysed mathematically by the principles of wave mechanics. Such a treatment leads to a picture of the electron as a charge-cloud of varying density. This varying density represents the probability of finding the electron at any particular point, the probability being highest where the cloud is densest. The probability density may be contoured in such a way as to enclose a series of volume increments within which the probability of finding the electron progressively falls off. The calculated charge-clouds (or probability-density contours) assume the shape of spheres around, or lobes extending out from, the nucleus. The name *electronic orbital* is given to each such volume within which an electron probably resides at any moment, and it may be pictured by drawing a selected probability-density contour outside which the proportion of charge is small (say less than 10 per cent) as a boundary surface of the orbital.

Fig. 1 is an electron-density map, derived from measuring the intensities of diffracted X-ray beams from a crystal, for it is the electron envelopes of the atoms that chiefly diffract X-rays. Fig. 2 shows the electron distribution in a radial direction from the nucleus in the alkali metals, and expresses the arrangement in principal quantum shells. Notice that the heavier atoms are larger, but that as the nuclear charge increases the inner shells move closer to the nucleus. Fig. 3 shows 'boundary surfaces' of an *s*-orbital, spherically symmetrical; three *p*-orbitals with a directional character which enables the *p* group to be subdivided into p_x, p_y, and p_z; whilst *d*-orbitals, of which one example is

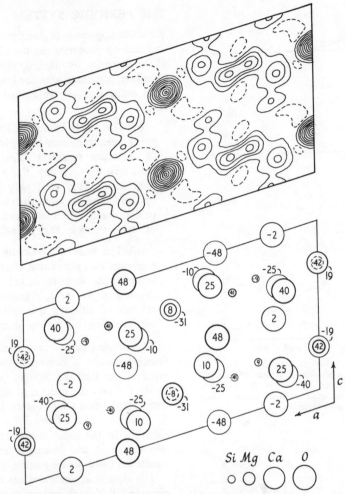

Fig. 1. Electron-density map of the mineral diopside, $CaMg[SiO_3]_2$, projected on the plane (010), and the atomic positions inferred from this with the aid of similar maps projected upon other crystal planes. (After BRAGG, W. L., 1929, p. 488.)

shown, may have five different orientations, and f-orbitals seven.

ENERGY CONSIDERATIONS

The principal quantum number, that is the shell in which the electron lies, chiefly determines its energy. The more distant shells have higher energy associated with them than the inner ones and, as mentioned already, it is the infall of electrons from an outer to an inner shell that leads to the radiation of light or X-rays. The outer shells are correspondingly less stable than inner ones, and this places a limit to the growth of heavier and heavier elements.

Within any quantum shell the subsidiary orbitals have energies in the order $s < p < d$, but the s-orbital of a particular shell may have roughly the same energy as the d-orbital of the next shell within. The usual order of energies is

$$1s < 2s < 2p < 3s < 3p < 3d$$
$$\sim 4s < 4p < 4d \sim 5s < 5p < 4f$$

As for electrons of the same orbital but opposite spin, these have the same energies, and this is true also of the energies of the orbitals of different orientation, the three in group p being equivalent, the five d arrangements equal to one another, and the seven f-orbitals likewise, except under the influence of magnetic fields, as already noted.

The energy of each type of electronic orbital may be calculated, and represents roughly the energy required to remove that electron—that is, the ionization potential of the electron.

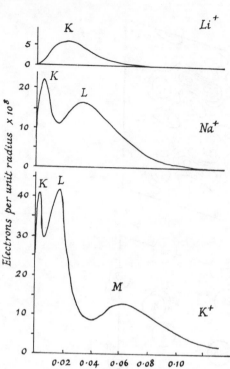

Fig. 2. Radial electron densities of the ions of the alkali metals (after COULSON, C. A., 1961, p. 38). Note the contraction of the inner quantum shells as the nuclear charge increases.

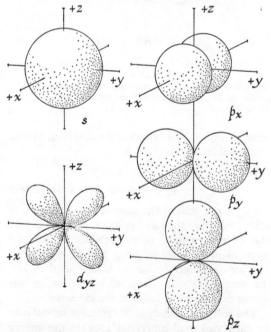

Fig. 3. Electron distribution, calculated from the Wave Equation, for the s-, p-, and one of the d-orbitals. (After COULSON, 1961, p. 25.) Cf. Fig. 7.

THE PERIODIC SYSTEM

With this apparatus of description we may proceed to build up progressively the electronic structures of the elements, by filling up the energy levels with electrons, beginning with the lowest level 1s.

In doing this we are guided by two important rules.

(i) The Pauli Exclusion Principle states that no two electrons may have precisely the same set of quantum numbers. This means that two electrons may occupy the same orbital space (e.g. 1s or $2p_x$, etc.) only if their spins are opposed; and no more than two can occupy one orbital space. When two electrons do occupy one orbital in this way, they are said to be *paired*.

(ii) The Rule of Maximum Multiplicity. This empirical rule states that electrons tend to avoid being in the same space orbital, or cell, so that in each of the p, d, f groups as many of the orbitals of differing orientation as possible are occupied first, before any pairing takes place.

The building up of atomic configurations on these principles leads, of course, to the familiar arrangement of the Periodic System (see Appendix IV). Table 1 shows the process of filling up the energy levels for the first ten elements. In this table we see the operation of the two rules given above, in that electrons in the same orbital space, or cell, have opposed spins, and passing from B to N the three p cells are filled before pairing begins.

The electronic configurations of the elements are given in full (though without distinction of magnetic quantum number) in Table 9 (p. 27). In following that table through, it will be noticed that the maximum number of electrons in a shell is $2n^2$, where n is the principal quantum number. It is important to notice, further, that the M shell becomes temporarily filled with 8 electrons at Ar (element 18), and the N shell starts to fill; but then the M shell starts to expand again at Sc (atomic number $Z = 21$). At Cu ($Z = 29$) it has achieved its full complement of 18 electrons, and after Zn ($Z = 30$) the N shell starts to expand again. The series Sc–Zn are transition elements. A comparable phenomenon occurs in the N shell at Y ($Z = 39$), whilst the O shell expands at La, and the N shell again at Ce to reach its ultimate maximum of 32 electrons at Yb. A final series of transition elements begins at Ac ($Z = 89$).

Because of the dominating influence of the outer electron orbitals on the chemical combining properties of an element, these transition series

Table 1. The electronic structure of the first ten elements.

Shell	K	L			
Quantum Group	$1s$	$2s$	$2p_x$	$2p_y$	$2p_z$
Element					
H	↿				
He	↿⇂				
Li	↿⇂	↿			
Be	↿⇂	↿⇂			
B	↿⇂	↿⇂	↿		
C	↿⇂	↿⇂	↿	↿	
N	↿⇂	↿⇂	↿	↿	↿
O	↿⇂	↿⇂	↿⇂	↿	↿
F	↿⇂	↿⇂	↿⇂	↿⇂	↿
Ne	↿⇂	↿⇂	↿⇂	↿⇂	↿⇂

comprise elements which, though increasing in atomic number, have very similar chemical properties. This is especially true of the Rare Earths whose chemical separation presents many difficulties.

THE FORMATION OF CHEMICAL BONDS

The interaction between the outer electron orbitals of atoms and the transfer, or partial transfer, of electrons from the control of the nucleus of one atom to that of another produces the forces that bind the atoms together. These forces may be of an electrostatic character, or they may arise by the reduction of the total energy of the outer electron orbitals, as a result of sharing, with consequent increases in the stability of the system.

For purposes of description we may divide chemical bonds into six types:

    (*i*) The metallic bond

    (*ii*) The covalent bond

    (*iii*) The ionic (electrovalent) bond

    (*iv*) The dative bond

    (*v*) The residual (van der Waals) bond

    (*vi*) The hydrogen bond

THE METALLIC BOND

Metals are conductors of electricity, and this implies that electrons must be free to flow through the crystal structure under the influence of even a small potential difference. Moreover, a pure metal is

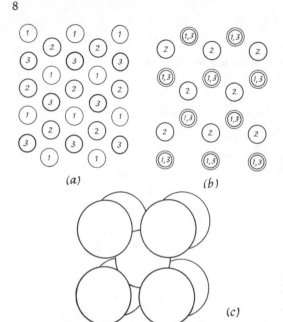

(a)   (b)

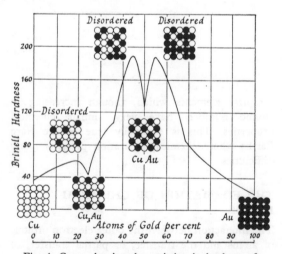

(c)

Fig. 3. (a) The three close-packed arrangements of equal
spheres. In the cubic face-centred and hexagonal arrangements
the circles are numbered to indicate successive layers. They
mark the centres of the spheres only: spheres in each layer
will in fact touch one another.   The cubic face-centred
arrangement repeats after every third layer, the hexagonal
arrangement after every second layer.

The cubic body-centred arrangement is a little less closely
packed than the other two, and the spheres do not touch all
their neighbours. In this drawing the spheres are shown at
their full size.

composed of atoms all of the same kind, hence all
the same size, which arrange themselves in one or
other of the three possible close-packings of equal
spheres (Fig. 3a). They exhibit a malleability and
ductility not found in heteroatomic compounds,
which may be explained partly by the large number
of planes of easy gliding, parallel with sheets of
uniformly-sized atoms that traverse the close-
packed structure; but which also implies a lack of
rigidity and fixed directional character in the bonds.

The metallic bond is regarded as involving a
yielding-up of outer shell electrons by each atom, to
a continuum of electrons pervading the whole
structure. The same idea may be expressed by
saying that the electron orbitals overlap, or the
individual orbital charge-clouds merge. The atoms
are held together by attraction between the atomic
cores (that is nuclei *plus* inner shell electrons) and
the ambient electrons held in common between near
neighbours, so that the neutrality of the atoms is
preserved. The lines of bonding force from a par-

ticular atom may be regarded as evenly distributed
all round it. The detached electrons have sometimes
been spoken of as forming an electron 'gas'. The
details of the establishment of the continuum of
electrons, and the existence of 'energy bands' within
which conduction occurs, cannot be dealt with
here.*

The entry of foreign ions of a different size,
which disturbs the regularity of the close-packed
structure and destroys the planes of easy gliding,
leads to a great increase in the hardness, tensile
strength, and sometimes the brittleness, of the metal,
as in the case of well-known alloys. In alloys in
general, random sharing of electrons and the lack
of necessity for a balance of electronic charges in the
immediate neighbourhood of an atom, means that
no fixed proportions between atoms of different
kinds need be maintained, and alloying takes place
in widely varying proportions. In certain cases,
when the proportion of foreign atoms reaches a
particular value, an intermetallic compound may
form in which the different kinds of atoms occupy
regularly arranged positions in the lattice. When
regularity is restored in this way, the compound may
be almost as soft as the pure metal, as is illustrated
by the disordered and ordered alloys of Cu with Au
(Fig. 4).

Fig. 4. Curve showing the variation in hardness of
alloys of copper and gold, and the relationship between
hardness and regularity or disorder in the atomic
arrangement. (Data of KURNAKOW N., ZEMCZUZNY, S.,
ZASEDATELEV, M., *J. Inst. Metals*, **15**, p. 325, 1916.)

* For further details consult COULSON, C. A. *Valence*,
p. 320 (Oxford Univ. Press, 1961), or EVANS, R. C. *An
introduction to crystal chemistry*, p. 79 (Cambridge Univ.
Press, 1964).

THE COVALENT BOND

The covalent (or homopolar) bond is formed when two atoms, which may be of different elements, share a pair of electrons, one electron being supplied by each atom. For example, in the HCl and Cl_2 molecules the bond may be formulated

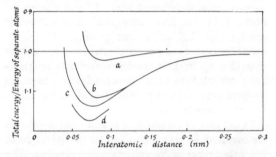

The number of covalent links that an atom may form, depends upon the number of electrons it requires to complete its outer electron shell to the stable configuration of the next inert gas in the periodic system; this number will be $8 - n$, where n is the number of electrons in its outermost quantum shell.

The nature of the covalent bond is best explained with reference to the formation of the hydrogen molecule. Imagine two hydrogen atoms some distance apart, so that they do not influence one another. Each will have the energy of the H atom ($-13 \cdot 60$ eV). Now imagine that we move them gradually together, so that the electron of each will begin to be influenced by the nucleus of the other. With the aid of the mathematical principles of wave mechanics it is possible to calculate the energy change as they move together and interact. The calculation involves the attraction of nucleus A for the charge cloud around B, and vice versa, as well as forces of repulsion between the two nuclei and the two charge clouds. The curve of energy against distance apart has a minimum (curve a, Fig. 5) where the attraction exceeds the repulsion, so that a bond will be formed, but this has much less energy than the known energy of the H_2 molecular bond.

If the two electrons are next regarded as indistinguishable, so that at close approach either may circulate about either nucleus, and the new combined wave function evaluated, we have a minimum energy (curve b, Fig. 5) much nearer the measured bond energy. When allowance is made for the screening of the nuclei one from the other, so that the electron orbitals are contracted into a smaller volume, there is a further improvement (curve c, Fig. 5); and another when we allow for polarization (i.e. distortion) of the two charge clouds. The

next stage is to allow for the possibility that the two electrons may at some time both be near the same nucleus—that is, the bond may partake of ionic character in some degree. This is spoken of as 'resonance' between a covalent and ionic state of the bond. Finally, by continuing such progressive refinement (and by using electronic computers) it has been possible to calculate an energy value for the bond of $4 \cdot 7467$ eV, which may be compared with an experimental value of $4 \cdot 7466 \pm 0 \cdot 0007$ eV. (curve d, Fig. 5).

This treatment of the problem is called the valency-bond theory, and is one of two mathematical approaches to the problem (the other being the 'theory of molecular orbitals'). It satisfactorily shows why pairing of electrons supplied by two atoms should produce a chemical linkage, and may be extended to atoms more complicated than hydrogen.

Fig. 5. The energy of the bond in the hydrogen molecule, calculated with progressive refinement (see text). (After Coulson, 1964, p. 116.)

In crystals, covalent bonding involves the completion of pairing in the outer electron orbitals (or cells) of atoms which, in their unbonded state, have unpaired electrons, and reference to Table 1 will show how the valencies of the first ten elements are explained on this basis.

But the question at once arises, how it is that carbon, which has two unpaired outer electrons, combines with two oxygen atoms to form CO_2, or with four adjacent C atoms in the structure of diamond (Fig. 6). The answer to this lies in the concept of *hybridization of orbitals*. Again the matter is treated mathematically by wave mechanics; but, briefly, Pauling here introduced the idea that one of the s-electrons is promoted by excitation (requiring about 400 kilojoules per g mol) to the unoccupied p cell (Table 1), so providing four unpaired electrons to form bonds. The regular arrangement of the atoms in diamond, for example, and other evidence, shows that these four bonds do

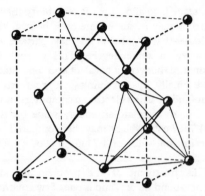

Fig. 6. The structure of diamond, showing tetrahedral co-ordination of carbon atoms. One tetrahedral group is outlined.

of the accurately calculated probability density contours for one of the four sp^3 hybrid orbitals in carbon, and the complete picture of the four bonds is shown schematically in Fig. 8.

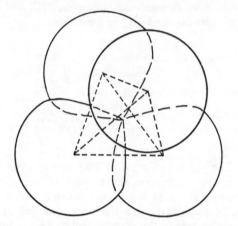

Fig. 8. Representation of the tetrahedral charge-cloud of the sp^3 hybrid orbitals of carbon, composed of four units of the type of Fig. 7.

not consist of three p-bonds, with the directional character of p_x, p_y, and p_z orbitals at 90° to one another, plus a different kind of bond due to the s-orbital, but are four equivalent bonds inclined at equal angles (109°28′) to one another—that is, towards the corners of a tetrahedron drawn around the atom (Fig. 6). In short, the differences between the s- and p-orbitals have been submerged, and a resultant uniform type of bond produced. This is *hybridization of electron orbitals*.

In the case of carbon, one s- and three p-orbitals are involved, so that it is called an sp^3 *hybrid bond*. But other types of hybridization are possible. The interesting feature about them is that, when analysed mathematically they are found to possess definite geometrical shapes—that is, the charge clouds radiate out from the central atom at definite angles to one another. An example is given in Fig. 7

The shapes of some other important hybrids are given in Table 2.

Table 2. The shapes of some hybrid bonds.

Hybrid	Geometrical arrangement	Coordination No.*
sp	linear	2
d^3sp^3	tetrahedral	4
dsp^2	square	4
d^2sp^3	octahedral	6

* The number of atoms linked to the central atom.

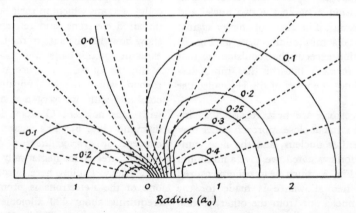

Radius (a_0)

Fig. 7. Contours of the wave function, ψ, for one orbital of the tetrahedral sp^3 hybrid of carbon. The pattern must be rotated around the base-line of the diagram to sweep out the shape of the orbital. This type of graph is the basis for representing bonding orbitals as lobes extending from the atomic core. (After COULSON, 1961, p. 199.) (Each unit of radius = 0·0529 nm)

The characteristic features of the covalent bond are its rigidly directional properties and its demand for satisfaction of electron requirements between immediate neighbours. In the first of these features it is the antithesis of the metallic bond. The second feature is an important difference from the ionic bond, where some latitude is possible in balancing charges, provided these do not accumulate over long distances in the structure.

THE DATIVE (COORDINATE) BOND

In conjunction with the covalent bond we must mention a bond which, like the covalent bond, results from the sharing of a pair of electrons between the atoms; but both shared electrons come from one atom. Such a bond is found in many complex ions in organic chemistry, and chemists call compounds involving such ions 'coordination compounds'. In mineralogy, however, we use the word coordination in a wider sense, so we shall speak of this as the dative bond, signifying that one atom gives two electrons to be shared between it and its partner. It is usually two paired outer shell electrons that do not take part in other bonds (a so-called *lone pair*) that are donated in the dative bond.

COVALENT AND DATIVE BONDS IN MINERALS

Amongst minerals, dominantly covalent bonds are found amongst the native non-metallic elements (diamond, sulphur) and the sulphides. There are many cases, however, where a *degree of covalency* exists in the bonding, and this has important effects upon the physical properties of minerals. We have seen above how this concept of the degree of ionicity or covalency enters into the wave equations. It is now a widely accepted idea in the interpretation of mineral structures, and a fruitful field of investigation lies ahead in determining the nature and extent of this control over mineral properties. The concept is discussed further on p. 17.

The carbonate, nitrate, sulphate, phosphate and silicate ionic groups, which possess a definite shape and do not dissociate in solution, are also to be formulated in terms of covalent or covalent and dative bonds. CO_3^{2-} and NO_3^- are plane triangular groups and SO_4^{2-}, PO_4^{3-} and SiO_4^{4-} are tetrahedral. In all cases the interatomic distances are intermediate between those appropriate to a double and a single bond between the atoms and suggest resonance between different structures as indicated in Fig. 9.

Because of its function in the formation of complex ions, the dative bond is very probably of great geochemical importance in the transport of elements in mineral-forming solutions, but investigation of these matters is as yet in its infancy.

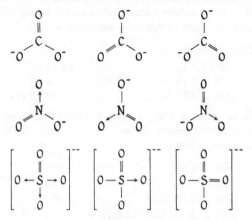

Fig. 9. Three "shaped" molecules whose bonding is in considerable part covalent in character. The bond arrangement in each case resonates between the different arrangements shown (and others in addition in the case of $[SO_4]^{2-}$), all the bonds in each molecule being the same, though compounded out of the structural types shown. The arrow indicates a dative bond.

THE IONIC (ELECTROVALENT) BOND

In this type of bond one of the atoms (the cation) surrenders one or more of its outer shell electrons to the other atom (the anion). By so doing both donor and recipient are said to become *ionized*, and each achieves a stable electronic configuration, which is that of the inert gas next preceding it (in the case of the cation) or next following it (in the case of the anion) in the periodic table. The *valency* of the atom is given by the number of electrons it can give or receive in assuming this stable configuration. In parting with electrons the cation acquires a positive charge, due to the excess of protons in the nucleus over the remaining complement of electrons, whilst the anion similarly acquires an excess of negative charge. It is the electrostatic attraction between these charged ions that constitutes the bond between them.

A simple example is provided by NaCl, where the electron transfer may be represented by

$$Na^{\times} + {}_{\circ}^{\circ}Cl_{\circ}^{\circ} \rightarrow [Na]^+ \begin{bmatrix} {}_{\circ}^{\circ} \times Cl_{\circ}^{\circ} \\ {}_{\circ\circ} \end{bmatrix}^-$$

THE 18-ELECTRON GROUP

In addition to the stable 8-electron configuration of the outermost shell, the 18-electron configuration has a relatively high degree of stability, and elements following transition series expansions to an 18-electron population may form ions by losing electrons to revert to this 18-electron outer shell.

Examples of this are

Zn	2.8.18.2	Zn^{2+}	2.8.18
Cd	2.8.18.18.2	Cd^{2+}	2.8.18.18
Hg	2.8.18.32.18.2	Hg^{2+}	2.8.18.32.18

In the cases of Cu, Ag and Au, of the immediately preceding Periodic Group IB, we find variable valencies as follows

Cu	2.8.18.1	Cu^+	2.8.18
Ag	2.8.18.18.1	Ag^+	2.8.18.18
Au	2.8.18.32.18.1	Au^+	2.8.18.32.18

$$Cu^{2+} \quad 2.8.17$$
$$Ag^{2+} \quad 2.8.18.17$$
$$Au^{3+} \quad 2.8.18.32.16$$

They thus link with the preceding transition series of each period, in which variable valencies are characteristic, and indicate some instability of the 18-electron shell no doubt related to the small energy differences between the electrons of the outermost shell and the most energetic levels of the preceding shell. As the nuclear charge increases, in Zn, Cd and Hg, this instability disappears.

IONIC CRYSTALS

Many minerals have been interpreted as ionic crystals, and this interpretation has been very successful in explaining their chemical constitution. In ionic crystals two major factors govern the type of structure that may form:

(*i*)   the relative sizes of the ions involved,

(*ii*)   the balance of charges between the ions.

THE CONCEPT OF IONIC SIZE

The 'size' or radius of an atom or ion is not a fixed quantity. When two oppositely charged ions attract each other they approach one another until, at a certain distance, forces of repulsion between them (arising either by partial overlapping of adjacent electronic charge-clouds, or by repulsion between the nuclei whose like charges are insufficiently screened at this distance by the electron envelope) balance the forces of attraction, and they remain at a definite mean distance apart. This distance of closest approach is the sum of the radii of the two ions concerned. What fraction of this distance represents the radius of the one and what fraction the radius of the other is determined by consideration of the interatomic distances in the pure elements (when the radius is half the distance of closest approach) and of the interionic distances in compounds. When the radius of one element in a compound is known, the distance of closest approach gives the radius of the other directly. Each new determination extends the range of possible inferences.

In certain cases other arguments can be used, for example in the case of the oxides, sulphides and selenides of magnesium and manganese (Fig. 10).

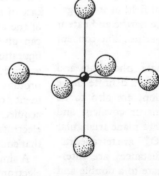 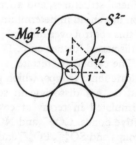

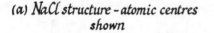

(a) NaCl structure – atomic centres shown   (b) 6-fold coordination of a single atom   (c) Central section of (b) – ionic boundaries shown

Fig. 10. Derivation of the radius of the S^{2-} ion.

All these compounds have NaCl structure. Their interatomic distances are

MgO	0·210 nm	MnO	0·224 nm
MgS	0·260 nm	MnS	0·259 nm
MgSe	0·273 nm	MnSe	0·273 nm

Here MgO and MnO have different interatomic distances, while in the other two pairs of compounds the distances are nearly the same. From this Goldschmidt argued that Mg and Mn ions make significant contributions to the interatomic distances in MgO and MnO; but in the sulphides and selenides the S and Se ions make up the whole of the interatomic distances—that is, in these cases the anions are mutually in contact with Mg or Mn lying in the interstices between them. Hence we may determine the radius of the S and Se ions as $0·26 \times \sqrt{2}/2$ ($= 0·184$ nm) and $0·273 \times \sqrt{2}/2$ ($= 0·193$ nm) respectively (Fig. 10).

V. M. Goldschmidt made extensive use of interatomic distances in this way to compile a table of ionic sizes that has long been used by mineralogists and crystallographers. In building up his scale Goldschmidt used the radii $O^{2-} = 0·132$ nm, and $F^- = 0·133$ nm as initial values. Chemists have preferred to use a scale due to Pauling, built up in a less empirical way by considering the regularities of ionic size in isoelectronic sequences, that is, sequences of elements with the same residual electron configuration after removal of valency electrons (corresponding to parts of the periods of the elements undisturbed by transition series.) The two sets of radii show reasonably good agreement, but Pauling gives values of $O^{2-} = 0·140$ nm and $F^- = 0·136$ nm. Revised Pauling values* are adopted here (Table 3).

Factors controlling ionic radii

The following factors influence ionic size in a regular way.

(*a*) *The state of ionization.* As the atomic number increases, the electron levels fill up and the atoms increase in size. We have already noticed, however, that the increase in nuclear charge at the same time draws the inner quantum shells closer to the nucleus (Fig. 2) so that, even in ions of the same valency, the 'core' of electrons inside the valency levels is contracted.

Upon ionization, cations, having lost electrons, tend to be small. Anions, having gained electrons, tend to be larger. But the higher the charge of the ions, whether positive or negative, the more strongly

* AHRENS, L. H. 1952. *Geochimica cosmochim. Acta*, 2, 168.

Table 3. **The ionic radii of the elements for 6-fold co-ordination (nanometres). (After Ahrens, 1952.)**

Ac³⁺	0·118	Hf⁴⁺	0·078	Pt²⁺	0·080
Ag⁺	0·126	Hg²⁺	0·110	Pt⁴⁺	0·065
Ag²⁺	0·089	Ho³⁺	0·091	Pu³⁺	0·108
Al³⁺	0·051	I⁻	0·216	Pu⁴⁺	0·093
Am³⁺	0·107	In³⁺	0·081	Ra²⁺	0·143
Am⁴⁺	0·092	Ir⁴⁺	0·068	Rb⁺	0·147
As³⁺	0·058	K⁺	0·133	Re⁴⁺	0·072
As⁵⁺	0·046	La³⁺	0·114	Re⁷⁺	0·056
At⁷⁺	0·062	Li⁺	0·068	Rh³⁺	0·068
Au⁺	0·137	Lu³⁺	0·085	Ru⁴⁺	0·067
Au³⁺	0·085	Mg²⁺	0·066	S²⁻	0·184
B³⁺	0·023	Mn²⁺	0·080	S⁶⁺	0·030
Ba²⁺	0·134	Mn³⁺	0·066	Sb³⁺	0·076
Be²⁺	0·035	Mn⁴⁺	0·060	Sb⁵⁺	0·062
Bi³⁺	0·096	Mn⁷⁺	0·046	Sc³⁺	0·081
Bi⁵⁺	0·074	Mo⁴⁺	0·070	Se²⁻	0·198
Br⁻	0·195	Mo⁶⁺	0·062	Se⁶⁺	0·042
C⁴⁻	0·260	N³⁻	0·171	Si⁴⁺	0·042
C⁴⁺	0·016	N³⁺	0·016	Sm³⁺	0·100
Ca²⁺	0·099	N⁵⁺	0·013	Sn²⁺	0·093
Cd²⁺	0·097	Na⁺	0·097	Sn⁴⁺	0·071
Ce³⁺	0·107	Nb⁴⁺	0·074	Sr²⁺	0·112
Ce⁴⁺	0·094	Nb⁵⁺	0·069	Ta⁵⁺	0·068
Cl⁻	0·181	Nd³⁺	0·104	Tb³⁺	0·093
Co²⁺	0·072	Ni²⁺	0·069	Tb⁴⁺	0·081
Co³⁺	0·063	Np³⁺	0·110	Tc⁷⁺	0·056
Cr³⁺	0·063	Np⁴⁺	0·095	Te²⁻	0·221
Cr⁶⁺	0·052	Np⁷⁺	0·071	Te⁶⁺	0·056
Cs⁺	0·167	O²⁻	0·140	Th⁴⁺	0·102
Cu⁺	0·096	Os⁶⁺	0·069	Ti³⁺	0·076
Cu²⁺	0·072	P³⁻	0·212	Ti⁴⁺	0·068
Dy³⁺	0·092	P⁵⁺	0·035	Tl⁺	0·147
Er³⁺	0·089	Pa³⁺	0·113	Tl³⁺	0·095
Eu³⁺	0·098	Pa⁴⁺	0·098	Tm³⁺	0·087
F⁻	0·136	Pa⁵⁺	0·089	U⁴⁺	0·097
Fe²⁺	0·074	Pb²⁺	0·120	U⁶⁺	0·080
Fe³⁺	0·064	Pb⁴⁺	0·084	V²⁺	0·088
Fr⁺	0·180	Pd²⁺	0·080	V³⁺	0·074
Ga³⁺	0·062	Pd⁴⁺	0·065	V⁴⁺	0·063
Gd³⁺	0·097	Pm³⁺	0·106	V⁵⁺	0·059
Ge²⁺	0·073	Po⁶⁺	0·067	W⁴⁺	0·070
Ge⁴⁺	0·053	Pr³⁺	0·106	W⁶⁺	0·062
		Pr⁴⁺	0·092	Y³⁺	0·092
				Yb³⁺	0·086
				Zn²⁺	0·074
				Zr⁴⁺	0·079

they tend to be drawn together in the crystal structure. This second effect tends to accentuate the smallness of cations, but somewhat to offset the increase of size of anions. The relative sizes of cations and anions in the first and second short periods are shown in Fig. 11. The important point to note is that *cations are usually small compared with anions.*

(*b*) *The state of coordination.* The coordination number of an ion is the number of immediately surrounding ions to which it is linked (see, for example, Fig. 10). This depends to some extent on its valency, and is in this way connected with the state of ioniza-

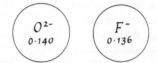

Electron gain decreasing : charge decreasing

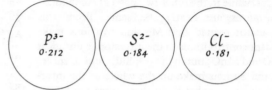

Electron loss increasing : charge increasing

Fig. 11. Ionic radii of typical anions and cations. Note the sizes of Si^{4+}, Al^{3+}, Mg^{2+}, and Na^+, respectively second, third, seventh and fourth in abundance (by number of atoms) in the earth's crust, compared to O^{2-} the most abundant element of the crust.

tion. Where two valency states are possible, the co-ordination state will be different in the two cases. Sometimes a particular ion, although possessing only one valency state, may occur in different co-ordinations, e.g. Al^{3+}, which may be found in 6-fold or 4-fold coordination.

The coordination of an ion is in large measure determined by its size, as we shall see (p. 15), but its size may also be modified (within limits) by its state of coordination, because the more bonds there are radiating from it the less will be their individual strength, and the less strongly will the ionic neighbours be drawn together. In comparing ionic sizes allowance has to be made for this, and the tabulated ionic radii (Table 3) are corrected to a standard state of 6-fold coordination, i.e. linkage to six neighbouring ions as in the NaCl structure (Fig. 10).

(c) *Polarization of the ion.* In general, the forces of attraction towards its neighbours acting on a given ion are not the same in all directions. They depend upon the valencies and electronic configurations of those neighbours, and the distances at which the neighbours lie. The tendency of variation in these forces in different directions is to distort, or *polarize*, the ion on which they act. The higher the charge on a cation, for example, and the closer its approach, the greater will be its distorting force

upon a neighbouring anion. Goldschmidt used the quantity Ze/r^2, where Ze is the charge and r the radius, as a measure of the polarizing power of an ion. Also, the smaller the coordination number of an ion, the less symmetrical will be the field of force in which it lies, and the greater the tendency to polarization.

Many atoms and ions (e.g. those of the inert gases, or ions with inert gas configurations) have a spherical symmetry of electron distribution. These have less distorting influence than less symmetrical ions like Cu^+. The resistance of different ions to distortion or polarization by their neighbours varies. Compact, highly charged cations are the least polarizable, whilst anions are more readily distorted. F^- and O^{2-} have by far the greatest rigidity amongst anions, however, and polarization effects upon them are usually neglected in fluorides and oxides.

Polarization of ions is in some respects akin to the assumption of a covalent character in the bond, and a certain amount of overlapping of the two concepts is to be found in the literature.* The concept of covalency is, however, considerably more developed than that of polarization, and the question of degree of ionicity or covalency of bonding is dealt with on p. 17.

* e.g. AHRENS, L. H., 1953. *Geochim. cosmochim. Acta*, 3, 1.

Radius ratio

Having defined what we mean by the size of an ion and seen the factors that influence it, we may consider its effect upon the types of crystal structure that may form. In doing so we shall speak of an ion as though it were a sphere with definite boundaries, although we know from the foregoing that it is actually a locally effective field of force subject to outside influences. It is, however, convenient and often adequate to regard it as a finite sphere.

When a cation attracts to itself a number of anions, these anions strive to come into contact with the cation. The number that can simultaneously do so depends upon the relative sizes of anions and cation. We have already seen that cations are relatively small, anions relatively large. If the cation is very small relative to the anions, only two can make contact with it. As its relative size increases three anions may touch it at once, and then four and so on. Whatever the actual sizes of the ions, it is the ratio of their radii that determines how many anions can touch the cation. This is called the *radius ratio*.

An arrangement where the anions are held away from contact with the cation because of their impingement on one another is unstable. Work is required against the forces of attraction between the oppositely charged ions and the potential energy of the system is increased. In other words the cation must at least nearly fill the hole between the anions; otherwise the arrangement will collapse to one in which fewer anions are grouped around the cation. The number of anions grouped around the cation is its *coordination number*. The possible regular arrangements of coordination, with the limiting values of

the radius ratio R_C/R_A between them, are given in Table 4. The derivation of these values is shown in Fig. 12.

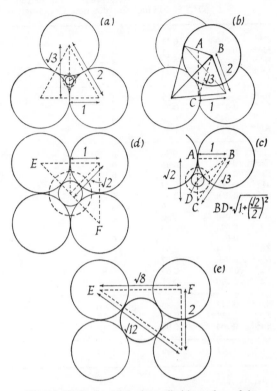

Fig. 12. Derivation of the lower limiting values of the radius ratio for the arrangements listed in Table 4. (*a*) Equilateral triangle. (*b*), (*c*) Tetrahedron. (*d*) Square and octahedron. (*e*) Cube. The theoretical upper limit of cation size for each arrangement is shown as a broken circle. [Note that *A*, *B*, *C* in (*c*) correspond to the points so lettered in (*b*). *E*, *F* in (*e*) correspond to the points so lettered in (*d*), (*e*) representing a section containing the face-diagonals of a cube.]

In very many minerals the important anion is O^{2-} and in Table 5 the radius ratios of some of the commoner cations with respect to O^{2-} are listed, with the expected and observed coordination numbers. The limiting values of the radius ratio are generally respected in crystal structures; but quite a number of exceptions occur, especially when we consider examples with anions other than O^{2-} and include artificial crystalline substances as well as minerals. These discrepancies point to departures from ionic bonding which are discussed below (p. 17).

The idea of ions linked into *coordination polyhedra* of the kinds listed in Table 4 (often some-

Table 4. Limiting values of the radius ratio for different coordinations.

Coordination Number	Arrangement of Anions	Radius Ratios at which transition occurs
2	Opposite one another	
		0.155 $(2\sqrt{3}/3 - 1)$
3	Corners of an equilateral triangle	
		0.225 $(\sqrt{1.5} - 1)$
4	Corners of a tetrahedron	
		0.414 $(\sqrt{2} - 1)$
4	Corners of a square ⎫	
6	Corners of an octahedron ⎬	
		0.732 $(\sqrt{12}/2 - 1)$
8	Corners of a cube	
		1.00
12	Closest packing of equal spheres	

Table 5. **Radius ratio of some common cations relative to oxygen.**

	$R/R_{O^{2-}}$ ($O^{2-} = 0.14$ nm)	Observed Coordination	Theoretical Coordination
C^{4+}	0·016	0·11	3	
N^{3+}	0·016	0·11	3	2
B^{3+}	0·023	0·16	3 or 4	3
S^{6+}	0·030	0·21	4	
Be^{2+}	0·035	0·25	4	
P^{5+}	0·035	0·25	4	4
Si^{4+}	0·042	0·30	4	
Al^{3+}	0·051	0·36	4 or 6	
Ga^{3+}	0·062	0·44	6	
Cr^{3+}	0·063	0·45	6	6
Fe^{3+}	0·064	0·46	6	
Mg^{2+}	0·066	0·47	6	
Li^{+}	0·068	0·49	6	
Ti^{4+}	0·068	0·49	6	
W^{4+}	0·070	0·50	4	
Fe^{2+}	0·074	0·53	6	
V^{3+}	0·074	0·53	6	
Zn^{2+}	0·074	0·53	6	
Mn^{2+}	0·080	0·57	6	
Na^{+}	0·097	0·69	6–8	
Ca^{2+}	0·099	0·71	8	
Sr^{2+}	0·112	0·80	8	
K^{+}	0·133	0·95	8–12	8
Ba^{2+}	0·134	0·96	8–12	
Rb^{+}	0·147	1·05	8–12	12

what distorted in actual structures) greatly simplifies the description of mineral structures. The polyhedra are linked in various ways, either through intermediate metal ions, or by sharing corners or edges.

The balance of charges between ions.

Pauling's Rules. The following empirical rules, formulated by Pauling, summarize the way in which the electrostatic forces that bind together the ions in ionic crystals operate.

(*i*) If the total charge on a cation be divided by the number of immediately surrounding anions (i.e. by its coordination number) the resulting fraction is the amount of charge it contributes to satisfying each anion.

(*ii*) The fractions of charge received by an anion from neighbouring cations must equal, or approximately equal, its valency: i.e. the charges holding the structure together must be balanced between each ion and its immediate neighbours, or if not with its immediate neighbours, at least within a short distance through the structure.

(*iii*) Common edges, and especially common faces, between coordination polyhedra diminish stability. This is because such arrangements bring the cations at the centres of the polyhedra nearer together, producing forces of electrical repulsion.

(*iv*) In a crystal containing different cations, those with high valency and low coordination number tend not to share polyhedral elements, the repulsive forces between such energetic ions being large.

(*v*) The number of essentially different constituents in a crystal structure tends to be small (Principle of Parsimony).

THE HYDROGEN BOND

In certain compounds of hydrogen, notably water and hydrogen fluoride, the individual molecules are associated; that is, they are linked together in chains, so that the hydrogen must be forming two bonds, although it has only one electron. The notable feature of this additional bond of hydrogen is that it is relatively weak, with energy of the order of 8 to 33 kilojoules per g mol compared with 200 to 400 kilojoules for most covalent bonds.

In ice, at sufficiently low temperatures, the H_2O molecules are almost all linked up in the fashion shown in Fig. 13 to give a tetrahedral arrangement of H atoms around the oxygens. In melting, about half these bonds are broken and the destruction of the framework allows the molecules to pack more closely together, producing the well-known increase in density of the liquid compared with the solid.

The elucidation of the hydrogen bond has presented particular difficulty. Even the position of the H atom has had to be determined by refined experiments (neutron diffraction) as it does not scatter X-rays sufficiently to be located with their aid. It turns out that the H atom is generally not placed midway between the two anions, as might have been expected, but is closer to one of them, as shown in Fig. 13.

Without going into details, we may say that the bond is regarded as dominantly electrostatic, the H atom being to some extent ionized in its bond with one anion, and the positive charge so produced interacting with the lone pair of electrons on another anion. Its formation seems to be dependent upon the small size of the hydrogen (or deuterium) atom and the absence of inner shell electrons, allowing close approach of the adjacent atoms without introducing repulsive forces. It is formed only between H and the most electronegative elements* (N, O,

*See p. 17.

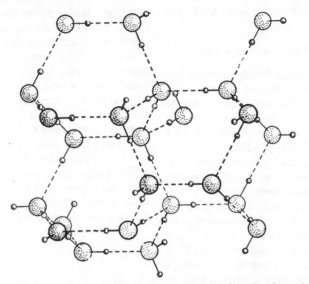

Fig. 13. The linkage of water molecules in ice. Each oxygen atom is joined to four others surrounding it tetrahedrally, and along each link lies an atom of hydrogen nearer to one oxygen than to the other. Each oxygen has two close hydrogens, joined to it by full lines in the diagram. The longer "hydrogen bonds" are shown by broken lines.

F, and S), a fact that tends to confirm its electro-static character.

The presence of hydrogen bonding is invoked in a number of minerals (boehmite, lepidocrocite, some layer silicates), especially those produced by weathering and other low temperature processes, and its existence in water, together with its low energy of formation and rupture, give it great significance in natural chemical processes taking place at normal temperatures. In particular, the making and breaking of hydrogen bonds is regarded as of outstanding importance in biological processes, especially in the formation of complex protein molecules that are the basis of the reproducibility of living structures.*

THE VAN DER WAALS BOND

The deviation of behaviour of gases from that predicted by the perfect gas laws is in part due to forces of attraction between the gas molecules, and Van der Waals, who first took account of this fact, has given his name to the weak bonding force involved. When a diatomic gas is condensed to a solid the bonds within the molecule are little changed, and the molecules are weakly held together by residual bonding forces of a different character from the covalent bonds within the molecule.

* See PAULING, L., 1960. *The nature of the chemical bond* (3rd edn), New York, Cornell Univ. Press, p. 498 ff.

Van der Waals forces are invoked to explain the weak cohesion between 'molecular' units in crystals of graphite and the weak retention of certain molecules in otherwise largely ionic crystal structures, as for example between the layers of micas and clay minerals.

ELECTRONEGATIVITY AND THE INTERMEDIATE IONIC-COVALENT CHARACTER OF BONDS

Although chemical bonds have been discussed above as though they were of distinct types, in fact it is found that most bonds have an intermediate character between that of the purely covalent and purely ionic bond. The study of bond energies, chiefly by measuring heats of formation and of dissociation, discloses that the energy of an actual bond between unlike atoms is usually greater than the calculated energy of a normal covalent bond between them. The additional energy is due to a degree of ionic character in the bond.

The normal covalent energy, just mentioned, is calculated approximately as the arithmetic mean* of the energies of the bonds between like atoms of the two elements involved. Thus, for two elements A and B the bond energy, D, is given by

$$D = \tfrac{1}{2}\{D(A-A) + D(B-B)\}$$

* In some cases the geometric mean, Δ', is used.

and if the observed bond energy is $D(A-B)$ the additional ionic energy, Δ, will be

$$\Delta = D(A-B) - \tfrac{1}{2}\{D(A-A) + D(B-B)\}.$$

The bond energies of single bonds between like atoms are known from the heats of dissociation of the molecules of the elements into atoms, using the empirical postulate that the dissociation energy of a polyatomic molecule may be equally divided between the number of bonds present.

It is found that the value of Δ in the above equation increases as the two atoms A and B become more and more different in their *electronegativity*, that is their power to draw electrons to themselves. This capacity to attract electrons is different from two other properties, the ionization potential and the electron affinity, but is related to them. [Thus the mean of the first ionization potential, (the energy needed to remove the first valency electron from an atom, and the energy of the reaction $A^+ + e^- \rightarrow A$) and the electron affinity (energy of the reaction $A + e^- \rightarrow A^-$) has been suggested as a measure of the electronegativity of the neutral atom.]

From a study of the values of Δ (additional ionic bond energy) for compounds of fourteen elements whose single bond energies are known, Pauling* has built up a complete scale of electronegativity of the elements with an arbitrary origin at $H = 2 \cdot 05$ (Table 6). On this scale, elements with electro-negativities greater than about 2 comprise the non-metals, separated by a heavy line in Table 6. Silicon and arsenic are placed with the non-metals because they, also, like boron, carbon, nitrogen, phosphorus and sulphur, tend to form discrete anionic groups with oxygen in mineral structures.

Because the ionic contribution to bond energy increases with the electronegativity difference between the elements involved, it is possible to make an approximate estimate of the percentage ionic character of any bond, from the electronegativities of the bonded elements. Fig. 14 shows the approximate relationship calculated.

We may express the idea of electronegativity in another way by saying that when an ionic bond is formed between two atoms the electron surrender may not be complete. In most bonds some part of the electron charge cloud remains associated with the donor. This is shown, for example, in X-ray

* PAULING, L. 1960. *The nature of the chemical bond.* New York, Cornell Univ. Press. Other scales are by MULLIKEN, R. S. 1934. *J. chem. Phys.*, **2**, p. 782, and POVARENNYKH, A. S. 1955. *Mem. All Union Miner. Soc. Acad. Sci. USSR*, **84** (4), 469–92.

diffraction study, where the X-rays are scattered chiefly by the electrons. In determining crystal structure, electron-density maps are prepared (Fig. 1, p. 5) by measuring the intensity of the diffracted beams, and in NaCl, for example, it is found from the total area under the radial distribution curve of electrons (of the type of Fig. 2) that there are $17 \cdot 85$ electrons around the Cl sites, whereas fully ionized Cl^- would have 18. The Na^+ thus retains a small part of its valency electron. NaCl may be regarded, however, as effectively purely ionic: in other cases there is a greater reluctance to make the surrender of electrons complete, and the bond becomes intermediate in character between the ionic and the covalent.

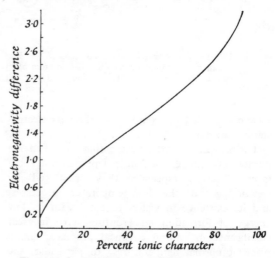

Fig. 14. Curve relating the difference in electro-negativity of two atoms to the degree of ionicity of the bond between them. (Data from PAULING, L., 1960, p. 98.)

ISOMORPHISM AND ATOMIC SUBSTITUTION

STRUCTURAL CONTROL OF MINERAL COMPOSITION

The geometrical factor of the radius ratio, and the necessity for a balance of charges between neighbouring ions, impose strict limitations on the types of structure that are stable in ionic, or largely ionic crystals. And these rules are sufficiently closely obeyed in nature for us to be able to say that the two properties of ionic size and valency determine the possibility of entry of any cation into a particular structural position. This has had a very important

Table 6. The electronegativities of the elements. (After Pauling, 1960.) Note that the electronegativity varies with the valency state of the atom. Thus V^{3+} 1·4, V^{4+} 1·6, V^{5+} ~ 1·8; Cr^{2+} 1·6, Cr^{3+} 1·5, Cr^{4+} ~ 2·1; Mn^{2+} 1·4, Mn^{3+} ~ 1·5, Mn^{7+} ~ 2·3; Fe^{2+} 1·8, Fe^{3+} 1·9.

IA	IIA	IIIB	IVB	VB	VIB	VIIB	VIII	VIII	VIII	IB	IIB	IIIA	IVA	VA	VIA	VIIA	0
1 H 2·1																	2 He
3 Li 1·0	4 Be 1·5											5 B 2·0	6 C 2·5	7 N 3·0	8 O 3·5	9 F 4·0	10 Ne
11 Na 0·9	12 Mg 1·2											13 Al 1·5	14 Si 1·8	15 P 2·1	16 S 2·5	17 Cl 3·0	18 A
19 K 0·8	20 Ca 1·0	21 Sc 1·3	22 Ti 1·5	23 V 1·6	24 Cr 1·6	25 Mn 1·5	26 Fe 1·8	27 Co 1·8	28 Ni 1·8	29 Cu 1·9	30 Zn 1·6	31 Ga 1·6	32 Ge 1·8	33 As 2·0	34 Se 2·4	35 Br 2·8	36 Kr
37 Rb 0·8	38 Sr 1·0	39 Y 1·2	40 Zr 1·4	41 Nb 1·6	42 Mo 1·8	43 Tc 1·9	44 Ru 2·2	45 Rh 2·2	46 Pd 2·2	47 Ag 1·9	48 Cd 1·7	49 In 1·7	50 Sn 1·8	51 Sb 1·9	52 Te 2·1	53 I 2·5	54 Xe
55 Cs 0·7	56 Ba 0·9	57 La 1·1	72 Hf 1·3	73 Ta 1·5	74 W 1·7	75 Re 1·9	76 Os 2·2	77 Ir 2·2	78 Pt 2·2	79 Au 2·4	80 Hg 1·9	81 Tl 1·8	82 Pb 1·8	83 Bi 1·9	84 Po 2·0	85 At 2·2	86 Rn
87 Fr 0·7	88 Ra 0·9	89 Ac 1·1															

58 Ce	59 Pr	60 Nd	61 Pm	62 Sm	63 Eu	64 Gd	65 Tb	66 Dy	67 Ho	68 Er	69 Tm	70 Yb 1·3	71 Lu 1·2
90 Th 1·3	91 Pa 1·5	92 U 1·7	93 Np 1·3	94 Pu	95 Am	96 Cm	97 Bk	98 Cf	99 E	100 Fm	101 Mv	102 No 1·3	

influence upon our thinking about the chemical compositions of minerals. We now see a given mineral as a *structure* with a number (usually a small number) of different types of structural position, or atomic site, into which suitable atomic tenants may enter. To qualify for entry, an atom must simply have the right size and charge, so that the atoms in each type of site in a crystal may be of more than one element, depending upon what was available in the environment in which the crystal grew.

No longer do we see a mineral as a chemical combination between two or more elements obeying the Law of Constant Proportions and combining in fixed proportions by weight, as is proper for chemical compounds in classical chemical theory. We shall be quite prepared to find, on chemical analysis, a composition that gives us fractional numbers of atoms of the elements involved, provided that the numbers of atoms of different elements, grouped according to their entry qualifications for the sites available, add up to integral numbers.

AN EXAMPLE OF CATION DISTRIBUTION

As an example we may quote the analysis of an augite, whose atoms must be divided amongst the number of sites shown in the diagram of the structure of diopside, $CaMg[SiO_3]_2$, in Fig. 1. (Reference may also be made to Fig. 227 p. 257, and the description of the structure there given.)

The steps in the calculation are:

(*i*) Divide the weight per cent of each oxide by its molecular weight to convert the analysis to molecular proportions.

(*ii*) Convert these to cation and anion (oxygen) proportions by multiplying by the number of each in the molecule.

(*iii*) Bring the number of oxygens to 24, this being the number belonging wholly to the unit cell of Fig. 1 (when we allow for sharing of boundary atoms between adjacent cells).

(*iv*) Adjust the cations proportionately to amounts relative to 24 O^{2-}

(*v*) Consider the cation radii and charges, and assign them to the available types of site in accordance with these properties.

In some cases there may be discussion centred around step (*v*), but usually there is not much room for disagreement. In our case, we have a deficiency of Si^{4+} for the sites so marked in Fig. 1, and we take some Al^{3+} as the most promising substitute. This leads to a surplus of negative charge on the oxygens linked to this trivalent ion, which is compensated for by the presence of Fe^{3+}, Al^{3+}, and Ti^{4+} in place of divalent ions outside the Si^{4+}—O^{2-} chains. Thus we have *substitution* of one atom for another in the basic formula of the mineral group, as shown at the bottom of the table.

Table 7.   Calculation of the formula of augite from Garbh Eilean, Shiant Is., Scotland analysed by R. J Murray, 1954. (*Geol. Mag.*, **91**, p. 22).

Oxides (wt %)		M.W.	Mol. Props.	Cations	Anions	Cations to 24 O^{2-}	Cation Grouping		
							Ion	Radius	Groups
SiO$_2$	50·29	60	0·838	0·838	1·676	7·572	Si^{4+}	0·042	7·572 ⎤
TiO$_2$	1·33	80	0·016	0·016	0·032	0·145	Al^{3+}	0·051	0·428 ⎬ 8·000
Al$_2$O$_3$	2·82	102	0·027	0·054	0·081	0·488	Al^{3+}	0·051	0·060 ⎤
Fe$_2$O$_3$	1·76	160	0·011	0·022	0·033	0·199	Ti^{4+}	0·068	0·145 ⎟
FeO	9·75	72	0·136	0·136	0·136	1·229	Fe^{3+}	0·064	0·199 ⎬ 3·997
MnO	0·26	71	0·004	0·004	0·004	0·036	Mg^{2+}	0·066	2·819 ⎟
MgO	12·49	40	0·312	0·312	0·312	2·819	Fe^{2+}	0·074	0·774 ⎦
CaO	21·14	56	0·376	0·376	0·376	3·398	Fe^{2+}	0·074	0·455 ⎤
Na$_2$O	0·44	62	0·006	0·012	0·006	0·108	Mn^{2+}	0·080	0·036 ⎟
K$_2$O	trace						Ca^{2+}	0·099	3·398 ⎬ 3·997
H$_2$O	—						Na$^+$	0·097	0·108 ⎦
Total	100·28				2·656				

Factor for cation proportions
$$\frac{24}{2\cdot656} = 9\cdot036$$

Basic formula per unit cell:
$$4[Ca(Mg, Fe)[Si_2O_6]]$$

expands to:
$$4[(Na, Ca, Mn, Fe^{2+})(Mg, Fe^{2+}, Fe^{3+}, Al, Ti)[(Al, Si)_2O_6]]$$

INCIDENCE OF ATOMIC SUBSTITUTION (DIADOCHY)

The concept of atomic substitution, or *diadochy*, was the result of the investigation of mineral structures by X-rays, and its introduction cleared away a cloud of obscurity that had surrounded the problem of the formulae of minerals (especially silicates) during the earlier phase of unassisted chemical analysis. It is now clear that diadochy occurs in most minerals, and that many fall into the category of *isomorphous series*, or solid solution series, with a continuous reciprocal variation in the proportions of a pair of elements, or of more than one pair, in their chemical constitution. In the *olivine series*, for example, there is continuous variation from forsterite, Mg_2SiO_4, to fayalite, Fe_2SiO_4, with the common olivines having a formula near to $Mg_{1.8}Fe_{0.2}SiO_4$. In the important group of *plagioclase feldspars* we have continuous substitution of $(Na^+ + Si^{4+})$ for $(Ca^2 + Al^3)$ from anorthite, $Ca_2Al_2Si_2O_8$, to albite, $Na\,AlSi_3O_8$.

The existence of this possibility explains the range of variation in composition of isomorphous mineral series with respect to major elements. Thus Mg^{2+} (radius 0·066 nm) and Fe^{2+} (0·074 nm) of similar size and the same charge, can replace each other with great freedom in most structures. Fe^{3+} (0·064 nm), Al^{3+} (0·051 nm) and Cr^{3+} (0·063 nm) can also substitute freely for one another, whilst Na^+ (0·097 nm) and K^+ (0·133 nm) do so with somewhat less facility at low temperatures.

The phenomenon of diadochy also explains, to a large extent, the location of minor, or trace, elements, occurring in amounts from 0·01 weight per cent down to a few hundred parts per million in minerals chiefly made up of major elements. For example, the distribution of V^{3+} (0·074 nm), as a trace element, closely follows that of the major element Fe^{3+} (0·064 nm), Rb^+ (0·147 nm) that of K^+ (0·133 nm), Sr^{2+} (0·112 nm) that of Ca^{2+} (0·099 nm), Ge^{4+} (0·053 nm) that of Si^{4+} (0·042 nm), Ga^{3+} (0·062 nm) that of Al^3 (0·051 nm), and so on.

FACTORS GOVERNING ATOMIC SUBSTITUTION

GOLDSCHMIDT'S RULES

Some variation is permissible in the properties of interchangeable ions in mineral structures, and the limits of tolerance were empirically defined by V. M. Goldschmidt.

Ionic size

A difference in ionic size amounting to not more than 15 per cent of the radius of the larger ion permits a wide range of substitution at ordinary temperatures. There is, in general, much more tolerance of the entry of foreign ions into a structure at high temperatures, when all the ions are in a high state of thermal vibration about their mean positions, and the geometrical control (sizes of available spaces) is less rigid. The structure is then in a state of relative disorder. With a fall in temperature this latitude diminishes, and foreign ions tend to be thrown out of the structure—exsolved—as separate phases of different composition, with definite boundaries between them and their host. The result is often the formation of regular lamellae of the exsolved substance along certain crystal planes in the host; that is, regularly related to its atomic structure. Common examples of such structures are seen in perthitic feldspar (where the inhomogeneity gives rise to a play of colours in the gem variety called moonstone), in certain pyroxenes and some magnetite.

Ionic charge

Of two suitably-sized ions available to occupy a given site, the more highly-charged one will generally be accepted first.

The two factors of ionic size and charge lead to three possible cases.

Camouflage. The examples of diadochic ions given above illustrate the case where a minor or foreign element has the same valency as, and a radius similar to, that of the major element it replaces. This was characterized by Goldschmidt as *camouflage* of the minor element.

Capture. An ion of higher charge than, and similar radius to a major element may be readily accepted or *captured* by a mineral structure in preference to the corresponding major element. Because of the high valency, such a substituent element will be strongly attracted to the structure and firmly held there when it has entered. Such ions tend to be concentrated in the early-precipitated crystals of the mineral within a given environment of crystallization. Similarly, an ion of the same charge as, but smaller radius than the major element will be preferentially incorporated into the structure. Here again, the applicant with the higher *ionic potential*, measured by the charge divided by the radius (Z/r), is the more successful.

Admission. An ion of similar radius and lower charge, or the same charge but somewhat greater radius than the corresponding major element may be *admitted* to the structure, but is not preferentially selected; it may tend to occur in late-formed crystals, when its concentration in the environs, relative to the major element, has been increased.

THE INFLUENCE OF ELECTRONEGATIVITY

A further factor in diadochy is the relative electronegativities of the atoms involved. We have seen already that difference in electronegativity between two atoms may be used to estimate the degree of ionicity or covalency of the bond between them. Cations being the electron donors, the higher the electronegativity of a cation the more covalent, and hence the weaker, its bonds will become. In its application to diadochy, it is found that, where substitution is possible between cations of appreciably different electronegativities (a difference of about 0·1 or more), the one of smaller electronegativity will be preferred. In the operation of this factor, as in Goldschmidt's Rules, we see the tendency of the crystal to incorporate preferentially those ions that make the greatest contribution of bond energy to the structure.

A number of examples of the influence of electronegativity have been cited.* A simple case is that of V^{3+}, which accompanies Fe^{3+} as noted above, but is also concentrated in the early-formed Fe^{3+}-bearing magnetite crystals of igneous rocks. The radius of V^{3+}, 0·074 nm, being greater than that of Fe^{3+}, 0·064 nm, we should not expect to find this concentration in early magnetite; but when we consider its lower electronegativity, 1·4, compared with Fe^{3+}, 1·9, its preferential entry into the structure is explained. The case of Cr^{3+} and Fe^{3+} is similar, though here the Cr radius, 0·063 nm is in any case a little less than Fe^{3+}. Its concentration in early crystals is, however, very marked, and this is explained by the cooperation of the electronegativity factor.

Successful as Goldschmidt's Rules have been in explaining most cases of element substitution in minerals, they will undoubtedly be further improved and refined as knowledge increases. Lately S. R. Nockolds has suggested† a new function,

*RINGWOOD, A. E. 1955. *Geochim. cosmochim. Acta*, 7, 189–202.

†NOCKOLDS, S.R., 1966. The behaviour of some elements during fractional crystallization. *Geochim. cosmochim. Acta*, 30, 267–78.

relative total bonding energy, taking into account charge, radius and electronegativity, which he uses with success to explain observed substitutions and order of entry of elements in crystallization from melts.

POLYMORPHISM

In the preceding section we considered the variations in chemical composition that may take place in any one mineral structure. Most minerals exhibit isomorphism of that kind, and usually members of isomorphous series are included under a single mineral name, as for example olivine, or plagioclase. Some isomorphous series (including the examples just given) are subdivided arbitrarily and different names applied to sections of the series; but often this is not done, and even where it is, greater convenience and precision may result from giving a percentage composition in terms of one of the end-member 'molecules', e.g. plagioclase An_{60} (= plagioclase with 60 per cent of the anorthite component, $Ca_2 Al_2 Si_2 O_8$).

The converse case must now be considered, where cne chemical composition may crystallize with different structures under different conditions, thus exhibiting *polymorphism*. In this case, each structure usually receives a separate species name, which emphasizes once more the fundamental rôle of crystal structure in mineralogy.

Polymorphism is of the greatest importance, because the different crystal structures assumed by one compound are controlled chiefly by the prevailing temperature and pressure of its crystallization. We may sometimes succeed in producing *inversions* from one form to another in the laboratory, and in determining the range of conditions under which each is stable, so enabling the conditions under which the mineral formed to be defined. Even if this is not possible, we may use the form of the polymorphic compound to compare one mineral association and set of mineral-forming conditions (*mineral paragenesis*) with another.

Some polymorphs have long been known, e.g. those of carbon (graphite and diamond), of SiO_2 (low and high quartz, tridymite, cristobalite), of $CaCO_3$ (calcite, aragonite), and of Al_2SiO_5 (andalusite, kyanite, sillimanite); but it is especially due to the high-pressure experiments of P. W. Bridgman at Harvard University that the widespread possibility of polymorphism became known. It now appears likely that nearly all crystalline substances are capable of inversions of this kind.

Table 8. Physical and optical properties of some polymorphs of silica.

Mineral	Crystal System	Cell sides (nm)	Cell Vol. (nm^3)	SiO$_2$ units per cell	Vol. per SiO$_2$ unit (nm^3)	D	Refractive indices
Lechatelierite	Glass					2·19	n 1·46 ¶
Quartz (low)	Trigonal	a 0·4913 c 5·5405	0·1305	3	0·0435	2·65	ε 1·553 ω 1·544
High temperature polymorphs:							
Tridymite (high)	Hexagonal	a 0·5046 c 0·8256	0·1820	4	0·0455	2·27★	α 1·471★ γ 1·483★
Cristobalite (high)	Cubic	a 0·7138	0·3637	8	0·0455	2·33★	ε 1·484★ ω 1·487★
High pressure polymorphs:							
Coesite	Monoclinic (pseudohexagonal)	a 0·714 b 1·237 c 0·714	0·6306	16	0·0394	2·92	α 1·594 γ 1·599
Stishovite	Tetragonal	a 0·4179 c 0·2665	0·0465	2	0·0233	4·3	ε 1·845 ω 1·800

★ These values pertain to the low modification (orthorhombic tridymite; tetragonal (?) cristobalite).

Silica, SiO$_2$, provides a good example of polymorphism, and one where striking new discoveries have recently been made, even though it might be thought to be a well-studied compound.

In Table 8 are listed some of the forms of SiO$_2$, and their physical properties and the relationships between them are shown in Fig. 15. All of these minerals, except stishovite, are arrangements of Si^{4+} ions in tetrahedral coordination with O^{2-}, and Fig. 16 shows the ranges of temperature and pressure within which each is stable.

DISPLACIVE TRANSFORMATIONS

Consider first the transition from low quartz to high quartz (Fig. 16), which takes place at 573°C at ordinary pressure, and at only slightly higher temperatures as the pressure increases. The arrangement of Si–O tetrahedra in high quartz is shown in Fig. 17a, and b, where they are seen to be joined in interlinked spirals, whilst in Fig. 17c is shown the difference in arrangement between high and low quartz. It will be seen that in low quartz the tetrahedral groups are slightly twisted with respect to their disposition in high quartz, in such a way as to reduce the six-fold symmetry of the hexagonal spaces in high quartz to three-fold symmetry. This change, however, does not involve any breaking of bonds, but simply a distortion of the high quartz structure. It therefore takes place easily, as the crystal passes through the transition temperature on

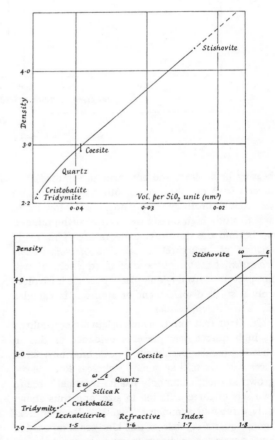

Fig. 15. Relationships between the physical properties of the silica polymorphs.

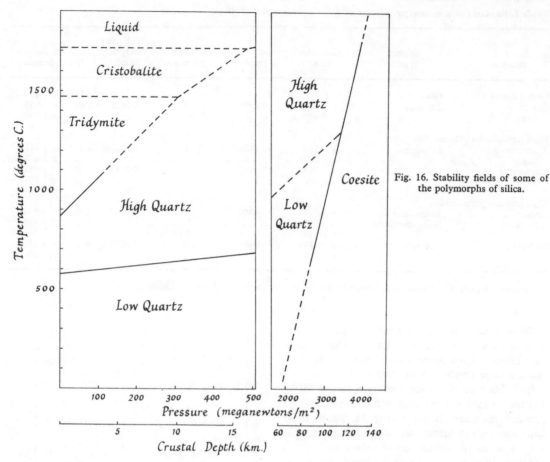

Fig. 16. Stability fields of some of the polymorphs of silica.

heating or cooling, and the structure immediately inverts from one form to the other. All the quartz we study at ordinary temperatures is the low form, and to study high quartz we must heat the mineral up beyond the transition temperature. This ready change of structure is called a *non-quenchable transition*, because quick cooling (quenching) will not prevent it from taking place. Because it involves only a small displacement of atoms it is called a *displacive transformation*.

It is true that quartz which originally crystallized as high quartz may preserve evidence of this in showing regular evenly-sized hexagonal bipyramid faces with six-fold symmetry, whereas low quartz, grown as such, commonly has its terminal faces in two sets of three differing in size, and thus shows only three-fold symmetry (Fig. 18). Regular hexagonal quartz crystals of this kind are composed of low quartz at ordinary temperatures, but are spoken of as *paramorphs* of high quartz, because

they retain their original symmetry. Accidents of growth may, however, make this distinction based upon form unreliable, for original high quartz may have grown distorted, whilst low quartz may grow with evenly matched terminal faces. But in all cases, the shapes of etch pits on these terminal faces will separate them into two sets of three, different in character, at ordinary temperatures.

Very refined measurements of the *exact* temperature of the high–low inversion, demanding special apparatus, may serve to show in which form a piece of quartz originally grew, for it has been found that the inversion may take place at up to about 0·6°C above or below 573°C, and that this variation is dependent upon the thermal history of the crystal.

Both tridymite and cristobalite exist in high and low modifications, related by displacive transformations of this kind, also, but these are not shown on the stability diagram.

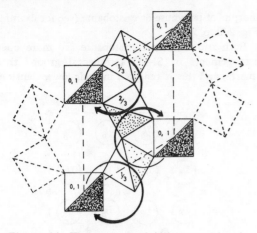

Fig. 17 (*a*). The structure of high-quartz, viewed along the *c*-axis.

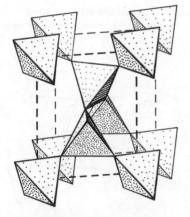

Fig. 17 (*b*). The linkage of Si—O tetrahedra in high-quartz.

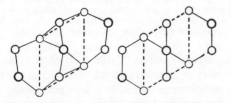

Fig. 17 (*c*). Displacement of tetrahedra in the inversion from high-quartz (left) to low-quartz. Silicon atoms only depicted.

Fig. 18 (*a*). Hexagonal bipyramidal crystal of quartz, showing six-fold symmetry.

Fig. 18 (*b*). Crystal of quartz terminated by two sets of rhombohedra, *r* and *z*, exhibiting three-fold symmetry.

RECONSTRUCTIVE TRANSFORMATIONS

If we now compare the structures of tridymite and cristobalite (Figs 19 and 20) with each other, and with quartz, we see that these three are all different ways of linking tetrahedral Si^{4+}—O^{2-} groups together. In order to convert one into another, linkages between tetrahedral groups must be broken and the groups rearranged. This is a *reconstructive transformation*, and proceeds much more sluggishly than a displacive transformation. It requires time for its completion, and if the higher temperature polymorph is cooled fairly rapidly through the inversion temperature the change will not take place; the high temperature form will then persist *metastably* at ordinary temperatures, where it is not the truly stable form. In this way we may handle and study tridymite (stable above 867°C) and cristobalite (stable above 1470°C) in the laboratory, at ordinary temperatures, whereas we cannot do this with high quartz (stable above 573°C). Tridymite and cristobalite occur naturally in lavas initially poured out at high temperatures under low pressures. With the passage of time each will tend to invert to quartz (especially if there is any rise of temperature to facilitate atomic rearrangement) but the

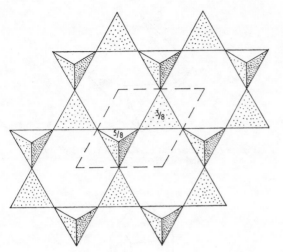

morphs of tridymite or cristobalite (see for example Fig. 242, p. 281).

Both tridymite and cristobalite are more open arrangements of Si—O tetrahedral groups than quartz, and this is reflected in their lower densities

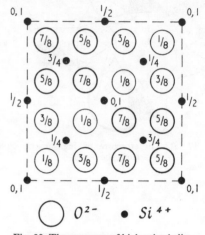

Fig. 19. The structure of high-tridymite.
(*a*) Plan of the structure viewed along the *c*-axis. The unit cell is outlined.

Fig. 20. The structure of high-cristobalite.
(*a*) Plan of the unit cell contents projected upon the cube face.

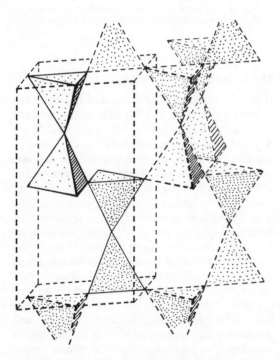

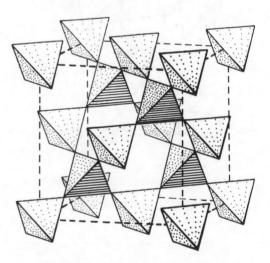

(*b*) The relatively open arrangement of Si—O tetrahedra. Only the line-shaded tetrahedra belong entirely to the unit cell illustrated. The others surround Si atoms at face-centres and corners of the cube.

(*b*) The arrangement of the Si—O tetrahedral groups. Only those shown in full outline belong to the unit cell outlined.

characteristic crystal shape of the original mineral tends to be retained so that we have quartz para-

and refractive indices (Table 8). High temperature polymorphs, in general, tend to be less dense than low temperature forms. It is also noteworthy that low temperature forms belong to crystal systems of lower symmetry than the high forms, because at high temperatures less strict orderliness of atoms is per-

Table 9. Electronic configurations of the first 92 elements.

Electron shells: K (1s); L (2s, 2p); M (3s, 3p, 3d); N (4s, 4p, 4d, 4f); O (5s, 5p)

Z	Element	1s	2s	2p	3s	3p	3d	4s	4p	4d	4f	5s	5p
1	H	1											
2	He	2											
3	Li	complete	1										
4	Be	complete	2										
5	B	complete	complete	1									
6	C	complete	complete	2									
7	N	complete	complete	3									
8	O	complete	complete	4									
9	F	complete	complete	5									
10	Ne	complete	complete	6									
11	Na			complete	1								
12	Mg			complete	2								
13	Al			complete	complete	1							
14	Si			complete	complete	2							
15	P			complete	complete	3							
16	S			complete	complete	4							
17	Cl			complete	complete	5							
18	Ar			complete	complete	6							
19	K				complete	complete	..	1					
20	Ca				complete	complete	..	2					
21	Sc				complete	complete	1	2					
22	Ti				complete	complete	2	2					
23	V				complete	complete	3	2					
24	Cr				complete	complete	5	1					
25	Mn				complete	complete	5	2					
26	Fe				complete	complete	6	2					
27	Co				complete	complete	7	2					
28	Ni				complete	complete	8	2					
29	Cu				complete	complete	10	1					
30	Zn				complete	complete	10	2					
31	Ga					complete	complete	complete	1				
32	Ge					complete	complete	complete	2				
33	As					complete	complete	complete	3				
34	Se					complete	complete	complete	4				
35	Br					complete	complete	complete	5				
36	Kr					complete	complete	complete	6				
37	Rb							complete	complete	..	..	1	
38	Sr							complete	complete	..	..	2	
39	Y							complete	complete	1	..	2	
40	Zr							complete	complete	2	..	2	
41	Nb							complete	complete	4	..	1	
42	Mo							complete	complete	5	..	1	
43	Tc							complete	complete	6	..	1	
44	Ru							complete	complete	7	..	1	
45	Rh							complete	complete	8	..	1	
46	Pd							complete	complete	10	..	..	
47	Ag							complete	complete		..	1	
48	Cd							complete	complete		..	2	
49	In							complete	complete		..	complete	1
50	Sn							complete	complete		..	complete	2
51	Sb							complete	complete		..	complete	3
52	Te							complete	complete		..	complete	4
53	I							complete	complete		..	complete	5
54	Xe							complete	complete		..	complete	6

Electron shells: N (4d, 4f); O (5s, 5p, 5d, 5f); P (6s, 6p, 6d, 6f); Q (7s)

Z	Element	4d	4f	5s	5p	5d	5f	6s	6p	6d	6f	7s
55	Cs	10	..	2	6	..	..	1				
56	Ba		..			..	..	2				
57	La		..			1	..	2				
58	Ce	complete	2	complete	complete	..	..	2				
59	Pr	complete	3	complete	complete	..	..	2				
60	Nd	complete	4	complete	complete	..	..	2				
61	Pm	complete	5	complete	complete	..	..	2				
62	Sm	complete	6	complete	complete	..	..	2				
63	Eu	complete	7	complete	complete	..	..	2				
64	Gd	complete	7	complete	complete	1	..	2				
65	Tb	complete	9	complete	complete	..	..	2				
66	Dy	complete	10	complete	complete	..	..	2				
67	Ho	complete	11	complete	complete	..	..	2				
68	Er	complete	12	complete	complete	..	..	2				
69	Tm	complete	13	complete	complete	..	..	2				
70	Yb	complete	14	complete	complete	..	..	2				
71	Lu		complete			1	..	2				
72	Hf		complete			2	..	2				
73	Ta		complete			3	..	2				
74	W		complete			4	..	2				
75	Re		complete			5	..	2				
76	Os		complete			6	..	2				
77	Ir		complete			9	..	0				
78	Pt		complete			9	..	1				
79	Au		complete			10	..	1				
80	Hg		complete			10	..	2				
81	Tl					complete	..	complete	1			
82	Pb					complete	..	complete	2			
83	Bi					complete	..	complete	3			
84	Po					complete	..	complete	4			
85	At					complete	..	complete	5			
86	Rn					complete	..	complete	6			
87	Fr						..	complete	complete	..	..	1
88	Ra						..	complete	complete	..	..	2
89	Ac						..	complete	complete	1	..	2
90	Th						..	complete	complete	2	..	2
91	Pa						2	complete	complete	1	..	2
92	U						3	complete	complete	1	..	2

missible, the atoms being in a higher state of thermal vibration about their mean positions (leading ultimately, of course to liquefaction, where all crystalline order is lost). The high forms are thus more nearly random or isotropic in structure, and this means more nearly the same in all directions, or more highly symmetrical.

STABILIZATION OF POLYMORPHS BY FOREIGN IONS

Tridymite is interesting in that it is always found, on analysis, to contain small amounts of other elements, as impurities; and when absolutely pure SiO_2 is crystallized under various conditions tridymite fails to form, quartz and cristobalite only being found. The presence of foreign ions (Al^{3+} replacing Si^{4+} and atoms like Na^+ and Ca^{2+} lying in the interstices of the structure) appears to be essential to stabilize the very open atomic framework.*

* MASON, B. 1953. Tridymite and christensenite. *Am. Mineral.*, **38**, 866–67.

HIGH PRESSURE POLYMORPHS

Both coesite and stishovite were discovered (in 1953 and 1961 respectively) by laboratory experiments at very high pressures. The transition curve between quartz and coesite has been partly located. Stishovite has been synthesised at 13 000 MN/m^2 and 1000°C by Akimoto, who gives the slope of the transition curve between coesite and stishovite as $P(MN/m^2)$ = 6700 + 0.028T (°C). Since they were prepared artificially, both have been found at Meteor Crater, Arizona, in sandstone altered at the tremendously high pressures generated by the impact and explosion of a very large meteorite.

Coesite might be expected to be the stable form of SiO_2 at 60–100 km depth within the earth, and stishovite at 400–500 km, if free SiO_2 exists there. Their high densities reflect the conditions of their formation. Stishovite is unique in having Si^{4+} in six-fold (octahedral) coordination with O^{2-}, instead of the usual four-fold tetrahedral arrangement.

Further Reading

BRAGG, W. L. and CLARINGBULL, G. F. 1965. *Crystal structures of minerals*. London, G. Bell & Sons.

EVANS, R. C. 1964. *An introduction to crystal chemistry*. Cambridge University Press.

GOLDSCHMIDT, V. M. (ed. A. Muir). 1958. *Geochemistry*. Oxford, Clarendon Press. (Chapter 6)

SEEL, F. (trans. N. N. Greenwood & H. Stadler). 1963. *Atomic structure and chemical bonding*. London, Methuen & Co.

PAULING, L. 1960. *The nature of the chemical bond*. New York, Cornell Univ. Press.

2

CRYSTAL DESCRIPTION

CRYSTAL SYMMETRY

It is a simple experiment to grow a crystal of copper sulphate, by suspending a "seed crystal" of this substance by a thread in a saturated solution of the salt. In the course of a week or so, with slow evaporation of the solution, the seed crystal will grow by the addition of copper and sulphate ions from the solution. These ions are not added to the surface at random, but in a regular way governed by the principles of structural cohesion given in the last chapter. Growing in this way, with the ions added in a regular geometrical pattern, the crystal takes on a regular geometrical shape, bounded by plane faces. Some of the faces are bigger than others, because the material is added at different rates in different directions, and this confers what is called a characteristic *habit* on the crystal. Different substances crystallize with different habits, such as the prismatic, the tabular or the equant habit (Fig. 21).

Crystals of the same substance may crystallize with different habits under different conditions, for example calcite ($CaCO_3$) as dog-tooth spar, nail-head spar, or (rarely) as prisms (Fig. 22).

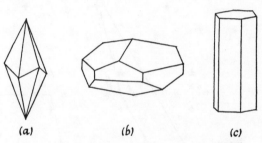

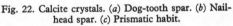

Fig. 22. Calcite crystals. (*a*) Dog-tooth spar. (*b*) Nail-head spar. (*c*) Prismatic habit.

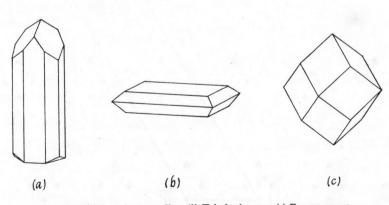

Fig. 21. (*a*) Prismatic tourmaline. (*b*) Tabular barytes. (*c*) Equant garnet.

THE UNIT CELL

The bounding planes, or faces, of a crystal bear a definite relation to the structural pattern in which the atoms are linked. Like any pattern, this internal structural pattern consists of units, each of which, in the case of a crystal, is a group of linked atoms in a fixed spatial relationship to one another, repeated over and over again to build the crystal. The smallest unit which, when repeated in three dimensions, will build the crystal, is called the *unit cell* of the structure.

The idea of the unit cell is most easily illustrated from a two-dimensional pattern, such as a wallpaper for example, and in Fig. 23 an example is given in which several possible unit cells are outlined.

SPACE LATTICES

The corners of the cells have the property that each is a point which has identical surroundings in the same orientation. Naturally there is a choice of corner points (e.g. abcd; efgh; kjlm, etc. in Fig. 23) but to fulfil the requirements of a unit cell they must have identical surroundings in the same orientation. Such an array of points is called a *lattice* and in the three-dimensional case a *space-lattice*.

If we wish to define the structural pattern of a crystal we may do so by stating, first of all, the dimensions (lengths of edges) and the angles between the edges of a selected unit cell, which is usually the smallest one that can be chosen, consistent with displaying the full symmetry. These edge-lengths

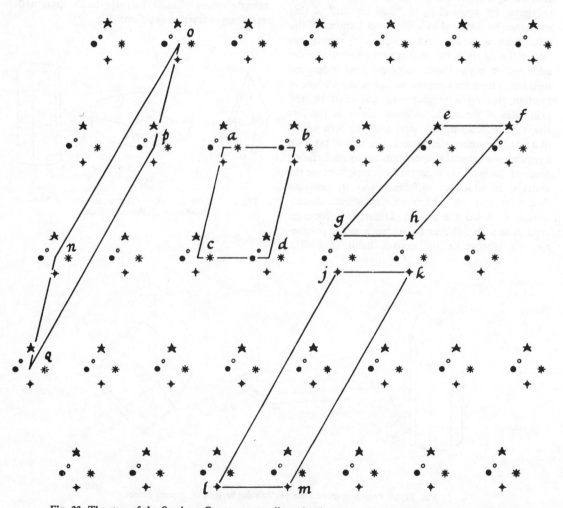

Fig. 23. The stars of the Southern Cross as a two-dimensional pattern, with possible unit cells outlined.

and angles are called the parameters of the unit cell, or *lattice parameters*. Secondly, we must state the *contents* of the unit cell in terms of atoms or atomic groups of the chemical elements making up the substance, and locate the atoms by co-ordinates along the cell edges.

For example, we can specify the pattern of Fig. 23 as follows (Fig. 24):

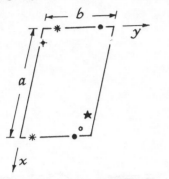

Fig. 24. One of the unit cells of Fig. 23.

$$x \wedge y = 76°; \quad a = 30 \text{ mm}, b = 20 \text{ mm}$$

with the stars of the Southern Cross as under

	co-ordinates in mm	co-ordinates in terms of unit lengths a and b
α-Crucis	24, 18	0·8, 0·9
β	0, 4	0·0, 0·2
γ	4·5, 0·8	0·15, 0·04
δ	0, 16	0·0, 0·8
ε	28·5, 16·8	0·95, 0·84

Now clearly, considering the whole field of crystalline substances, the edge-lengths and angles between them are infinitely variable, and the contents of atoms may be regarded as infinitely variable, too. But it turns out, by experiment with different arrays of points, that the number of basically different types of lattice is limited. Two-dimensional lattices may be based on a square pattern, on rectangles, on parallelograms, or on a hexagonal arrangement, which can be alternatively described in terms of a 60° rhombic cell (Fig. 25).*

In the case of three-dimensional space-lattices, there are only 14 basic types, as was established by the French crystallographer Auguste Bravais. One

* Notice that uncentred hexagons do not form a lattice, as defined above, because the surroundings of every point are not the same.

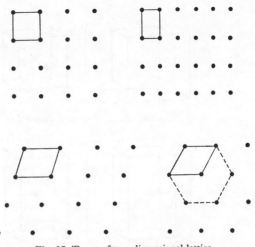

Fig. 25. Types of two-dimensional lattice.

unit cell of each of these 14 *Bravais lattices* is illustrated in Fig. 26. Each unit cell is a brick of a particular shape which can be stacked with others of its kind (and size) to form an indefinitely extended pile without any gaps between the bricks. In any particular mineral the bricks (the unit cells) may be considered identical (though they do show limited variation in the way described on p. 148). Different mineral species have unit cells of different types, sizes and (where the angle between edges is not fixed at 90° or 120°) different angles as well.

The Bravais lattices fall into three types (Fig. 26):

(*i*) Primitive, or P lattices in which the unit cell has a lattice point at each corner only. There are seven P lattices.

(*ii*) Body-centred or I lattices (from the German word innencentrierte) in which there is also a lattice point at the centre of the cell.

(*iii*) Face-centred lattices with points at the centres of all the faces (F lattices), or at the centres of one pair of faces (denoted A, B, or C lattices in different cases).

The fact that there are only 14 lattice types may be confirmed by trial; to demonstrate some of the trials involved, Fig. 27 shows the possibilities in the Tetragonal system that prove either to be the same as lattices already included in the 14, or turn out not to produce a true lattice array conforming with our definition.

In the description of a crystal structure mentioned above, the lattice is defined as one of these 14 types, with dimensions, and angles where necessary, added.

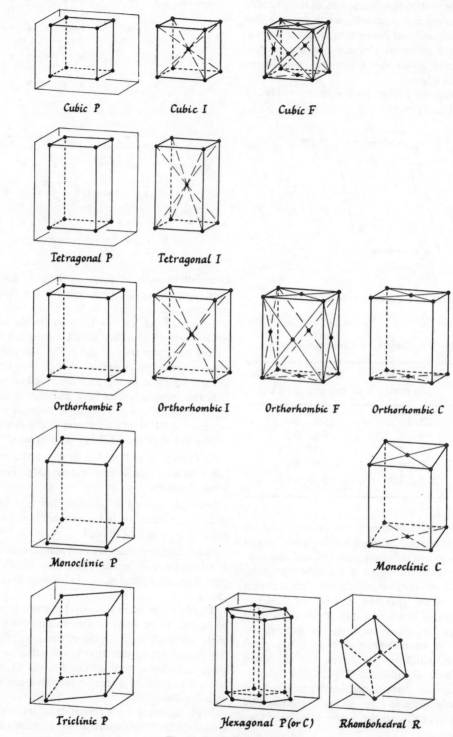

Cubic P   Cubic I   Cubic F

Tetragonal P   Tetragonal I

Orthorhombic P   Orthorhombic I   Orthorhombic F   Orthorhombic C

Monoclinic P   Monoclinic C

Triclinic P   Hexagonal P (or C)   Rhombohedral R

Fig. 26. The 14 Bravais lattices.

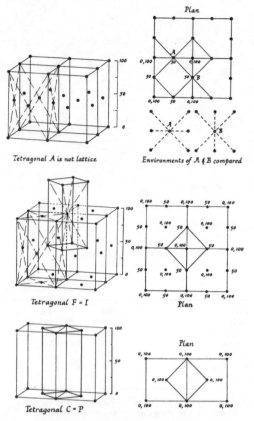

Plan

A
0,100 0,100
50 50 B

Environments of A & B compared

Tetragonal A is not lattice

Tetragonal F = I

Plan

Tetragonal C = P

Plan

Fig. 27. Trials of additional lattices that might be postulated in the tetragonal system.

The unit cell of a mineral is, of course, very small; it is on the atomic scale of size with edge-lengths of 0·3 to 1·5 nm. Let us now see how these tiny building units express themselves in the outward appearance of crystals we can touch and see.

THE EXTERNAL SYMMETRY ELEMENTS

When we handle a well-developed crystal we see that it possesses a certain symmetry, and to describe this we make use of the elements of symmetry defined below.

(*i*) *Rotation axis*. If an axis can be chosen through an object, rotation about which gives the observer exactly the same view more than once in a single rotation, the object has an axis of rotational symmetry. In crystals the axis may be two-fold (diad), three-fold (triad), four-fold (tetrad), or six-fold (hexad) depending on the number of times the same view is seen in one rotation (Fig. 28).

(*ii*) *Plane of symmetry* (*mirror plane*). If a plane can be chosen through the object such that each feature on one side of the plane has an exactly equivalent feature on the other, in mirror-image relationship (that is, the same distance from the plane as its opposite and capable of being joined to its opposite by a normal through the plane) then this is a plane of symmetry (Fig. 29).

(*iii*) *Centre of symmetry* (see also p. 90). When each feature on one side of an object can be joined by an imaginary line through the object's centre to an exactly similar feature the same distance from the centre on the other side, the object has a centre of symmetry. In a crystal model this may be tested by laying the model on the table and observing whether each solid angle (coign) resting on the table is represented by a like coign in reversed position on the opposite side of the upper surface (Fig. 30).

(*iv*) *Axis of rotary inversion or inversion axis*. This element is a little more subtle than the others. It is present when a rotation followed by an inversion

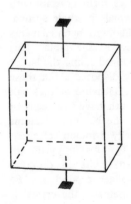

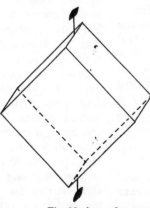

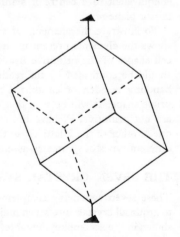

Fig. 28. Axes of symmetry.

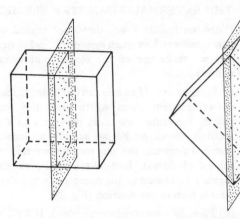

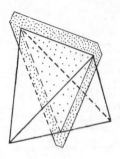

Fig. 29. Planes of symmetry.

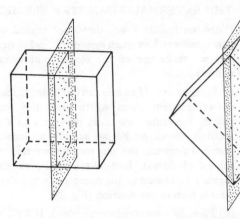

Fig. 30. A crystal (of the triclinic system) showing a centre of symmetry only (equivalent to $\bar{1}$).

Table 10. The major crystal systems.

System	Symmetry of the Holosymmetric Class				
Cubic	$3A^{IV}$	$4A^{III}$	$6A^{II}$	$9P$	C
Tetragonal	$1A^{IV}$	$4A^{II}$	$5P$	C	
Hexagonal	$1A^{VI}$	$6A^{II}$	$7P$	C	
Trigonal (Rhombohedral)	$1A^{III}$	$3A^{II}$	$3P$	C	
Orthorhombic	$3A^{II}$	$3P$	C		
Monoclinic	$1A^{II}$	$1P$	C		
Triclinic	C				

A indicates an axis of symmetry and the Roman superscript the degree of the rotational symmetry; P indicates a plane of symmetry; and C indicates a centre of symmetry.

across a centre brings one to a new position, the process being repeated until one has returned to the starting point (Fig. 31). Inversion axes may be one-fold, two-fold (inversion diad), triad, tetrad or hexad, written $\bar{1}, \bar{2}, \bar{3}, \bar{4}, \bar{6}$ respectively. The operation of an inversion axis may be described in terms of combinations of the other symmetry elements already mentioned, but the element is convenient to have in addition to the others, to build up the scheme shown in Fig. 68. Note particularly that $\bar{1}$ is equivalent to a centre of symmetry and $\bar{2}$ to a mirror plane m.

To illustrate the elements of symmetry, Fig. 32 shows the elements present in the seven basic unit cell shapes of the primitive Bravais lattices. These should be confirmed by the reader, if possible, by handling wooden or cardboard models of appropriate shape. It will be seen that those lattices that are not primitive, P, have the same symmetry as one or other of the P lattices, so that we have seven different types of symmetry-assemblage only.

THE SEVEN CRYSTAL SYSTEMS

These seven symmetry complements can be used to group all crystals into seven major crystal systems shown in the accompanying table.

All crystals can be placed in one of these seven systems simply by determining their external symmetry elements. In the great majority of cases the crystal will possess the full complement of symmetry shown in the table for its system. However, not all crystals in any system have the full ('normal' or holosymmetric) symmetry of that system. Some mineral species form crystals with less symmetry than the normal for their system, but with a greater complement of symmetry elements than the next system below in the table. This happens when the atoms arranged round the lattice points are not symmetrically disposed about those points. For example in Fig. 33 halite or rock salt (NaCl) and pyrite (FeS_2) both have unit cells of cubic F pattern; but while the Na and Cl atoms can be regarded as spherically symmetrical about their lattice points, the S_2 groups in pyrite are elongated (dumb-bell-shaped) units in different orientations along the cell-edges around any Fe atom, so that the total symmetry of the crystal is lowered. Pyrite belongs to class $m3$ ($= 2/m3$) of

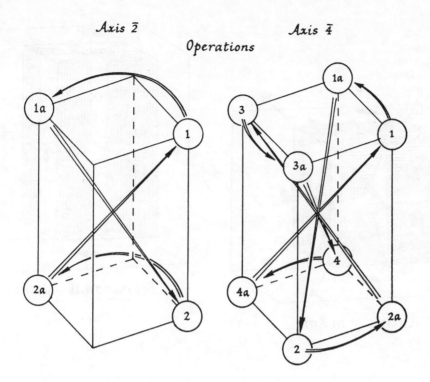

Axis $\bar{2}$ Operations   Axis $\bar{4}$

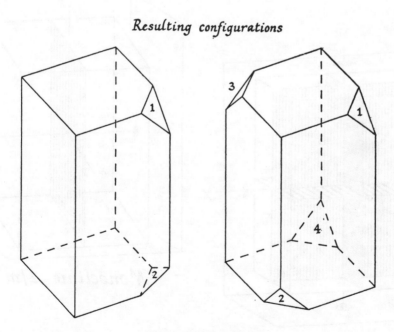

Resulting configurations

Fig. 31. Inversion axes.

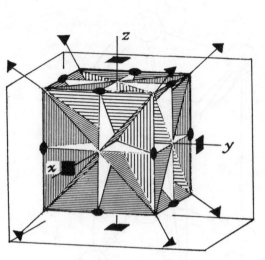

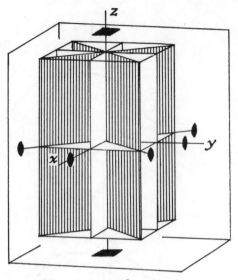

Cubic   m3m

Tetragonal 4/mmm

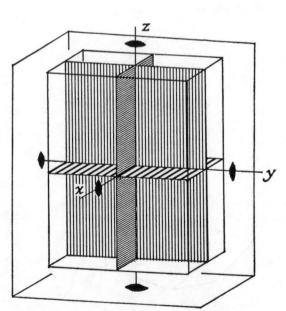

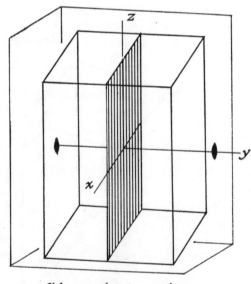

Orthorhombic mmm

Monoclinic 2/m

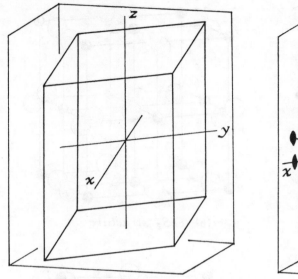

Triclinic $\bar{1}$

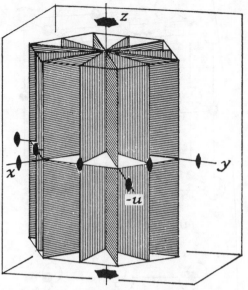

Hexagonal 6/mmm

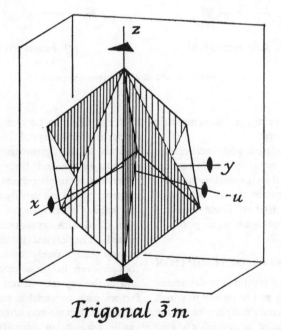

Trigonal $\bar{3}m$

Fig. 32. Elements of symmetry of the holosymmetric classes of the seven crystal systems. The symbols below the drawings are explained on p. 63.

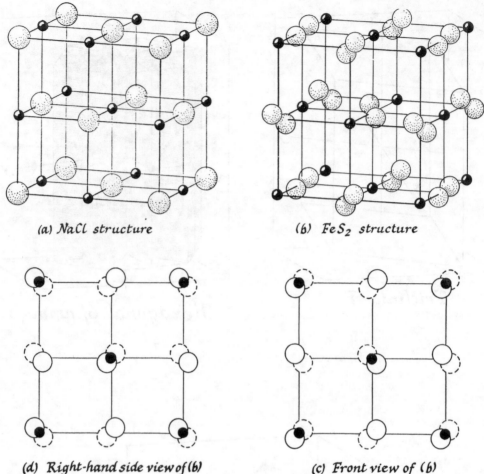

(a) *NaCl structure*   (b) *FeS₂ structure*

(d) *Right-hand side view of (b)*   (c) *Front view of (b)*

Fig. 33. Structures of halite and pyrite compared.

Fig. 68, while halite belongs to the holosymmetric class, *m3m* of the Cubic system.

When this factor is considered, additional classes must be made under each major system to accommodate crystals of lower than normal symmetry, giving, with the holosymmetric classes, a total of 32 sets of symmetry combinations. These are summarized in Fig. 68 with an explanation on p. 63.

PROCEDURE IN CRYSTAL DESCRIPTION

The first step in classifying a crystal is to determine what symmetry it possesses, and so decide to which crystal system it belongs. In the example taken at the beginning of this chapter, of a crystal growing freely suspended in a saturated solution, the sym-metry is likely to be clearly discernible by simple inspection of the crystal. In rocks and ores, however, crystals usually grow interlocked with others of various kinds, and their symmetrical growth is prevented by impingement on other crystals. Good natural crystals are usually the result of growth from the walls of a vein-fissure, or cavity, which has not been filled with crystalline material before growth ceased. The crystal is thus able to develop its natural faces freely and display its symmetry, though even in these circumstances the attached end of the crystal will not show good faces, whilst factors such as variable supply of solution lead to distortion. But we may often recognize the symmetry easily enough to identify well-known minerals simply by inspection.

THE LAW OF CONSTANCY OF INTERFACIAL ANGLE

To be certain of the symmetry in the case of un-known minerals, or to describe fully all the faces on known ones, the angles between the faces must be accurately measured. This is the procedure known as goniometry, which is fundamental in crystallo-graphy, and which has established, by repeated measurement of many examples, the *Law of Constancy of Interfacial Angle*. This Law states that, in any two crystals of the same substance the angles between corresponding faces, measured in a plane normal to the edge between them, is always the same. It is the angles between the faces, not the relative sizes of the faces which is fundamental in crystal measurement.*

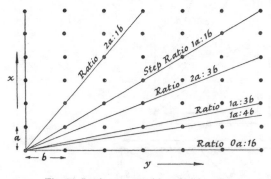

Fig. 34. Lattice array and interfacial angle.

The reason for this constancy of angle between corresponding faces of the same substance is readily appreciated from the considerations (*i*) that the unit cell shape and size are the same for all crystals of the compound, being determined by the chemical elements present and the angles between their bonds, and (*ii*) that each bounding face of a crystal is parallel to planes of equivalent atoms or atom groups in the lattice array. There is an infinite number of such planes, but in some the atoms will be sparse whilst in others they will be densely spaced. It is the planes with greater density of atom sites that tend to form the common crystal faces. This prin-ciple, known as the *Bravais Rule*, is illustrated in two dimensions in Fig. 34, which shows a lattice array with equivalent points connected by lines with various ratios of steps, of length equal to the unit

* One result of the Law of Constancy of Interfacial Angle is that we may identify many crystalline compounds solely by measurement of their interfacial angles. The Barker Index of Crystals provides the data to do this. With the rarity of suitable crystals for measurement, however, it is a method not often used for identifying natural minerals.

cell edge, along x and y. The simple ratios have more points per unit of length.

If, in Fig. 34, the lines of different step-ratio are taken as the traces of crystal faces, it is clear that the angle between any one of them and x or y depends upon the lengths of the cell sides a and b, so that, for a given set of lattice dimensions, the interfacial angles must be constant.

THE LAW OF RATIONAL RATIOS OF INTERCEPTS

To show how this works in three dimensions, Fig. 35 shows a crystal built to our own specifica-tion. The drawings 1, 2 and 3 show three views of a lattice of Orthorhombic P type, with the repeat distances a, b and c along the axes x, y and z in the ratio $a:b:c = 0.75:1:1.5$.

On these lattice sections are drawn the traces of planes with various simple gradients, or step-ratios, in the units a, b and c along the co-ordinates, e.g. 1 in 1, 1 in 2, etc. The ones chosen all have a relatively high density of lattice points along their length. Drawings 1A, 2A, 3A are enlargements of one unit cell of the lattice with the traces of the same planes marked. Two things are at once clear.

(*i*) Once the *ratio* of the repeat distances a, b and c is established (whatever the absolute values of their lengths) the *slopes* of the various planes are fixed. They will be the same for all crystals of the same substance (which always has the same lattice), in accordance with the Law of Constancy of Inter-facial Angle. This introduces the idea of the *axial ratio*, the ratio of the repeat distances (here $0.75:1:1.5$), which is fixed for any particular substance.

(*ii*) In any given substance, the slopes of all possible faces will bear a simple relationship to one another, in terms of gradients, or step-ratios, along the axes x, y, z, which will always be in multiples of a fixed unit of length along each axis. This is the *Law of Rational Ratios of Intercepts*, which intro-duces the idea of *unit length* along each axis. This unit length is fundamentally the lattice repeat distance along the axis; but some multiple of this would serve equally well as unit length. Once we define unit lengths along the axes by nominating one face as a standard or reference plane (the para-metral plane mentioned on p. 51), we shall find that the slopes of all other faces of the crystal will bear a simple relation to the reference plane, and can be defined by intercepts on the axes that are simple multiples or sub-multiples of the unit lengths.

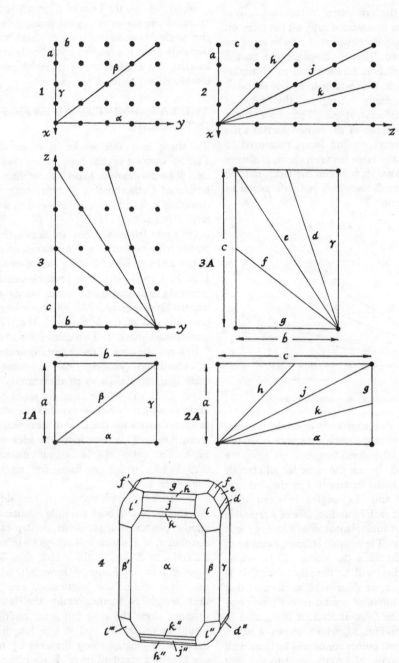

Fig. 35. Slopes of faces related to lattice sections.

Drawing 4 in Fig. 35 shows a crystal bounded by the planes whose traces are shown on the lattice sections. It also has a set of faces l, l', l'', l''' which cut all three axes in a step-ratio $x:y:z = 1:1:1$. The student may like to draw a lattice net, from the data given above, on a diagonal plane of the unit cell that will contain z and draw the trace of this face on it.

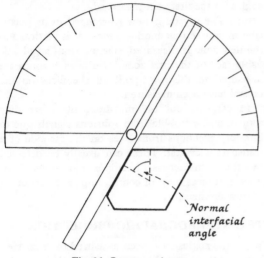

Fig. 36. Contact goniometer.

In Fig. 35 we have built up a crystal on a lattice of our own choice, and we can quote angles between its bounding faces, based on the step-ratios in a lattice of known symmetry and axial ratio $a:b:c$. The list is given in Table 11, using the *normal interfacial angle*, shown in Fig. 36, which is the one always recorded in crystal measurement.

What the morphological crystallographer does is the reverse of this. He measures the angles between the faces present on a crystal, determines the symmetry, and works out the axial ratio. When he has done this, and defined the slopes of the faces

Table 11. Interfacial angles of the crystal of Fig. 35.

$\alpha \wedge \beta$	36°52'	the angle whose tangent is $a/b = 0.75/1$ written $\tan^{-1} 0.75$
$\beta \wedge \gamma$	53°07'	$\tan^{-1} b/a = 1/0.75 = 1.33$
$\alpha \wedge \gamma$	90°00'	
$\alpha \wedge k$	14°02'	$\tan^{-1} a/2c = 0.75/3 = 0.25$
$\alpha \wedge j$	26°34'	$\tan^{-1} a/c = 0.75/1.5 = 0.5$
$a \wedge h$	45°00'	$\tan^{-1} 2a/c = 1.5/1.5 = 1.0$
$\alpha \wedge g$	90°00'	
$\gamma \wedge d$	18°16'	$\tan^{-1} b/2c = 1/3 = 0.33$
$\gamma \wedge e$	33°26'	$\tan^{-1} b/c = 1/1.5 = 0.66$
$\gamma \wedge f$	53°07'	$\tan^{-1} 2b/c = 2/1.5 = 1.33$
$\gamma \wedge g$	90°00'	
$\beta \wedge l$	21°48'	$\tan^{-1}\left(\dfrac{a}{\sqrt{(a^2+b^2)}} \cdot \dfrac{1}{c}\right) = 0.5625/1.5 = 0.375$
$\beta \wedge g$	90°00'	

actually present on his crystal, with a note on their relative sizes, imperfections or other special features, his description of the crystal morphology is complete.

We shall now explain how this is done.

MEASUREMENT OF INTERFACIAL ANGLES: GONIOMETRY

To determine the symmetry, interfacial angles are measured either with a *contact goniometer* (Fig. 36) or a *one-circle reflecting goniometer* like that of Fig. 37. More elaborate two-circle goniometers are also available, but they will not be described here.

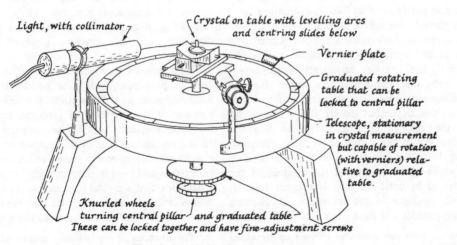

Fig. 37. One-circle reflecting goniometer.

The use of the contact goniometer will be obvious from the figure. The only precaution to be emphasized is that the plane in which the measurement is made should be normal to the line of intersection of the pair of faces being measured. A few moments experiment on a wooden model will show the reason for this.

USE OF THE ONE-CIRCLE REFLECTING GONIOMETER

The crystal to be measured should preferably be small (say 5 mm across) and have smooth perfect faces.* It is first sketched and its faces numbered on the sketch and identified by their shape and any cracks or marks on them. Then proceed as follows:

(*i*) Set the crystal in plasticine on the table surmounting the central pillar of the instrument, with some prominent set of parallel interfacial edges vertical.

(*ii*) Place the telescope at a position about 90° from the lamp collimator.

(*iii*) Unclamp the central pillar and turn it (and the crystal) until the reflection of the signal from the collimator by one of the crystal faces is seen in the telescope. It is convenient if the chosen face is roughly parallel to one of the two arcs that serve to tilt the crystal.

(*iv*) By means of the arcs, centre the signal on the crosswires of the telescope.

(*v*) Turn the crystal until the reflection from a second face is seen in the telescope, and centre this on the cross wires. The first signal may now require a small readjustment.

(*vi*) All the faces whose intersecting edges are members of the parallel set that was initially placed vertical should now reflect signals to the centre of the telescope as the crystal is rotated. Clamp the central pillar to the graduated circle, centre the signal from the first face accurately on the crosswires, using the fine-adjustment screw for the final adjustment, read the graduated circle at two points on its circumference by the verniers and record the angles. Turning the pillar and the graduated circle (to which it is now clamped) together, bring the signal from each face in turn on to the cross-wires and record the reading of the circle for each, until the first face again reflects. Then all the faces in the set whose intersecting edges are parallel will have been measured. Such a

---

* Imperfect faces may sometimes be used by cementing chips of microscope cover-glass to them.

set of faces is known as a *zone* of the crystal, and the common direction of edge is the *zone-axis*. In Fig. 35 (4) the following groups of faces forms parts of zones: β', α, β, γ with zone-axis parallel to z-axis; g, h, j, k, α, k'', h'', j'' with zone-axis parallel to y-axis; g, l', β', l'''; e, l, α, l'''; l', α, l''.

The differences between the averages of each pair of vernier readings give the normal interfacial angles for the first zone measured.

(*vii*) The crystal is now rearranged in its plasticine mount so that another zone-axis is vertical and the new zone is measured, starting from an identifiable face of the first zone.* Successive zones are measured in this way until all the faces on the crystal have been measured.

In carrying out the measurements there may sometimes be trouble from spurious signals due to interior reflections from faces on the far side of a transparent crystal. These can usually be detected by their diffuseness. Multiple reflections from imperfect faces may, of course, impair the accuracy of measurement.

THE STEREOGRAPHIC PROJECTION

When the angles have been measured the next step is to plot them on a diagram to exhibit their angular relationships and the symmetry of the crystal. For this purpose we use the stereographic projection, the most important property of which, for our purpose, is that on it angular relationships are preserved—in a word, it is *angle-true*.

It is worth noting that, because of this property, besides its use in crystallography the stereographic projection is of great use in many other geological problems involving the angular relationships of planes and directions, e.g. those of dip and strike of bedding, plunge of fold-axes or linear structures, directions of mine tunnels, inclined drillholes, and so on.

The principles on which this projection is drawn, and its properties, will now be described. The following section, as far as p. 50, is a self-contained account of this method of plotting, and can be tackled as a separate topic by the student if he wishes. In the succeeding part of the chapter on crystallography, however, stereograms will constantly be referred to and used in illustrations.

Imagine that a crystal is placed at the centre of a hollow sphere and that, from the centre of the sphere, lines are drawn normal to the crystal faces

---

* The advantage of the two-circle goniometer lies in the avoidance of this re-setting of the crystal for each new zone.

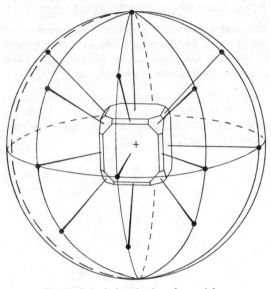

Fig. 38. Spherical projection of crystal faces.

To convert the spherical projection to a two-dimensional diagram, imagine a horizontal plane through the centre of the sphere and let all the poles in the upper hemisphere be joined by straight lines to the South Pole of the sphere. The points where these lines cut the central horizontal plane are the *poles of the faces in stereographic projection* (Fig. 39). Poles in the lower hemisphere, which would fall outside the equatorial circle if projected from the South Pole (e.g. D′ in Fig. 40) are joined to the North Pole and plot, as before, at the intersection of this join with the equatorial plane. Upper and lower face-poles on the spherical projection are distinguished on the stereographic projection by

(extended if necessary) and produced to intersect the sphere (Fig. 38). The points where the lines meet the sphere are called the *poles* of the faces in spherical projection. The angular relationships of the faces are reproduced in the latitude and longitude of these poles on the sphere, and the symmetry of their arrangement will be shown accurately in spite of any distortion in the original crystal.

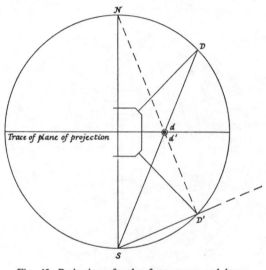

Fig. 40. Projection of poles from upper and lower hemispheres.

different symbols, a dot for an upper hemisphere face-pole and a circle for a lower hemisphere pole.

To locate any face-pole on the stereographic projection, or *stereogram*, we require two co-ordinates similar to the latitude and longitude of places on the earth's surface. An upper horizontal face of the crystal, as placed at the centre of the sphere, will plot at the centre of the stereogram while vertical faces will plot around its circumference, which is called the *primitive circle* (or simply *the primitive*).

By plotting one of the measured zones around the primitive circle we establish a basis for plotting all the other faces. In the crystal of Fig. 35 (4) the zone α, β, γ, ..., γ′ with its prominent direction of edge defining the long axis of the crystal is an obvious choice for this role. The poles of all the

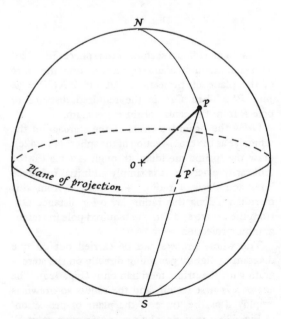

Fig. 39. The pole of a face in stereographic projection.

faces of one zone will lie on a great circle of the sphere.* Hence by locating the poles of faces common to two zones first, we can define the zonal great circles and plot the other poles along them.

The stereographic projection has the following valuable properties for mineralogical work:

(*i*) it is angle-true (as has been noted above),

(*ii*) great circles and small circles of the sphere both plot as circles, or arcs of circles, on the projection. They can therefore be drawn exactly with compasses.

In the next section the geometry of the projection and the basic constructions that it employs are considered. In practical use, however, plotting is usually done with the aid of a *Wulff net* (Fig. 41)

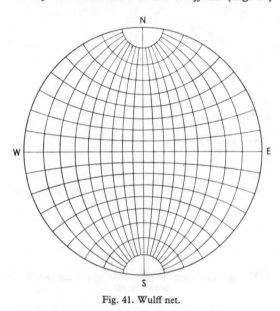

Fig. 41. Wulff net.

which is a set of arcs of great circles (analogous to longitude lines) and a set of small circles about a north and south pole of the primitive circle (analogous to latitude lines). These allow the location of face-poles by co-ordinates and the drawing of great circles passing through two or more points. Plotting is done on a sheet of tracing paper laid over the printed net and pivoted about a pin at the centre of the net. An example of the use of the net is given on p. 48.

*A *great circle* is the circle formed by the intersection of a plane passing through the centre of the sphere with the sphere's surface. The meridians of longitude on the globe of the earth are great circles, while the parallels of latitude, except for the equator, are *small circles* whose planes do not pass through the centre.

STEREOGRAPHIC CONSTRUCTIONS

It is useful, however, to be able to plot without the aid of a net, and learning to do so gives a greater insight into the properties of the projection. The following constructions show how this is done.

(*i*) To LOCATE A POLE AT AN ANGLE RHO (ρ) FROM THE NORTH POLE OF THE SPHERE AND (+) PHI (ϕ) FROM THE "PRIME MERIDIAN" OF POLE A ON THE PRIMITIVE CIRCLE (FIG. 42a).

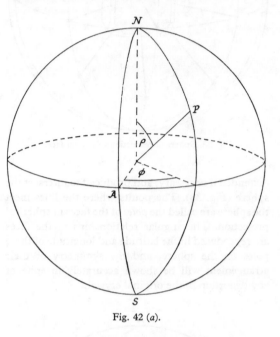

Fig. 42 (*a*).

Draw a meridional section of the sphere (Fig. 42b) with a horizontal diameter representing the trace of the plane of projection. Set off $\angle NOP = \rho$. Join PS. Then OP' is the required distance of pole P from the centre of the stereogram.

Draw the primitive circle of the stereogram the same size as the cross-section of the sphere (Fig. 42c), draw the 'prime meridian' through A (this can be arbitrarily placed as it is simply an initial reference line) and draw a radius $+ \phi$ degrees from this meridian. Along this radius lay off a distance OP' from the centre and P is the required pole in stereographic projection.

The whole process can be carried out on one diagram by lightly pencilling directly on the stereogram with the prime meridian on it (Fig. 42c). The angle ϕ is first set off, and the radius so drawn is employed as the 'trace of the plane of projection' of Fig. 42b, the angle ρ being set off from a point N'

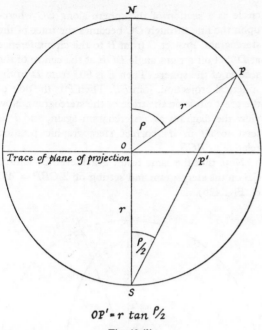

$$OP' = r \tan \rho/2$$

Fig. 42 (b).

normal to the radius and the join made to S' diametrically opposite, whence P' is located and the construction lines can be erased (dashed lines of Fig. 42c).

From Fig. 42b it can be seen that if the radius,

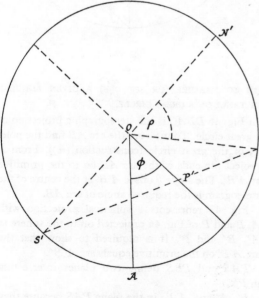

Fig. 42 (c).

r, of the stereogram is taken as unity, the length $OP' = \frac{1}{2} \tan \rho$, which is another way of locating P.

(ii) TO DRAW A SMALL CIRCLE ABOUT ANY POLE

Draw a section of the sphere through the pole P' about which the circle is to be drawn (Fig. 43). Mark off the angular radius of the small circle on either side of the pole (points Q' and R'). Join these two points to the south pole of the sphere. Bisect QR, the intercept between these joins on the trace of the projection plane, to find C the centre of the required circle, of radius CQ, on the stereogram. Note that the circle on the sphere is a circle on the stereogram; but its centre does not coincide with the stereographic location of pole P' at P.

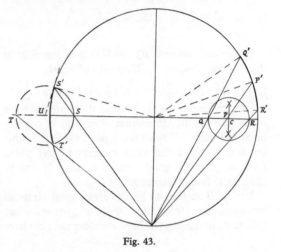

Fig. 43.

If the centre of the required small circle is on the primitive the same construction can be used, as shown in Fig. 43, but a tangent to the section of the sphere at one of the circumferential points of the small circle (e.g. S') drawn to intersect the trace of the plane of projection gives the required centre, U, more directly.

(iii) TO FIND THE CENTRE OF A GREAT CIRCLE AT A GIVEN ANGULAR DISTANCE FROM THE POLE OF THE SPHERICAL PROJECTION.

Draw a cross-section of the sphere (Fig. 44) and locate $A'B'$, the trace of a great circle ρ degrees from the north pole. Join $A'S$ by a line which cuts the trace of the stereogram at A. Join SB' and produce the join to cut the trace of the stereogram at B. Then the intercept AB represents a diameter of the great circle in projection, and its mid-point C is the required centre.

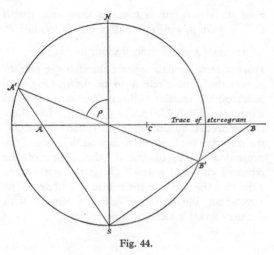

Fig. 44.

With constructions (*ii*) and (*iii*) it is possible to draw a stereographic net like that of Fig. 41.

(*iv*) TO FIND THE POLE OF A GIVEN GREAT CIRCLE.

A great circle may be regarded as the representation of a plane on the stereogram. The pole of a great circle is a point 90° from every point on its circumference. It bears the same relation to the plane represented by the great circle as a face-pole does to the crystal face it represents (*cf*. Fig. 38).

Let *ACB* in Fig. 45 be the given great circle in stereographic projection. Draw a perpendicular, *OC*, to its diameter *AB*. Now consider the primitive

circle as a section of the sphere along *OC* whereupon the line through *OC* becomes the trace of the stereogram. Project *C* from *B* to the circumference at *D*. Set off a right angle *DOE* at the centre of the section of the sphere. Then *E* is 90° from *D* in the spherical projection. Join *EB*. Then *P* is the pole of the great circle on the trace of the stereogram. Now view the diagram as a stereogram again, and *P* is seen to be in its correct stereographic position relative to *ACB*.

(Note that the same result is achieved by joining *CB* on the stereogram and setting off ∠*CBP* = 45° *cf*. Fig. 42b).

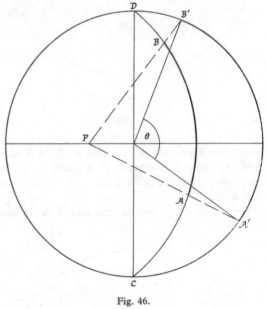

Fig. 46.

(*v*) TO MEASURE (OR SET OFF) A GIVEN ANGULAR DISTANCE ON A GREAT CIRCLE.

In Fig. 46 *DBAC* is the stereographic projection of a great circle. To measure the arc *AB* find the pole, *P*, of the great circle (construction (*iv*)). From *P* project the ends of the arc *AB* on to the primitive at *A'B'*. The angle *θ* of arc *A'B'* at the centre of the stereogram is the required angle of arc *AB*.

Fig. 47 represents a spherical projection with *A*, *B* and *P* of Fig. 46 projected on to the sphere at *A″*, *B″* and *P'*. It is required to show that the arc *A'B'* on the primitive equals arc *A″B″*.

PAS and *PBS* define two planes intersecting in *PS*.

P', *A″* and *A'* lie in the plane *PAS* because they lie on extensions of straight lines lying in that plane.

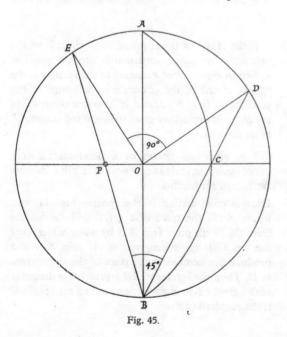

Fig. 45.

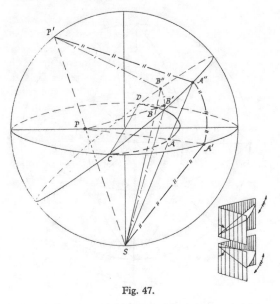

Fig. 47.

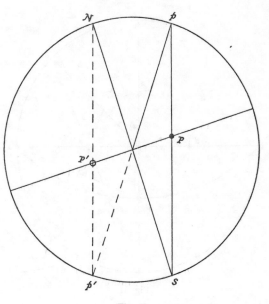

Fig. 49.

Similarly P', B'' and B' lie in plane PBS.

As S, the pole of the primitive is 90° from A' and B', so P', the pole of plane $CA''B''D$ is 90° from A'' and B''. Hence circles $CA'B'D$ and $CA''B''D$ are equally inclined to axis $P'S$, and the planes $P'A''A'S$ and $P'B''B'S$ intersecting in that axis cut off equal arcs $A'B'$ and $A''B''$ on the two circles. (This condition is displayed in the inset to Fig. 47).

(vi) TO FIND THE DIRECTION OF EDGE PRODUCED BY THE INTERSECTION OF TWO CRYSTAL FACES.

Let A and B in the stereogram of Fig. 48 be the poles of the two faces. On a diameter of the stereogram drawn through A set off A' 90° from A, and similarly B' 90° from B on a diameter through B. Construct the great circle $CA'D$ on diameter CD normal to AA' and great circle $EB'F$ on diameter EF normal to BB' (construction *(iv)* in reverse). These two great circles are stereographic representations of the planes of which A and B are the poles. The pole G in which these two great circles intersect gives the required direction of edge and OG is the stereographic projection of the line of intersection of the two planes (or faces).

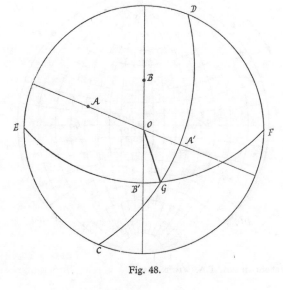

Fig. 48.

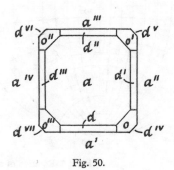

Fig. 50.

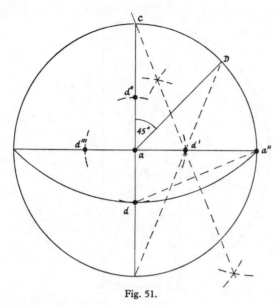

Fig. 51.

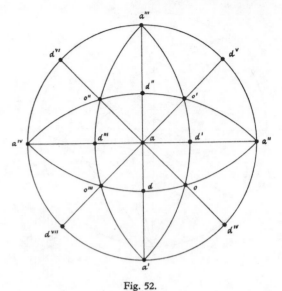

Fig. 52.

(vii) TO FIND THE OPPOSITE OF A POLE IN STEREO-GRAPHIC PROJECTION.

The opposite of a pole is a point 180° from it on the sphere.

In the stereogram Fig. 49 to find the opposite of pole P draw a diameter through P and another diameter NS normal thereto. Using the primitive as a section of the sphere, project P on to the sphere at p. Locate p' 180° from p at the opposite end of a diameter. Re-project p' back on to the temporary 'trace of the stereogram' by joining it to N thus

fixing P'. Regarding the diagram as a stereogram once more, P' is the stereographic opposite of P. P' is marked by an open circle because it is a pole in the lower hemisphere of the spherical projection, whereas P, marked by a dot, is an upper hemisphere pole.

PLOTTING A STEREOGRAM

Fig. 50 is a plan view of the crystal at the centre of the sphere in Fig. 38. The angle between faces of type a is 90°; between faces of types a and d the angle is 45° and between a and o 54° 44'.

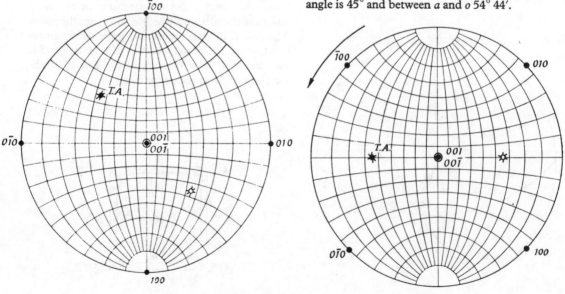

Fig. 53. Rotation of a stereogram about an axis. T.A. = twin axis.

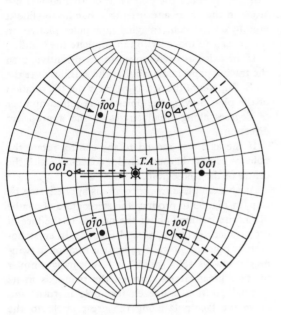

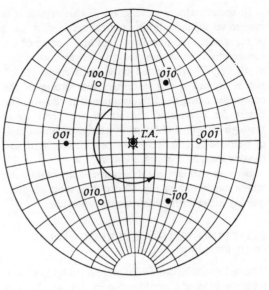

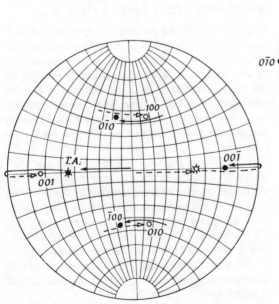

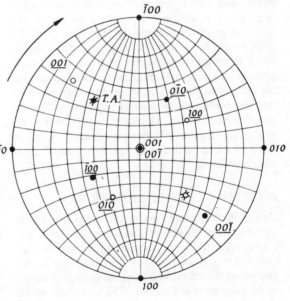

In constructing a stereogram of this crystal it is easy to set off faces in the vertical zone, a^i, d^{iv}, a^{ii}, d^v etc. around the primitive. In Fig. 51 some of the construction is shown for the other face-poles. d, d^i, d^{ii}, d^{iii} are located on a small circle of $45°$ radius around a at the centre, the setting of the compasses being fixed by locating d^i as shown. The centre for describing the arc of the great circle through a^{ii} and d is found by the intersection at C of perpendicular bisectors of two chords, $a^{ii}d$ and the one from a^{ii} through a and d^{iii}.

Because this is a highly symmetrical crystal, the whole stereogram of Fig. 52 could be built up quite quickly with relatively few construction lines.

PLOTTING WITH A STEREOGRAPHIC (WULFF) NET

The task could be done even more quickly, however, with the help of a net like that of Fig. 41. A sheet of tracing paper is laid over the net and pivots on a pin at its centre. The poles around the primitive are located from the peripheral graduations, and a is at the centre. Each vertical zone-circle such as $a^i aa^{iii}$ or $d^{iv}ad^{vi}$ is brought, in turn, over the N–S or E–W diameter of the net by pivoting the tracing paper, and the appropriate stereographic angles (e.g. $a \wedge d = 45°$; $a \wedge o = 54°\ 44'$) scaled off from the graduations along the diameter. Inclined great circles are drawn by bringing the face-poles along them into conjunction with an appropriate great circle between N and S of the net and tracing the arc to join them. The attempt to align poles in this way readily shows whether or not they lie on a common great circle. Moreover, two poles can be used in this fashion to define a great circle, whereas three would be needed to find its centre and draw it with compasses.

ROTATION OF THE STEREOGRAM

One of the important uses of the net is to rotate a whole set of poles (and/or planes), once plotted, into a new orientation, to compare their relationships with other sets of poles, as, for example, in the analysis of the effects of folding on geological strata, veins, ore-shoots and the like. The axis about which rotation is desired is brought over the E–W diameter of the net, by rotating the tracing paper. The axis is then moved along this diameter, either to the primitive or to the centre of the stereogram, the other points of the plot being moved through the same angle, along the small circles on which they lie,

in the same sense as the axis, having regard to the hemisphere in which they lie. (Think of the spherical projection, and imagine it as a ball with a cover or skin that slips about over the interior. The axis about which the plot is to be rotated is a pin stuck in the cover and used as a handle to slide it around on the ball. Imagine how poles plotted on the cover will move as you manipulate the handle.)

If the axis has been moved to the primitive, turn the tracing paper, now, so that the axis lies over the N–S diameter of the net. Now carry out the rotation of the plot, moving the poles the required number of degrees along small circles in the appropriate direction.

If the axis was moved to the centre, rotate the poles as required around circles concentric with the centre of the stereogram.

As an example, this procedure is applied, step by step, in Fig. 53 to produce the stereogram of a cube of fluorite twinned about the normal to the face of the octahedron (see p. 68 and Fig. 72). In crystallographic practice, however, since the rotation in twinning is always $180°$ (p. 68) it is normally simpler to trace great circles through the twin-axis in its original position and each face-pole in turn, and move the face-pole along the great circle to the twin-axis, and an equal number of degrees beyond it, to its twinned position. The effect is that of twirling the great circle, with its face-pole attached, through $180°$ about the twin-axis.

CRYSTALLOGRAPHIC REFERENCE AXES AND THE NAMING OF FACES

Measurement of the interfacial angles of a crystal, and their display on a stereogram, exhibits the symmetry of the crystal, which is a fundamental step in its description. With the symmetry entered on the stereogram, as is done for example in Fig. 54 for the crystal of Figs. 38 and 50, it becomes possible to see how another face-pole added to the diagram would have to be repeated in order to satisfy the requirements of the symmetry.

In Fig. 54 the pole of a face in the zone $a'od'o'a'''$ of Fig. 52 is added in the right lower quadrant, and the symmetry elements are inserted by conventional signs (square for a tetrad, triangle for a triad, ellipse for a diad, and heavy line for a mirror plane). By the operation of the triad through o the initial face-pole is repeated around o. The tetrads through a', a'' and a then ensure that it must be repeated in all quadrants of the stereogram, in both upper and lower hemispheres.

This concept of the symmetry *operating* on a given point to reproduce it a number of times in fulfilment of the 'commands' of the symmetry elements is a useful one. It is the basis on which the 32 classes of symmetry are built up in Fig. 68. When a single pole is entered on a stereogram of a crystal with a given symmetry complement, the symmetry elements—axes, mirror planes, and centre (if present)—*generate* further poles to build up a set of faces known collectively as a *form* (Fig. 54).

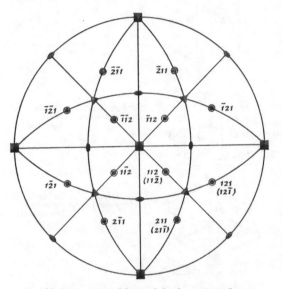

Fig. 54. Stereogram of faces of the form {211}. Lower faces are indexed only in the lower right-hand quadrant.

The next step in crystal description is to give designations to the various faces (grouped in these related sets, or forms) that may be present. In order to do this we establish a set of co-ordinates which will enable us to specify particular faces. These co-ordinates are called the crystallographic reference axes (or simply *crystallographic axes*) and are selected in accordance with agreed conventions. They are chosen to coincide with important symmetry axes and/or important directions of edge (zone axes) in the crystal: usually they are parallel with and have the same relative lengths as the edges of the unit cell.

It is important to be clear that symmetry axes are real entities, existing and recognisable in the crystal by anyone who knows what symmetry is, irrespective of systems of face nomenclature. Crystallographic axes, on the other hand, are a set of co-ordinates, selected on the basis of convenience,

for the purpose of identifying or specifying particular faces. In certain cases (hexagonal and trigonal systems) two different co-ordinate systems are available.

The usual crystallographic axes for the several systems are listed in Table 12. The sketches in the table show the positions in which a crystal, or crystal model, of each system is held for description, and the way the crystallographic reference axes of each are lettered, with the positive and negative directions of each. Note that in the monoclinic system $+x$ slopes downwards towards the observer, while in the hexagonal and trigonal systems the observer's line of sight bisects the angle between $+x$ and $-u$.

Two of the systems, the monoclinic and triclinic, are non-orthogonal—that is their axes are not all mutually at right angles or other fixed angle. Hence in describing them the inter-axial angle between x and z in the monoclinic system and all the inter-axial angles in the triclinic system must be stated. The sketches show the Greek letters by which these angles are named.

In the hexagonal and trigonal systems the angles between the horizontal axes are always $120°$, so they need not be specified.

Any face of the crystal, considered as an indefinitely extended plane, will cut one or more of the reference axes, and its angular position will be given by stating relative intercepts that it makes on these axes. We have seen (p. 39) in discussing the crystal lattice and the Law of Constancy of Interfacial Angle, that crystal faces are planes of rather high density of lattice points, and the Law of Rational Ratios of Intercepts expresses the fact that their angular relations to one another are determined by simple ratios of lengths along the unit cell edges.

THE PARAMETRAL PLANE

In the description of the external form of a crystal, when its symmetry is determined and the crystallographic reference axes chosen, a crystal face is selected that cuts all three reference axes (or all four in the hexagonal and trigonal systems). This face is then arbitrarily designated the *parametral plane* and is taken to define the unit of length, or simply *unit length* along each reference axis. The absolute lengths of these units (in millimetres, say) is irrelevant. All that matters are their relative lengths, which fix the slope of the crystal plane. This will be clear if it is remembered that a cube of say, fluorspar, is still a cube whether it be 5 mm or 5 cm along the edge, and the planes that

Table 12. Crystallographic (reference) axes for the seven crystal systems.

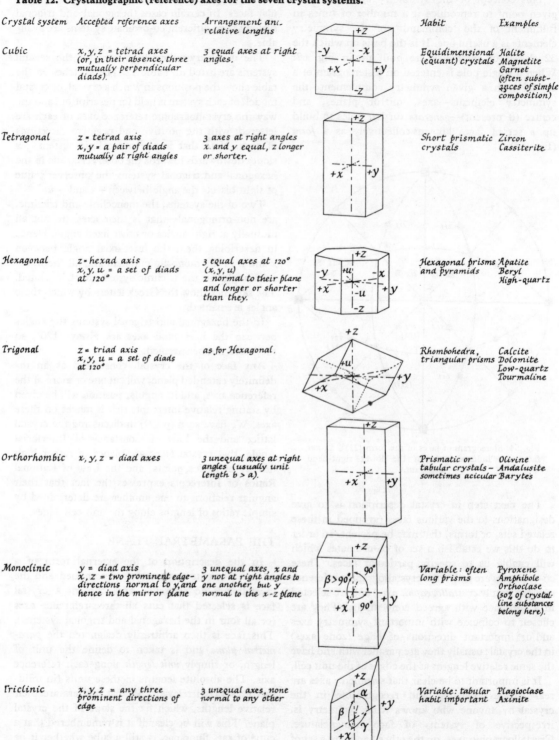

Crystal system	Accepted reference axes	Arrangement and relative lengths	Habit	Examples
Cubic	x, y, z = tetrad axes (or, in their absence, three mutually perpendicular diads).	3 equal axes at right angles.	Equidimensional (equant) crystals	Halite Magnetite Garnet (often substances of simple composition)
Tetragonal	z = tetrad axis x, y = a pair of diads mutually at right angles	3 axes at right angles. x and y equal, z longer or shorter.	Short prismatic crystals	Zircon Cassiterite
Hexagonal	z = hexad axis x, y, u = a set of diads at 120°	3 equal axes at 120° (x, y, u) z normal to their plane and longer or shorter than they.	Hexagonal prisms and pyramids	Apatite Beryl High-quartz
Trigonal	z = triad axis x, y, u = a set of diads at 120°	as for Hexagonal.	Rhombohedra, triangular prisms	Calcite Dolomite Low-quartz Tourmaline
Orthorhombic	x, y, z = diad axes	3 unequal axes at right angles (usually unit length $b > a$).	Prismatic or tabular crystals – sometimes acicular	Olivine Andalusite Barytes
Monoclinic	y = diad axis x, z = two prominent edge-directions normal to y, and hence in the mirror plane	3 unequal axes, x and y not at right angles to one another, but y normal to the x-z plane	Variable: often long prisms	Pyroxene Amphibole Orthoclase (50% of crystalline substances belong here).
Triclinic	x, y, z = any three prominent directions of edge	3 unequal axes, none normal to any other	Variable: tabular habit important	Plagioclase Axinite

bound it are still cube faces whether they be 2·5 mm or 2·5 cm from the origin of the crystallographic reference axes. In crystallography *parallel planes are the same*, and receive the same designation.

AXIAL RATIO

The choice of the parametral plane, then, defines the unit lengths of the reference axes and establishes the *axial ratio*, which is characteristic for each crystalline substance. The unit lengths in the directions x, y and z are called a, b and c respectively. We have seen (p. 39) that fundamentally the axial ratio $a:b:c$ expresses the relative edge-lengths of the unit cell. In the early days of crystallography this was as near as could be got to describing the fundamental constants of the crystal structure, and it could happen that the chosen parametral plane, which is simply a well-developed face cutting all three axes, might give a multiple of the relative edge-length of the unit cell as unit length along one axis. Nowadays, with the help of X-rays, we can measure the cell sides in absolute units, and so confirm the ratios of length determined by earlier crystallographers from the slopes of faces. Even if their chosen parametral plane should prove to

define a unit length that is a multiple of true cell edge-length in one direction, it could still function perfectly well as a basis for describing the faces present on the crystal. This is because all the recorded intercepts of crystal faces upon the axes are relative and can easily be transformed, if required, to intercepts in terms of the true cell edges, when these have been determined by X-ray methods.

The calculation of the axial ratio from interfacial angles is dealt with on p. 58.

MILLER SYMBOLS FOR CRYSTAL FACES

Having decided on a parametral plane, and relative units of length along the reference axes, the slopes of all other faces present can be described with reference to these units.

The Miller* system of indices provides a simple notation for describing the slopes of other crystal

* W. H. Miller, born in Wales in 1801, was Professor of Mineralogy in the University of Cambridge from 1832 until his death in 1880. Of a number of notations proposed for specifying crystal planes (in particular those of Weiss, Naumann and later Goldschmidt) the Miller system has become universally accepted amongst crystallographers today. For the hexagonal system we use here a modification of Miller's axes, suggested by Bravais.

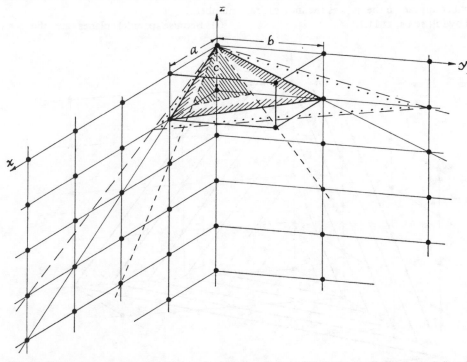

Fig. 55. The parametral plane (full line, partly shaded) and planes 231 (broken outline, fully shaded) and 324 (dots).

faces in terms of the unit lengths. The intercepts of a face, in terms of the unit lengths, *a*, *b* and *c* are first written down. In Fig. 55 for the parametral plane, which is outlined in full line, they will be

$$1a \quad 1b \quad 1c$$

and for the other planes shown

$$\tfrac{1}{2}a \quad \tfrac{1}{3}b \quad 1c$$

$$4/3a \quad 2b \quad 1c$$

Since the intercepts are always written in the order *a*, *b*, *c*, the letters themselves may be omitted. Then take the reciprocals of the intercepts, and multiply through to clear fractions, thus:

Intercepts			Reciprocals			Miller Indices
1	1	1	1/1	1/1	1/1	111
$\tfrac{1}{2}$	$\tfrac{1}{3}$	1	2/1	3/1	1/1	231
4/3	2	1	3/4	$\tfrac{1}{2}$	1/1	324

The three Miller indices together constitute the *Miller Symbol* of the face and define its relative slope.

When an intercept is on the negative end of an axis this is indicated by a 'bar' over the index figure, e.g. $3\bar{2}1$, said as: three, bar two, one. When a two-figure number appears in the indices the indices are separated by full stops, as 11.1.6.

Faces parallel to an axis cut that axis at infinity. Hence we have, for example, in Fig. 63

Intercepts	Reciprocals	Miller indices
1, ∞, ∞	1/1, 1/∞, 1/∞	100

Figure 56 shows part of the *family of planes* with intercepts $\tfrac{1}{2}a$, $\tfrac{1}{3}b$, 1*c* and Miller symbol 231. Planes of this slope have been drawn using successive lattice points as origin. (The figure illustrates, incidentally, the point that in crystallography parallel planes are the same and have the same symbol.) Note that the successive planes divide up the lengths *a*, *b* and *c* into a number of segments equal to the index figure for that axis. Another way of putting this is to say that if the intercepts are always expressed as fractions of unit length with the numerator unity, the denominators give the indices directly. This leads to a general statement:

If a family of planes divides the unit lengths *a*, *b* and *c* into *h*, *k* and *l* segments respectively, the intercepts are *a*/*h*, *b*/*k* and *c*/*l* and the indices are *hkl*.

The use of the symbol *hkl* to mean 'any plane', in a perfectly general sense, is common in crystallography.

There are three points to note about Miller index numbers:

(*i*) Because parallel planes are the same, the

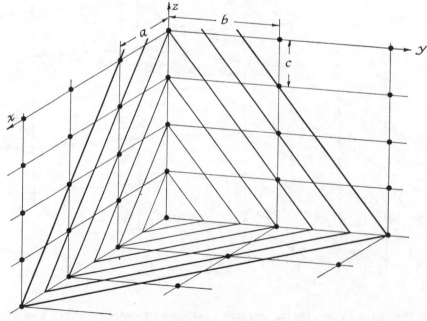

Fig. 56. The family of planes 231.

numbers in the symbol are always cancelled down to their lowest terms. In the description of external morphology the symbol 022, for example, will not occur, being equivalent to 011. In the study of the fine structure of crystals, however, 022 and like symbols have meaning, for, as study of Fig. 57

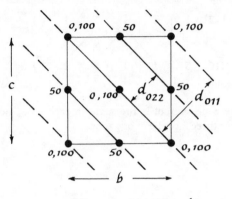

Plan on 100 of cube

Fig. 57. Difference in spacing of planes designated 011 and 022.

will show, there may be a *family of planes* that would be so designated, which, though parallel to 011 will have a *closer interplanar spacing, d,* than a family 011. If the lattice were not face-centred, family 022 would disappear. So a family characteristic that is lost when only interfacial angles are measured becomes apparent when we use X-rays to measure interplanar spacings.

(*ii*) Because of the introduction of reciprocals in writing Miller indices, a relatively large index figure denotes a relatively small intercept upon the axis to which it refers. The rule is: *the larger the number the smaller the distance.*

(*iii*) Notation. 110 denotes a set of planes of given slope (and spacing). (110) strictly denotes a single face, but the brackets are often omitted unless required for clarity. {110} denotes all the faces of a form (see p.57).

THE ZONE SYMBOL

Certain useful relationships obtain between the Miller symbols of faces and the symbols for *zone axes*. A zone axis (defined above, p. 42) may be specified by co-ordinates U, V, W, which are multipliers of the unit lengths on the crystallographic reference axes as shown in Fig. 58 where the heavy line is the zone axis parallel to the edge direction qr in which faces pqr and qrs meet. The

zone symbol $[UVW]$ is enclosed in square brackets to indicate that a *line* and not a plane is referred to. These relationships are beyond the scope of this book, but are dealt with in texts on crystallography (e.g. Phillips, F. C., 1946. *An introduction to crystallography.* London, Longmans, Green & Co., Ch. VIII).

Here we shall be content to mention two of them.

(*i*) The Addition Rule (a particular case of the Weiss Zone Law) states: the indices of two faces in the same zone always add up to the indices of a face lying between them, and bevelling the edge between them. This rule often permits the indexing of faces on a stereogram by inspection, when indices have been given to the main faces. It is

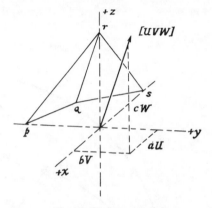

Fig. 58. The zone symbol.

particularly useful when the face to be indexed lies at the intersection of two zones. Examples of this relationship can be seen in Fig. 59.

(*ii*) To test whether a third face lies in a zone with two others, we derive the zone symbol of the two faces defining the zone as follows: write the indices of each face twice, those of the second below the first, and cross off the end numbers in each row thus:

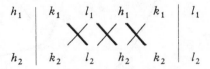

In each group linked by a cross, multiply together the numbers joined by a heavy line and subtract from this the product of the pair joined by the light line, working from left to right. This gives the zone symbol:

$$U = (k_1 l_2 - l_1 k_2)$$
$$V = (l_1 h_2 - h_1 l_2)$$
$$W = (h_1 k_2 - k_1 h_2)$$

Do the same with the indices of one of these two faces and the indices of the third face. If the third face is in a zone with the other two, the zone symbol should be the same as before.

For example, we shall show, in Fig. 70. p. 66, that face 51Ō1 is in a zone with 01Ī1 and 11Ž1. Note first that in the hexagonal and trigonal systems we omit the third index figure in setting down the indices—it is not needed to define the planes.

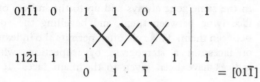

$$01\bar{1}1 \quad 0 \;\Big|\; 1 \quad 1 \quad 0 \quad 1 \;\Big|\; 1$$
$$\qquad\qquad \times \quad \times \quad \times$$
$$11\bar{2}1 \quad 1 \;\Big|\; 1 \quad 1 \quad 1 \quad 1 \;\Big|\; 1$$
$$\qquad\qquad\quad 0 \quad 1 \quad \bar{1} \qquad\qquad = [01\bar{1}]$$

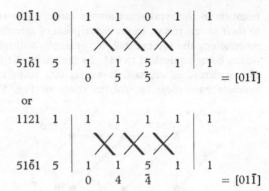

$$01\bar{1}1 \quad 0 \;\Big|\; 1 \quad 1 \quad 0 \quad 1 \;\Big|\; 1$$
$$\qquad\qquad \times \quad \times \quad \times$$
$$51\bar{6}1 \quad 5 \;\Big|\; 1 \quad 1 \quad 5 \quad 1 \;\Big|\; 1$$
$$\qquad\qquad\quad 0 \quad 5 \quad \bar{5} \qquad\qquad = [01\bar{1}]$$

or

$$1121 \quad 1 \;\Big|\; 1 \quad 1 \quad 1 \quad 1 \;\Big|\; 1$$
$$\qquad\qquad \times \quad \times \quad \times$$
$$51\bar{6}1 \quad 5 \;\Big|\; 1 \quad 1 \quad 5 \quad 1 \;\Big|\; 1$$
$$\qquad\qquad\quad 0 \quad 4 \quad \bar{4} \qquad\qquad = [01\bar{1}]$$

The zone symbol is the same in each case; hence these three faces lie in one zone.

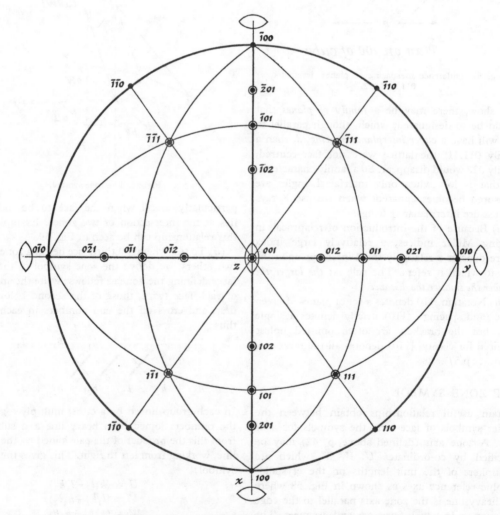

Fig. 59. Stereogram illustrating the Addition Rule for indices in the same zone.

CALCULATION OF AXIAL RATIOS

Returning to Fig. 35, p40, where a lattice was set up with selected cell dimensions, for demonstration purposes, we are now in a position to plot the poles to its principal faces on a stereogram and exhibit its symmetry. This is done in Fig. 59.

The diad axes must, by convention, be taken as crystallographic axes in the orthorhombic system. This gives us only one plane cutting x, y and z to serve as a parametral plane, and this is designated 111. The faces normal to the three diads become 100, 010, and 001.

All the faces in the zone between 001 and 100 will be normal to the xz plane and parallel to y, so they have 0 as the second index figure. Similarly, all poles in the yz plane have 0 in the first place, while those around the primitive, being parallel to z, have 0 in the last place. Planes represented by poles in the 001–111 zone will have an intercept ratio on x and y the same as that of the parametral plane, so they will have a symbol with the first two index numbers the same, e.g. 110, 221, etc.

In the zone 100, 111, $\bar{1}$11, $\bar{1}$00, the face truncating the edge between 111 and $\bar{1}$11 will have indices that are the sum of these two, that is, 022 = 011.

When the indexing has reached this stage, we can calculate the axial ratios of the crystal by plane trigonometry, making use of the interfacial angles between planes normal to the crystallographic axes (planes with indices 100, 010 or 001) and those that cut two axes and are parallel to the third (101, 011, 110). Where the required plane is not represented as a face on the crystal, its position can be obtained from the stereogram by drawing great circles representing two zones in which it must lie, and using their intersection to locate the pole required. The interfacial angle can then be measured (approximately) from the stereogram.

In Fig. 60 the trigonometrical ratios used for the several crystal systems (excluding the triclinic) are displayed. The application of the orthorhombic set to the stereogram of Fig. 59 leads us back, of course, to the axial ratio

$$a:b:c = 0.75:1:1.5$$

from which the interfacial angles were originally derived in Fig. 35 and the accompanying Table 11.

The calculation of axial ratios can be carried out also by the solution of spherical triangles, from the stereographic projection. This method will not be dealt with here: the reader is referred to F. C. Phillips, *An introduction to crystallography* (Long-mans, Green, 1946) for an account of the procedure. In particular, the proof of the formulae for triclinic axial ratios is best carried out by spherical trigonometry, because the proof by plane trigonometry is long and clumsy. The result is given below.

Formulae for Triclinic Axial Ratios:

$$a/b = \frac{\sin 100 \wedge 110 . \sin 010 \wedge 001}{\sin 010 \wedge 110 . \sin 100 \wedge 001}$$

$$c/b = \frac{\sin 011 \wedge 001 . \sin 100 \wedge 010}{\sin 011 \wedge 010 . \sin 100 \wedge 001}$$

Approximate values for the interaxial angles in the triclinic system may be found by measurement on the stereogram as in Fig. 61. For accurate values the spherical triangles must be solved.

CRYSTALLOGRAPHIC FORMS

In Fig. 62 all the faces of the *form* {111} of zircon are labelled. These planes 111, $\bar{1}$11, $\bar{1}\bar{1}$1, 1$\bar{1}$1, 11$\bar{1}$, $\bar{1}$1$\bar{1}$, $\bar{1}\bar{1}\bar{1}$, 1$\bar{1}\bar{1}$, are all generated by the operation of tetragonal symmetry on a single pole on the stereogram, and they all have the same slope relative to the crystallographic axes, though in different senses. Together they constitute the *form* {111} denoted by enclosing in curly brackets the Miller symbol for the face that cuts all the axes in their positive directions. The faces of the form {110} are also labelled on the drawing.

THE NAMING OF CRYSTAL FORMS

In the course of its development crystallography has accreted a superfluity of names. Some of the major systems have more than one name and the classes have three or four names used by different authors. Names have also been given to the forms developed in each class, and here, too, there is duplication. A few of the names are in common use, but most are little used because they convey no more information than the Miller symbol of the form. In this book they are consigned to Appendix I where they are available for reference if needed. But there are some important exceptions which follow.

UNIQUE, SPECIAL AND GENERAL FORMS

These are terms describing the relationship of faces to symmetry elements.

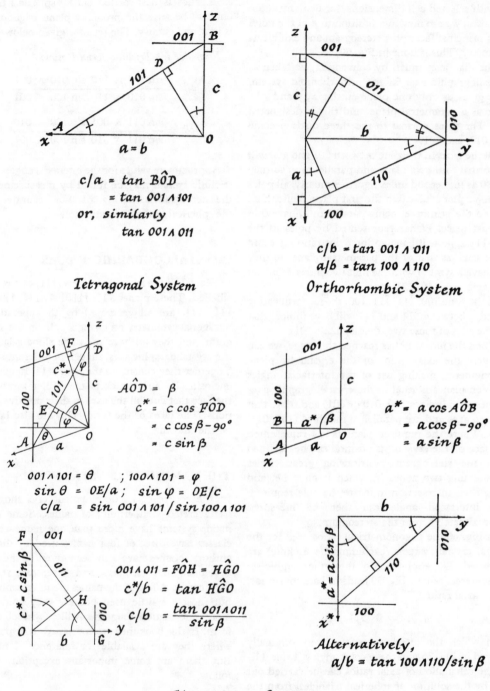

Tetragonal System

$c/a = tan\ B\hat{O}D$
$= tan\ 001 \wedge 101$
or, similarly
$tan\ 001 \wedge 011$

Orthorhombic System

$c/b = tan\ 001 \wedge 011$
$a/b = tan\ 100 \wedge 110$

Monoclinic System

$A\hat{O}D = \beta$
$c^* = c\ cos\ F\hat{O}D$
$= c\ cos\ \beta - 90°$
$= c\ sin\ \beta$

$001 \wedge 101 = \theta$; $100 \wedge 101 = \varphi$
$sin\ \theta = OE/a$; $sin\ \varphi = OE/c$
$c/a = sin\ 001 \wedge 101 / sin\ 100 \wedge 101$

$001 \wedge 011 = F\hat{O}H = H\hat{G}O$
$c^*/b = tan\ H\hat{G}O$
$c/b = \dfrac{tan\ 001 \wedge 011}{sin\ \beta}$

$a^* = a\ cos\ A\hat{O}B$
$= a\ cos\ \beta - 90°$
$= a\ sin\ \beta$

Alternatively,
$a/b = tan\ 100 \wedge 110 / sin\ \beta$

Fig. 60. Calculation of axial ratios; *continued opposite.*

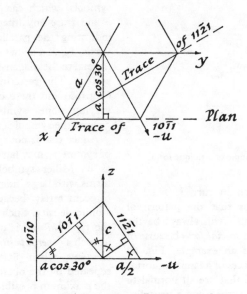

$$c/a = tan\ 0001 \wedge 10\bar{1}1 \times cos\ 30°$$
$$or\ c/a = tan\ 0001 \wedge 11\bar{2}1/2$$

Hexagonal System

Fig. 60. Calculation of axial ratios; *continued from previous page.*

A *unique form* is one developed from a plane normal to an axis of symmetry. Such a plane, being normal to a line with a particular direction in relation to the crystallographic reference axes, can itself have only one slope in relation to these axes. It has therefore, a unique Miller symbol (the same digits always, though of course the order in which they are written changes). An example of such a form is the cube (Fig. 63), normal to the tetrad axis in the cubic system, with the symbol 100 (or $\bar{1}00$, 010, etc.)

A face that is normal to a symmetry plane may have varying slopes relative to the crystallographic axes, but the restriction that it must always be

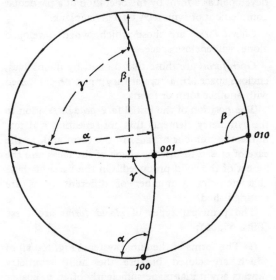

Fig. 61. Measurement of triclinic interaxial angles.

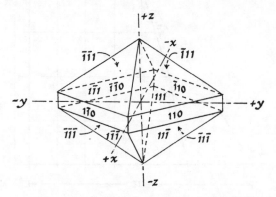

Fig. 62. Faces of the forms {111} and {110} in zircon.

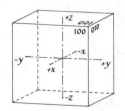

Fig. 63. The cube as an example of a unique form.

normal to the plane, which in turn is fixed in relation to the axes, means that the pattern of numbers in the Miller symbol will always be the same. Such a form is called a *special form* because of this special relationship. As an example, Fig. 64 shows in cross-section the traces of a family of forms in the orthorhombic system that are all normal to the symmetry plane containing the *x* and *z* crystallographic axes. Notice the common pattern of the

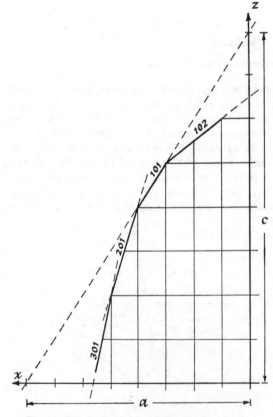

Fig. 64. The family of forms *h0l*, specially related to the *xz* plane.

symbols, which can be generalized as *h0l*, where *h* and *l* are any integers, but the 0 occurs in all, indicating the parallelism of all members to the *y* axis. Fig. 54 shows stereographically the form 211 normal to the diagonal symmetry planes of the cubic system. The resulting solid is the trapezohedron (Fig. 65) and there can be a family of these with varying slopes symbolized by 311, 322, etc. In general the symbol is *hll* (with $h > l$).

A face that is not specially related to any element of symmetry may have any combination of numbers as its Miller symbol, though, as we have seen, forms with large index numbers in the symbol tend to occur rarely, because they will be planes of low atomic density. Such a form will have a symbol of the type *hkl* and is called *the general form*. A pole in such a *general position* in relation to the symmetry elements, when plotted on a stereogram, will be repeated by the operation of the symmetry to give the maximum possible number of faces of any form in that crystal class. Study of the general form of an unknown crystal will disclose the full symmetry present; whereas special or unique forms may appear, with precisely the same aspect, in the holosymmetric class of a crystal system (Table 10) and in classes of lower symmetry in that system. Thus cubes of pyrite are common, though we have seen earlier in this chapter that pyrite has less than full cubic symmetry.

CLOSED AND OPEN FORMS

These terms describe the arrangement of the faces developed as a form by the operation of a particular complement of symmetry on a given plane.

Closed forms are those which, when developed alone, will enclose space.

Open forms are those that do not, by themselves, enclose space. In a crystal they must be combined with another form or forms.

The position of the initial face-pole in relation to the symmetry elements that are repeating it determines the number of faces in the form and, in the case of closed forms, the shape of the faces and the aspect of the solid produced. On the basis of these characteristics a number of different types are distinguished.

The principal types of *closed forms* are these (Figs. 65, 66):

(*i*) The forms of the cubic system (Fig. 65) all of which are closed because the high symmetry repeats any initial plane sufficiently often to ensure this.

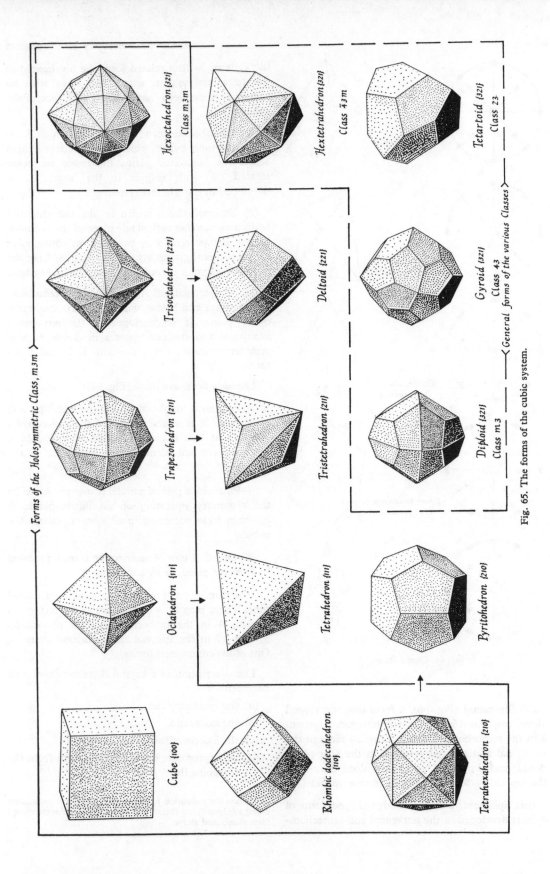

Fig. 65. The forms of the cubic system.

Cube {100}

Rhombic dodecahedron {110}

Tetrahexahedron {210}

Octahedron {111}

Tetrahedron {111}

Pyritohedron {210}

Trapezohedron {211}

Tristetrahedron {211}

Diploid {321}
Class m3

Trisoctahedron {221}

Deltoid {221}

Gyroid {321}
Class 43

Hexoctahedron {321}
Class m3m

Hextetrahedron {321}
Class 43m

Tetartoid {321}
Class 23

⟨ Forms of the Holosymmetric Class, m3m ⟩

⟨ General forms of the various Classes ⟩

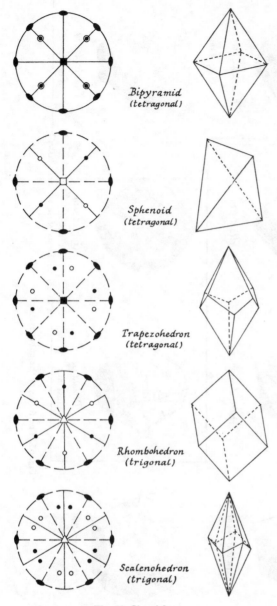

Bipyramid
(tetragonal)

Sphenoid
(tetragonal)

Trapezohedron
(tetragonal)

Rhombohedron
(trigonal)

Scalenohedron
(trigonal)

Fig. 66. Closed forms.

bic systems, with a related form the bisphenoid in the tetragonal system where two faces with an obtuse edge between them replace each face of the sphenoid.

(*iv*) Rhombohedron, a set of three pyramidal faces above and three below, which are not mirror images across the horizontal plane, the lower set being rotated 60° with respect to the upper. It is developed in the trigonal system.

(*v*) Scalenohedron, which is like the rhombohedron except that each single plane of the rhombohedron is represented by two with an obtuse edge between them, giving a total of 12 faces. Like the rhombohedron, it is confined to the trigonal system.

(*vi*) Trapezohedron. Again a set of pyramid-like planes above and below, without a horizontal symmetry plane. It is developed in the tetragonal, hexagonal and trigonal systems in classes lacking symmetry planes, but possessing full axial symmetry.

The *open forms* are these (Fig. 67):

(*i*) Pedion, a single plane not repeated by any symmetry. It is characteristic of the *hemimorphic* classes that lack a centre of symmetry and have different terminal planes at opposite ends of the crystal.

(*ii*) Pinacoid, a pair of parallel planes produced by the symmetry operating on an initial plane. A common form occurring in all systems except the cubic.

(*iii*) Dome. A pair of intersecting planes, repeated by a mirror plane, or by a diad axis.*

(*iv*) Prism. A set of planes forming an open-ended tube. There may be 4 or 8 planes in the tetragonal system, 4 in the orthorhombic and monoclinic, 3, 6 or 12 in the trigonal and hexagonal systems. One of the commonest forms.

The description of a crystal therefore involves a statement of

(*a*) the symmetry class,

(*b*) the axial ratio,

(*c*) the indices of the forms present,

(*d*) an appropriate name for each form from the foregoing list.

(*ii*) Bipyramid (Fig. 66), a form that, developed alone, has 8 or 16 faces in the tetragonal system, 8 in the orthorhombic and 6, 12 or 24 faces in the hexagonal and trigonal systems, in the shape of a double-ended pyramid with apices along the *z*-axis, the top and bottom halves being mirror images.

(*iii*) Sphenoid, a double-wedge-shaped form of 4 faces developed in the tetragonal and orthorhom-

---

* When the repetition is by a diad, the form is sometimes called a *sphenoid*; but this term is better reserved for the closed form mentioned above.

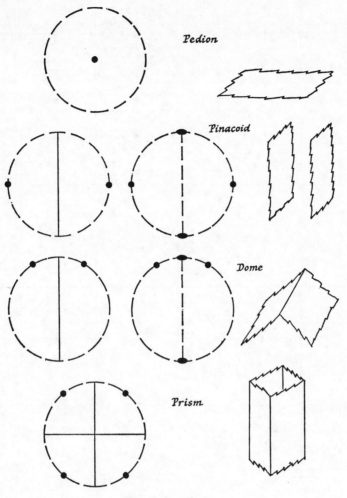

Fig. 67. Open forms.

THE 32 CRYSTAL CLASSES

In Fig. 68 the 32 possible classes of external crystal symmetry are developed by the process of building up the symmetry by successive additions of new elements.

The procedure is to draw a stereogram and place on it a face-pole in a *general* position (see. p. 60). Then elements of symmetry are added to the stereogram and the forms generated are recorded. In this process some symmetry elements generate others. In the International Notation (the Hermann-Mauguin notation), written at the bottom right-hand corner of each box, only the minimum symmetry required to define a class is stated, though other elements may be implied in this statement. In Fig. 68 the full complement of symmetry of each class is shown by symbols on the stereogram of the general form, in addition to the abbreviated notation written below. The symbols for symmetry axes are simple ones—ovals for diads, triangles for triads, squares for tetrads and hexagons for hexads, in solid black for rotation axes and unfilled for inversion axes. A full line on the stereograms denotes a symmetry plane: a broken line is used for the primitive circle or an axis when these are not coincident with mirror planes.

In the International Notation a rotation axis is indicated by a figure for the degree of rotation, 1, 2, 3, 4, or 6; an inversion axis by a number with a bar over it, $\bar{2}$, etc.; and a symmetry plane (mirror plane) by m. In Fig. 68 the column headings have these meanings:

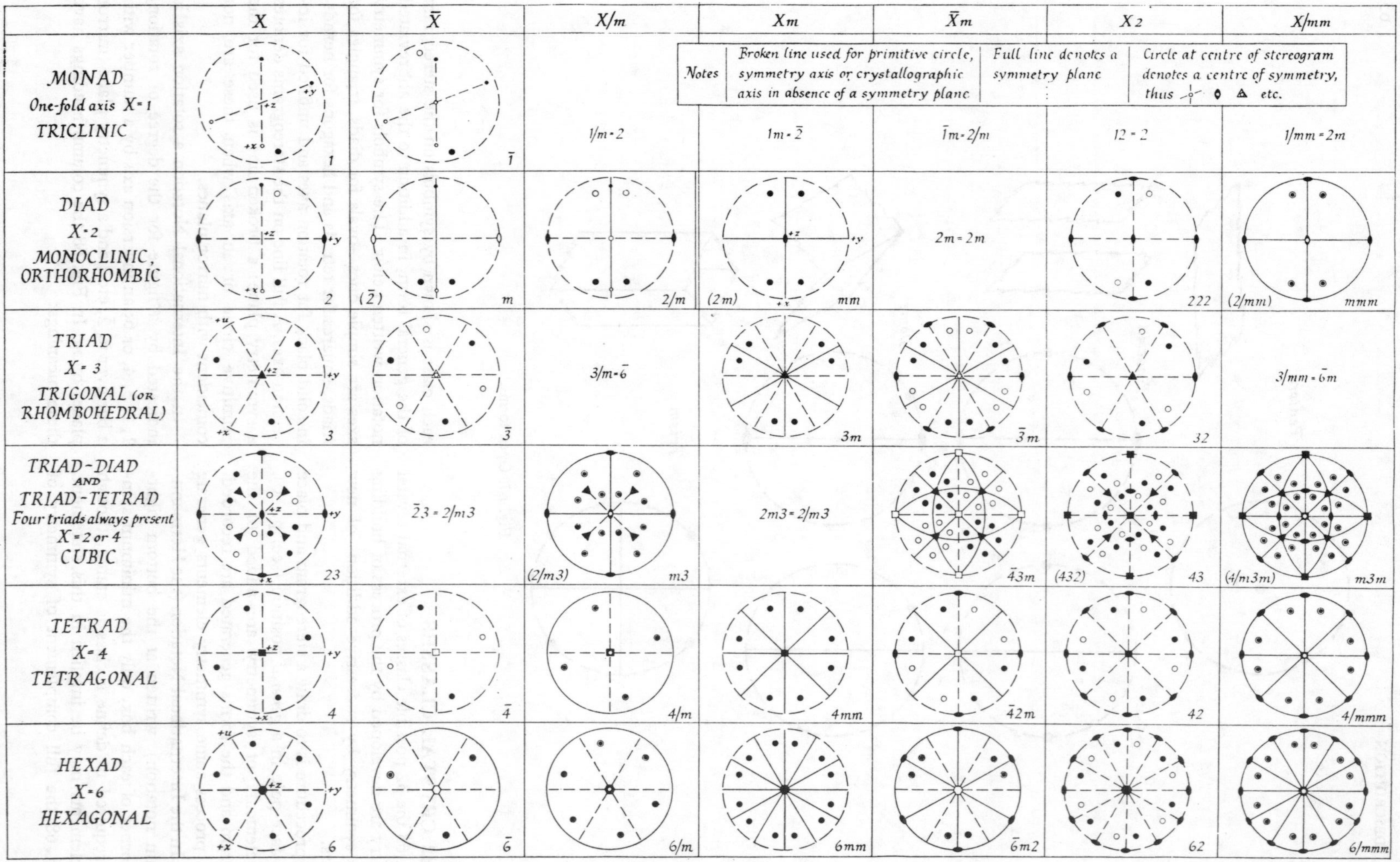

The 32 crystal classes table:

Notes	X	X̄	X/m	Xm	X̄m	X2	X/mm
MONAD One-fold axis X=1 TRICLINIC	1	1̄	1/m=2	1m=2̄	1̄m=2m	12=2	1/mm=2m
		(2̄)		(2m)			(2/mm)
DIAD X=2 MONOCLINIC, ORTHORHOMBIC	2	m	2/m	mm	2m=2m	222	mmm
TRIAD X=3 TRIGONAL (OR RHOMBOHEDRAL)	3	3̄	3/m=6̄	3m	3̄m	32	3/mm=6̄m
TRIAD-DIAD AND TRIAD-TETRAD Four triads always present X=2 or 4 CUBIC	23	2̄3=2/m3	(2/m)3 m3	2m3=2/m3	43m (43̄2)	43 (432)	(4/m3̄m) m3m
TETRAD X=4 TETRAGONAL	4	4̄	4/m	4mm	4̄2m	42	4/mmm
HEXAD X=6 HEXAGONAL	6	6̄	6/m	6mm	6̄m2	62	6/mmm

Notes column legend:
- Broken line used for primitive circle, symmetry axis or crystallographic axis in absence of a symmetry plane
- Full line denotes a symmetry plane
- Circle at centre of stereogram denotes a centre of symmetry, thus ⊕ ◯ △ etc.

Fig. 68. The 32 crystal classes.

X	a rotation axis alone
$\bar{X}$	an inversion axis alone
X/m	a rotation axis with a symmetry plane *normal* to it
Xm	a rotation axis with a symmetry plane that is not normal to it (usually a vertical symmetry plane)
$\bar{X}m$	an inversion axis with a symmetry plane not normal to it
$X2$	a rotation axis with a diad normal to it
X/mm	a rotation axis with a symmetry plane normal to it and another not so

In writing the symbol, the principal symmetry axis is placed first. (In the figure this axis is set at right angles to the plane of the paper, except in the monoclinic classes 2, m and 2/m, where it runs from left ot right so that $+z$ remains at the centre). Then comes a symmetry plane normal to this, if present, followed by secondary axes.

There are certain additional conventions. In the cubic system there are always four triads as secondary axes, and the convention is followed of putting m before the 3 when there is a symmetry plane containing two principal axes, but not a triad (the principal planes of the cube); and m after the 3 when the planes contain a triad and one principal axis (the diagonal planes of the cube in its conventional setting).

There are certain niceties, too, in the order of symbols in classes $\bar{3}m$, $\bar{4}2m$ and $\bar{6}m2$. If we wrote $\bar{3}2m$ this would imply that the mirror was not normal to the diad, which would be wrong. Some writers use $\bar{3}2/m$, but the International Notation just leaves out reference to the diads, which arise automatically anyway. In the other two symbols cited the order carries an implication of the relation of the horizontal elements to the chosen crystallographic axes which is of value in dealing with internal atomic arrangement.

SUPPLEMENTARY FORMS AND ENANTIOMORPHY

In studying the stereograms of Fig. 68 it will be noticed that, on some of them, if a different choice of position for the initial face-pole of the general form had been made, a different solid of similar but distinctive appearance would have been produced by the operation of the symmetry. There are two cases.

SUPPLEMENTARY FORMS

In class $m3$ $(= 2/m3)$, for example, the initial pole might be placed either at A or B in Fig. 69. The two resulting stereograms and the solids they represent are shown in the figure. It is easily seen from the stereograms that they are related by a rotation of $90°$. Together they would make up all the faces of the holosymmetric class $m3m$, and hence are called *supplementary forms*. There is a distinction between them because, for any given orientation of the internal atomic pattern, the relation of the faces of choice A to the atomic arrangement would be different from that of choice B.

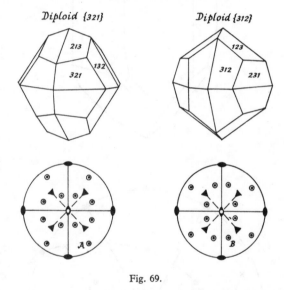

Supplementary Forms

Diploid {321}   Diploid {312}

Fig. 69.

The two choices are often distinguished as 'positive' and 'negative' forms, in this case 'positive' and 'negative' diploids. Sometimes the application of these terms is simple, the 'positive' form having the initial face-pole in the octant of the stereogram where all indices are positive. In other cases the assignment of the names is less obvious (the diploid is an example of this) and it is probably best simply to define the one meant by quoting its indices, e.g. diploid {321} to the left and diploid {312} to the right in Fig. 69. There is also the possibility of confusion with the positive and negative optical character of minerals (see Chap. 4), which is why the words are put in inverted commas here.

Enantiomorphy

Right-handed quartz

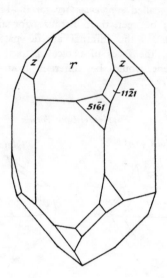
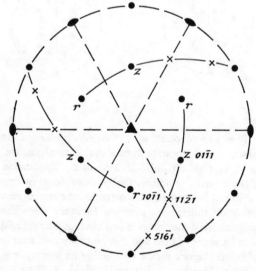

Left-handed quartz

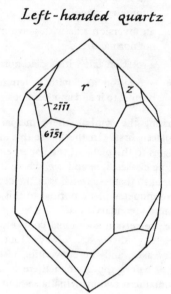
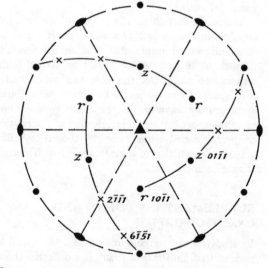

Fig. 70.

ENANTIOMORPHY

In class 32, for example, there is another kind of relationship, illustrated in Fig. 70. We have a choice in placing an initial face-pole of a general form (like {51$\bar{0}$1}) on the stereogram. These two options are illustrated in the figure. It will be seen that the left-hand stereogram cannot be re-orientated so as to be identical with the one on the right, and the same is true, of course, of the solids illustrated above the stereograms. The two are related in the same way as right and left gloves, and cannot be precisely superimposed. They are said to be *enantiomorphous*. Such crystals as these possess the property of rotating the plane of polarization of a beam of polarized light and are described as right- or left-handed, according to whether they rotate its plane to the left or right (as viewed by an observer looking through the crystal toward the light-source). Low-quartz, class 32, the example chosen for illustration, is the most notable case of enantiomorphy: it also shows supplementary forms (the rhombohedrons *r* and *z* of the figure) so that there are four members in the family of the general form, 'positive' and 'negative' right-handed and 'positive' and 'negative' left-handed crystals. The spiral linkage of SiO_4 tetrahedra in the structure of quartz, illustrated in Fig. 17 shows the relationship of the optical activity to the atomic arrangement.

Enantiomorphy is shown by many organic compounds and the property is retained by them even in solution, so that the dextro- or laevo-rotatory power of the solution helps to characterize the compound.*

Of its nature, enantiomorphy can only occur where there is no symmetry plane, nor any operation of inversion in the structure. It is therefore found only in classes in which rotation axes are the sole symmetry elements. These classes are:

1	Triclinic
2, 222	Orthorhombic
3, 32	Trigonal
23, 43	Cubic
4, 42	Tetragonal
6, 62	Hexagonal

It may happen that crystals belonging to one of the classes of lower symmetry than the holosymmetric do not show faces of the general form

* Dorothy Sayers and E. C. Bentley have written a good detective story with the optical activity of muscarine as the clue.

(as defined above) and therefore their true symmetry may not be recognizable from goniometric measurements. In such cases the possession of the power of rotating the plane of polarized light may be a useful clue to the class to which they belong.

ETCH FIGURES, PIEZOELECTRICITY AND PYROELECTRICITY

In the circumstances described in the last paragraph, the true symmetry may sometimes be shown by natural markings on the crystal faces. This is often the case with pyrite (class *m3*), already mentioned on p. 34, where cubes of the mineral may have striated faces, produced by the incipient development of planes other than {100}. The run of the striations groups the faces into three pairs of opposites, and indicates that the principal axes have two-fold, not four-fold symmetry (Fig. 71).

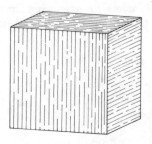

Fig. 71. Striated cube of pyrite.

In other cases groups of faces that appear to confer high symmetry may be differentiated into two or more distinct forms by the shapes and orientations of pits, called *etch-figures*, developed on them by suitable solvents such as mineral acids, caustic alkalis or hydrofluoric acid. The rate of solution often varies along different directions in the crystal, related to the internal atomic arrangement, and the orientation of the angular depressions formed may give clues to the symmetry, or may serve to distinguish right- and left-handed forms. Examples of these features are given in the descriptions of individual minerals.

Piezoelectricity and pyroelectricity are related phenomena in which stress and heat, respectively, cause the separation of electric charges on the surface of the crystal. These properties are more fully discussed in Chapter 3. The point of importance at the moment is that they are shown only by crystals that have no centre of symmetry. A simple test for piezoelectricity is mentioned on p. 90, and the possession of this property limits the class to which the crystal may belong to one of the non-centro-symmetrical classes.

TWINNING

One consequence of the symmetry of the internal structure of crystals is the possibility of growth of *twinned crystals*. A twinned crystal is a single crystal divided into two (or more) parts with one part in reversed structural orientation with respect to the next. Through a lattice there run planes across which the lattice inclination may be reversed in orientation without any lack of fit between the lattice rows at the plane of junction. When the atom groups (hung, as it were, upon these lattice points) are so arranged that they can equally well form part of the pattern on either side of the plane of reversal, twinning can occur. When, in the addition of atoms to a growing crystal, such a change in the orientation of the repeat directions

occurs, the succeeding atom layers will be added in the new orientation and a twinned crystal will result.

It will be clear at once that such a change of lattice orientation cannot take place across a plane of symmetry, because the twinning operation produces a condition of reflection symmetry across the lattice plane and if this existed already no change in orientation would occur. Likewise, the change is equivalent to a rotation of the lattice around a diad axis normal to the lattice plane concerned, so that no two-fold axis, or other rotation axis of even degree can be present normal to the plane. A triad axis normal to the plane is not excluded, however.

In the external form of the crystal, twinning is often indicated by the presence of re-entrant angles between faces (Fig. 72). These do not usually occur on simple crystal individuals.

The relation between the parts of a twin can always be described in terms either of a reflection across a plane which, with rare exceptions, is a possible crystal face and has rational indices, or as a 180° rotation about an axis which again is usually,

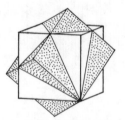

Fluorite: Cubic: interpenetrant cubes: twin-axis ⊥1̄1̄1̄

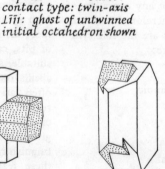

Spinel: Cubic: contact type: twin-axis 1̄1̄1̄: ghost of untwinned initial octahedron shown

Cassiterite: Tetragonal: knee-twin: contact type: twin-plane 011

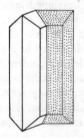

Aragonite: Orthorhombic: contact twin: twin-plane 1̄10

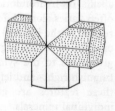

Staurolite: Ortho-rhombic: interpenetrant: twin-plane 032

Orthoclase: Monoclinic: twin-axis z: inter-penetrant across 010: the Carlsbad twin

Gypsum: Mono-clinic: swallow-tail twin: twin & composition-plane 100

Fig. 72. Examples of twinned crystals.

but not always simply related to the crystallographic axes. The plane that serves thus in description is called the *twin-plane*, and the axis the *twin-axis*. In centrosymmetrical crystals there will be both a twin-plane and a twin-axis normal to it, and the twin may be described in terms of either a reflection or a rotation. For stereographic work the description in terms of rotation is easier to use.

The description of the relation between the parts of a twin is called the *twin-law*, and a number of well-known laws are exhibited by the several crystal systems.

The plane by which the two parts of a twin are united is called the *composition-plane* and·is often the same as the twin-plane, but not always (cf. p. 284). Twins united along a well-defined plane are called *contact twins*. In some cases, however, close examination (perhaps by microscope in thin section) will show that portions of the structure with opposite orientations are intermingled in the two parts of the crystal, and in other cases the two parts of the crystal appear to grow through one another to produce *interpenetrant twins* (Fig. 72). The description of their relative orientation in terms of a twin-axis or twin-plane remains valid, but the composition plane is dissipated in a multitude of matched lattice planes throughout the volume of the body.

The relation between the twin-axis and the composition-plane in contact twins is made the basis of further classification:

Normal twins have the twin-axis normal to the composition-plane which is the twin-plane;

Parallel twins have as twin-axis a possible direction of a crystal edge (zone-axis), and the twin-axis lies in the composition-plane;

Complex twins have the twin-axis lying in the composition-plane and normal to a possible crystal edge.

Examples of these types are given in the article on the important group of feldspars (p. 284).

REPEATED TWINNING

Twinning may be repeated within one crystal, with one of two results. Where the twin-axes are parallel in the successive elements of the twin, or lie in the composition-plane, parallel lamellae of crystalline material with reversed orientation are formed, known as *polysynthetic twins*.

Where the twin-axes are not parallel in successive elements, the parts of the twin form various polygonal shapes. If the lattice elements of the crystal are close to that of a class of higher symmetry, repeated twinning may confer an apparently higher symmetry on the twinned individual. Such *mimetic twinning* is well illustrated by the example of aragonite ($CaCO_3$) shown in Fig. 73. Here, in an orthorhombic mineral, the angle $110 \wedge 1\bar{1}0$ is $63°48'$ which gives pseudohexagonal symmetry to the twin.

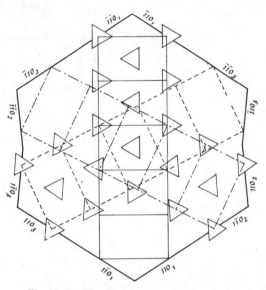

Fig. 73. A trilling in aragonite: mimetic twinning.

EXAMPLES OF TWINNING

To illustrate the main points of this description, Fig. 72 shows a few well-known twins, examples of which are readily available for study. Others will be found in the descriptions of individual minerals.

THE CAUSES OF TWINNING

From the point of view of origin, there are two main types of twinning, *growth twinning* and *secondary twinning*, the second being subdivided into *transformation twinning* and *deformation twinning*.

GROWTH TWINNING

In our description of twinning we have made use of the idea of rotation of one part of the twin relative to the other; but in most cases this is purely to explain the relation of the parts. In growth twinning the twinning has usually existed from the beginning of growth of the crystal, and no actual physical rotation is involved.

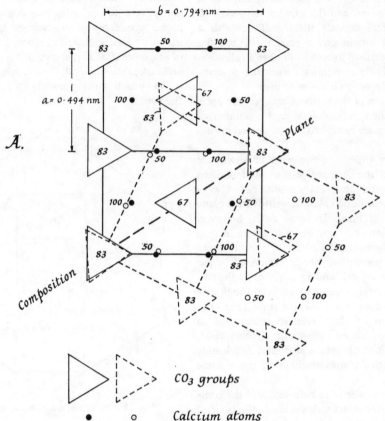

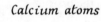

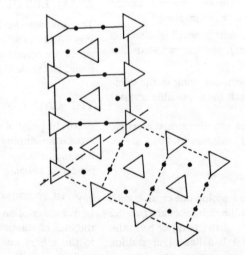

Fig. 74. Relation of twinning to structure in aragonite.

The underlying cause of growth twinning is not well understood. Crystals of some minerals are almost invariably twinned; other minerals may rarely or never show twinning.

First, of course, the symmetry must be conducive to its development. Twinning is rather rare, for example, in the hexagonal system, though frequent in the trigonal. In the former we have a large number of rotation axes of even degree, which may help to explain this difference. Where, in the atomic arrangement, there are planes or axes about which the atoms are nearly symmetrically disposed, twinning will be favoured, as is the case in aragonite, for example (Fig. 74). The diagram shows the unit cells superimposed with the 110 plane common, to illustrate the amount of mismatching involved in the twin. This is very small along the composition plane, but becomes quite large in the next rank of CO_3 groups on either side (diagram A). In Fig. 74B the overlap is omitted, and one has to sight carefully along the rows of triangles to see any misfit.

The classical theory of growth twinning emphasizes the importance of minimizing the internal lattice energy of the crystal. A perfect single crystal with an orderly arrangement of coordination polyhedra is presented as the lowest energy condition. Twinning, by disturbing this, will increase the lattice energy; but if the matching between the nearest neighbour atoms across the composition plane is close, so that the arrangement of anions around an individual cation is little disturbed, the reversal of atom groups more than one bond-length away from the plane does not greatly affect the total lattice energy, and twinning can occur. Apart from this, the initiation of twinning is ascribed to stacking mistakes in the very early stage of crystal growth, when the crystal energy field tending to retain atoms in correct orientation is said to be small. If the crystal is growing rapidly from a supersaturated solution the fault will be preserved as a twin.

Nevertheless, the initiation of growth twinning is not confined to the early stage of crystal nucleation as this would suggest. For example, in polysynthetic twinning of plagioclase new lamellae in reversed orientation continue to develop at intervals throughout the growth of the crystal. Also overgrowths of plagioclase on relatively large seed crystals in sediments show twinning independent of that in the core, so that the conditions that encourage reversals, and the susceptibility of the lattice to respond to them, are not entirely governed by weak lattice energy fields in crystal infancy. Moreover, the idea of crystal growth being primarily directed to minimizing lattice energy does little to explain why some crystals are almost invariably twinned.

The position seems to be that the probability of a stacking reversal depends upon there being a volume of the atomic pattern nearly common to the two orientations in the twin, and that the reversal can take place at any stage. Its preservation as a twin probably depends upon the rate of growth of the crystal and of the supply of ions to the two parts. Many common observations underline the importance of the environment of growth in the initiation of twins. There is more than a hint that in certain minerals in certain conditions twinning will inevitably occur. For example, plagioclase feldspars in igneous rocks are almost always twinned, while they are often untwinned in low-grade metamorphic rocks. Statistical study suggests that the frequency of the different twin laws is also different in igneous and metamorphic plagioclase. Composition plays a part, too, for the width of lamellae of one orientation is often much greater than those of the reversed orientation in the calcium-rich plagioclase of gabbros, while the two sets are more equal in more sodium-rich plagioclase. It must be remembered that the growth of crystals is not always an equilibrium process, but often takes place in response to a continuous change of the environment in one direction, as in the cooling of a melt or the evaporation of a solution. This means that conditions of growth, and particularly the concentration of other substances, will be changing, perhaps by a series of small steps. These changes may favour the occurrence of stacking faults. Moreover, growth mistakes will probably not be rectified, unless conditions change from those of precipitation to those of resorption.

Much work remains to be done in this field, but it is a study that seems capable of yielding useful information about conditions during crystal growth.

TRANSFORMATION TWINNING

Twinning may occur when a crystal inverts from one polymorph to another, by a displacive transformation. It provides a means of relieving strain built up by the small displacements of atoms involved in the change of symmetry.

An example is provided by the triclinic mineral anorthoclase, which inverts to a monoclinic form at temperatures up to 700°C (the temperature of inversion varying with the composition). In the triclinic condition it shows, in thin sections, cross-hatched (tartan pattern) twins on two laws, the Albite and Pericline Laws described on p. 285. This characteristic pattern may be seen, under the microscope, to disappear on heating and reappear on cooling, as the crystal passes through the inversion temperature. The scale of the twinning varies, too, from sub-microscopic (detectable by X-rays) through fine to coarser lamellae seen microscopically, this variation depending, presumably, on the rate of cooling, which governs the length of time the crystal is allowed to readjust its structure to the changing conditions—that is the *annealing time*.

Low-quartz often displays a transformation twin, called the Dauphiné twin, resulting from the high- to low-quartz inversion described on p. 24. The structural change is shown in Fig. 17(c). The twin-axis is crystallographic z (the triad axis) and the two parts of the twin (which are both of the same hand) interpenetrate in very irregular fashion. Because of the six-fold symmetry along this axis in high-quartz (class 62) the twin disappears when this polymorph forms at 575°C.

Transformation twinning in leucite is illustrated in Fig. 256, p. 293.

DEFORMATION TWINNING

If an elongated cleavage fragment of calcite a few millimetres long is held with one long obtuse edge resting on the table, and the opposite obtuse edge is pressed with a blunt knife, a small part of the crystal can be reversed in its obliquity to the long edges, as a result of twinning on the plane $1\bar{1}02$ (Fig. 75).

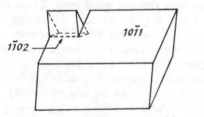

Calcite cleavage-piece twinned by pressure

Fig. 75.

Mechanically-induced twinning of this kind is closely related to *glide* in crystals. These two processes together are responsible for plastic flow of crystalline materials and are hence of the greatest importance in geology, metallurgy and engineering.

Glide consists of slipping of one part of the crystal past another. It takes place on lattice planes with low indices and the movement is in the direction of close-spaced atomic rows. The slip-plane, T, and the slip-direction, t, are inherent properties of the lattice, and independent of the direction of the applied load. Deformation depends upon a possible slip-plane being orientated favourably to the direction of load. The distance moved on each plane of movement must be an integral number of lattice repeat distances along t, and may vary on adjacent slip-planes (Fig. 76).

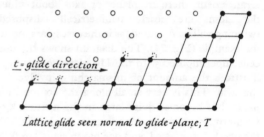

Lattice glide seen normal to glide-plane, T

Fig. 76. Lattice gliding.

In deformation twinning, a stressed portion of the crystal takes up a twinned orientation, during which the movement on each plane in the displaced part of the crystal, parallel to the composition-plane, is identical and is usually only a fraction of a lattice repeat in the direction of movement, while the total distance moved by any atomic plane is proportional to its distance from the twin-plane (Fig. 77). The result is *homogeneous strain*. If we take a plane at right-angles to the plane of movement and containing the direction of movement, and draw a circle on it before twinning takes place, the circle will be distorted by the twinning into an ellipse. Where this ellipse would cut the original circle there are two directions of *no resultant distortion*, symbolized as η_1 and η_2 which are the traces of planes of no distortion. The *first undistorted plane*, K_1, is the twin-plane, and the *second undistorted plane*, K_2, makes an angle with the first whose value defines the amount of movement or *shear*. It will be seen from Fig. 77 that, if the angle between these two undistorted planes is 2ϕ the shear, s, is given by

$$s = 2 \cot 2\phi.$$

In general any twin can be specified by defining K_1 and η_2, or K_2 and η_1; if K_1 is rational the twin may be described by reflection across that plane, while if η_1 is a rational direction it is described by a 180° rotation around the direction of shear. Frequently, all these parameters will be rational. From a knowledge of the crystal lattice it is possible

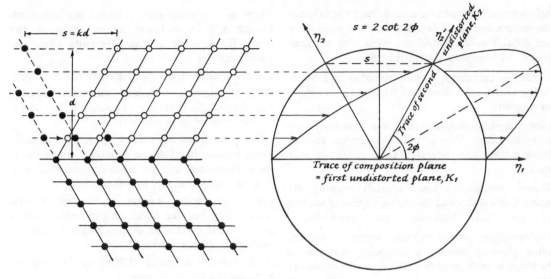

Fig. 77. Deformation twinning.

to calculate the amount of shear for a given crystal thickness, or conversely, if the undistorted planes can be recognized, as may be the case in deformed metals, the movement in the twin can be derived.

The propagation of slip in gliding is described in terms of a rapid sweeping movement of an initial dislocation, travelling at a speed of the order of 1/10th the velocity of sound, over the slip plane.

Physical theory is also developed, more tentatively, to show how the sweeping dislocation may, in addition, have a component normal to the initial plane and climb, in helical fashion, through successive lattice planes, to produce the homogeneous shear characteristic of deformation twins.

Both gliding and deformation twinning require the uptake of energy, supplied by the external deforming force, to move the atoms from their lowest energy positions, which they occupy in the perfect crystal structure. After gliding the atoms return to their lowest energy positions, but the internal energy of the system is increased at crystal imperfections at the terminations of the glide. In twinning the internal energy at the twin-plane will be increased. The possibility of gliding and deformation twinning depends on the energy characteristics of the atomic bond in particular directions in the crystal and, as has been explained earlier, these are favourable in the metallic bond, conferring on metals their ductility. Deformation twins form in metals, however, at stresses much less than those calculated from the absorption of elastic strain energy. The theory of propagation of glide and twinning by movement of initial dislocations is designed, in part, to explain this.

PARALLEL GROWTH OF CRYSTALS AND THE FORMS OF CRYSTAL AGGREGATES

There is a common tendency for groups of crystals to grow in approximately parallel orientation, which is a result of the precipitating atomic groups forming nuclei in a preferred orientation on the substrate, governed presumably by surface energy factors. To this is added the influence of the direction of supply of material from solution. The relation between the crystals does not obey any geometrical law, as it does in twinning, and the parallelism between them is often not exact.

Direction of supply of solutions and the conditions of deposition also govern the formation of distinctive aggregates. of crystals, which may be characteristic of particular substances. These aggregates have been given names, often derived from Latin or Greek roots, descriptive of the shape. Notes on the more common of these follow.

Botryoidal—like a bunch of grapes. A shape often assumed by substances precipitated as colloidal gels, subject to the influence of surface tension. Examples are malachite and psilomelane.

Compact—a mass in which individual crystals are so small as to be hardly visible. Subdivisible into microcrystalline, when the crystals can be seen under the microscope, and cryptocrystalline when they are difficult to see individually even with a microscope. An example is provided by flint.

Coralloid—branching, bluntly-rounded and interlacing forms, seen sometimes in deposits of calcite or aragonite.

Dendritic—branching growths producing miniature tree-like forms. Shown by native copper, and by black psilomelane infiltrated along cracks in host rock.

Drusy—close-packed small crystals growing inwards into a cavity and displaying a serrated surface of terminal faces. Often seen in quartz veinstone.

Fibrous—thin, parallel crystals closely in contact, often growing normal to opposing surfaces of bedding, as with gypsum in shale, or of veinlets, as in asbestos.

Filiform—slender, long, hair-like aggregates, as in millerite.

Granular—a widely-embracing term to cover more or less equidimensional grains spanning a wide range of coarse to fine grain-size, passing into the compact condition when the grains are no longer distinctly visible.

Lamellar—thin foils or scales as in the micas.

Massive—interlocked crystals, not displayed as distinct granules, but not fine enough to be called compact, as for example calcite in a marble. Often shown by sulphide minerals in mixed ores.

Mammillary—breast-like: rounded surfaces intersecting in open v-shaped grooves, often in fact more like irregular quilting. Hematite often provides examples.

Mossy—a diminutive form of the dendritic aggregate.

Nodular—forming separated ellipsoidal masses. Siderite nodules in shale give examples. The degree of regularity of shape varies widely.

Oolitic—small closely-packed spheroids or ellipsoids, looking like fish-roe. Exhibited by calcite in some limestones and by hematite, or other iron minerals in sedimentary iron-ores.

Pisolitic—a coarser variant of the oolitic with spheroids of pea size. Often seen in bauxite, where the colloidal conditions of precipitation probably produce it, *cf.* botryoidal.

Reniform—like a kidney. Displayed by hematite and closely allied to the mamillary.

Radiated—radially disposed needle-like crystals, or blades, e.g. gypsum or tourmaline in some of their crystallizations.

Reticulated—a mesh-work of crossing and re-crossing crystals.

Stalactitic—pendent, tapering masses, which may unite with deposit growing upwards from a cavity floor (stalagmite) to form columns. Formed by percolating waters depositing their dissolved material while dripping from a crack. Pre-eminently a calcite deposit of limestone caves, but other minerals may show the same formation.

Wiry—a form often shown by native silver and gold.

Further Reading

BISHOP, A. C. 1967. *An outline of crystal morphology*. London, Hutchinson & Co.

COTTRELL, A. H., and BILBY, B. A. 1951. A mechanism for the growth of deformation twins in crystals. *Phil. Mag.*, **42**, pp. 573–81.

PHILLIPS, F. C. 1946. *An introduction to crystallography*. London, Longman Green & Co. (Part 1).

HALL, E. O. 1954. *Twinning and diffusionless transformations in metals*. London, Butterworth & Co.

3

DENSITY, MECHANICAL PROPERTIES AND PROPERTIES OF LATTICE EXCITATION

DENSITY

DENSITY AND SPECIFIC GRAVITY

The specific gravity (S.G.) of a body is defined as the ratio of its weight to the weight of an equal volume of water. It is expressed simply as a number.

The density (D) of a body is the mass of unit volume and is expressed in g per cm^3 (or kg/m^3).

As we shall see, both are usually found by the measurement of displacement of water (either directly or indirectly). With the wide use of the gram and the cubic centimetre as the units of measurement, the difference between the two quantities comes down to the fact that in recording density the mass of the displaced water should be corrected to its mass at $4°C$. Strictly speaking, the volume of the body should also be corrected to a standard temperature, from a knowledge of its coefficient of thermal expansion, and the weighings of the body and liquid should be corrected for the mass of the air that they displace; but neither of these corrections is necessary in ordinary mineralogical work.

The correction for the temperature of the water displaced does, however, make an appreciable difference to the result, and it may be carried out as follows. Using tm for the temperature of the mineral, and tw for the temperature of the water, the density actually measured may be described as D_{tw}^{tm} in air. The value required is $D_{4°C}^{tm}$ in air. These two values are related by the equation

$$D_{4°C}^{tm} = D_{tw}^{tm} \times D_{tw} \text{ water}$$

Expressed as a correction this is

$$D_{tw}^{tm} - D_{4°C}^{tm} = D_{tw}^{tm}(1 - D_{tw} \text{ water})$$

The density of water at various temperatures is known so that the correction can be made from the following table.

Table 13. Correction factor $[1-D_{(abs)} \text{ water}]$ for density of water at various temperatures.

Formula: $D_{4°C}^{tm} = D_{tw}^{tm} - (D_{tw}^{tm} \times \text{table value} \equiv tw°C)$

Temp. $tw°C$		Temp. $tw°C$	
10	0·0003	20	0·0018
11	4	21	20
12	5	22	22
13	6	23	25
14	8	24	27
15	9	25	30
16	0·0011	26	32
17	12	27	35
18	14	28	38
19	16	29	41

ESTIMATION OF DENSITY

The mere fact that a mineral substance weighs heavy or light in the hand, for its size, is helpful in deciding what the mineral may be. Provided that the mineral constitutes most of the lump, and that the lump is, say, 30 g or more in weight, with practice a fair estimate of the density of a mineral may be made by tossing it gently in the hand. Minerals containing heavy atoms—Fe and above in the Periodic table—are naturally heavy.

More particularly, the relationship between density and colour may be helpful. Dark-coloured substances are very often relatively heavy and light-coloured substances light in weight. One's attention is at once arrested when a dark substance like graphite (C, $D = 2·23$) proves to be light to pick up, while barytes ($BaSO_4$, $D = 4·5$) is unexpectedly heavy in relation to its light colour.

This qualitative estimation of density is very useful in the practical identification of minerals.

MEASUREMENT OF DENSITY

The density of a rock (i.e. an aggregate of minerals) may mean one of three things:

(*i*) The density of the dry rock with its pore spaces filled with air. This is the density of the builder using stone in construction.

(*ii*) The density of the rock with its pores filled with water, as it would be in the upper layers of the earth's crust. This is the density of the geophysicist calculating the gravitational effect of near-surface rocks.

(*iii*) The density of the mineral matter making up the rock.

In all cases

$$D_{tw}^{tm} = \frac{w_1}{w_1 - w_2}$$

where w_1 is the weight in air and w_2 the weight suspended in water. The weight in water is less than that in air, the difference $w_1 - w_2$ being the weight of the water displaced which gives the volume of the body being determined.

The mineralogist is concerned with the density of solid homogeneous mineral grains unaffected by pore space or the presence of impurities. The problems of density measurement are (*i*) selection of a pure sample uncontaminated by other minerals

(a)   (b)

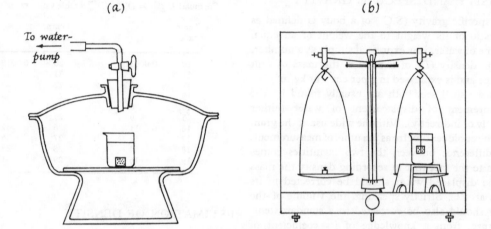

To water-pump

Fig. 78. (*a*) Evacuable desiccator for extracting air from the sample. (*b*) Arrangement for direct weighing in air and in water.

In (*i*) the dry rock is given a light coating of wax to seal the surface. It is now weighed first in air and then suspended in water, the wax preventing the water from entering the pores.

In (*ii*) the rock is saturated by placing it in a beaker of water inside an evacuable desiccator (Fig. 78a) the outlet of which is attached to a water pump. When the air is exhausted from the desiccator it is also drawn out of the pores of the rock and replaced there by water. The saturated piece of rock is then lightly dried on the surface and weighed first in air and then suspended in water.

In (*iii*) the dry rock is weighed in air. It is then placed in water and the air drawn out of the pores and replaced by water as in (*ii*). In this condition it is weighed suspended in water.

and (*ii*) avoiding trapped air in the sample when it is immersed in a fluid. The ways of solving these problems will be discussed first, and then available methods of measurement will be described.

PREPARATION OF THE SAMPLE

If large, homogeneous, transparent crystals without inclusions or cleavage cracks are available there is no problem. But in practice the amount of material available will often be small; it may be polycrystalline with pore space along grain boundaries; it may be fractured or have cleavage cracks with air in them; it may contain inclusions or occur as grains interlocked with other minerals; and if it is opaque there will be difficulty in finding out about its internal defects.

Porous material must be prepared by crushing it so that the displacement fluid can readily enter all intergranular spaces.

Physical separation of interlocked grains is best achieved by crushing the aggregate to particles smaller than the average size of the constituent mineral grains and selecting homogeneous grains. Mineral separation is an art in which density differences themselves are employed, as well as differences in magnetic and surface properties, in order to eliminate unwanted minerals. Using these methods on crushed material, or simply by hand-picking, a sample of homogeneous grains can be obtained. If the grain size is fairly small, say < 0·5 mm (30 to 60 mesh B.S.S.) the grains can readily be inspected for homogeneity under a microscope. Thoroughly opaque minerals still present a difficulty, however, and a sample of such grains may have to be polished for inspection under the reflected-light microscope.

Even if a large piece of the mineral is available, it may be better to use carefully inspected smaller pieces broken from it for density measurement, rather than the whole lump, to guard against possible inhomogeneity. On the other hand, provided it is certain that the material is homogeneous and free from open cracks, a sample of several grams weight will yield more accurate results than when the weighings are small. These rival considerations must be balanced for each particular case.

REMOVAL OF AIR

When a large fragment is weighed in water, air bubbles on the surface must be removed by touching them with a small brush. Their removal may be helped by adding a drop of a wetting agent to the water, which should not noticeably affect its density. The fragment should be suspended by a fine nylon thread (fine fishing cast is suitable) which does not displace much water, and does not absorb water.

Crushed samples must be freed from air by putting the vessel containing the grains immersed in water into an evacuable desiccator and exhausting the air, as described above.

Rather than water, a liquid of lower viscosity, the density of which is known (e.g. carbon tetrachloride), can be used as the displacement fluid. It has been suggested that this may be poured over the dry grains inside the desiccator by a suitable tilting arrangement, after the air has been exhausted. The low viscosity fluid more readily displaces air from crushed grains or fine cracks.

The logical extension of this idea is to use air itself as the displacement fluid, in the apparatus called a volumenometer, and for finely powdered samples this may be the only practicable method of volume measurement. Fine powders may be impossible to handle immersed in liquids because they tend to aggregate into spherical pellets that will not disperse.

With these principles in mind the available methods of measurement may be described.

DIRECT WEIGHING IN AIR AND IN LIQUID

This may be done on a chemical balance. If the fragment is large enough to be suspended in a nylon loop, it is hung from the hook on the balance beam and weighed. Then it is immersed in a beaker of water set on a bridge spanning the balance pan below (Fig. 78b), and weighed again. The uncorrected density is given by

$$D = \frac{w_1}{w_1 - w_2}$$

w_1 and w_2 being the weights in air and in water. The temperature at which the measurement was made should be recorded as a superscript and the water temperature, or the temperature to which it has been corrected, as a subscript to the symbol D, as discussed above. If a fluid other than water is used the density is given by

$$D = \frac{w_1 \times D_f}{w_1 - w_2}$$

where w_2 is the weight in this fluid and D_f is the density of the fluid.

For smaller fragments a glass fibre with two small pans along its length may be hung from the balance hook, one pan remaining above the fluid and the other immersed throughout the measurement. The mineral grains are transferred from the upper to the lower pan between weighings and the level of the liquid adjusted so that the length of glass fibre immersed is the same during each weighing. This eliminates the error due to the thread, and is more convenient. A sensitive torsion balance weighing to 50 mg and furnished with two such pans (Fig. 79) gives quick and accurate densities.

The limitations of the method are (i) the need for grains large enough to be handled, (ii) the sensitivity of the balance, taking into account surface tension effects of the fluid on the submerged pan and (iii) the need for removal of air bubbles.

RELATIVE WEIGHING IN AIR AND IN LIQUID

A counterpoised beam balance, or a spring balance,

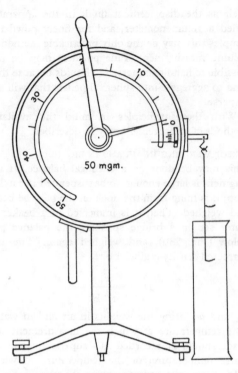

50 mgm.

Fig. 79. Torsion balance.

can give the ratio of weights in air and liquid without direct weighing.

This is shown in Fig. 80. A beam is balanced on a knife-edge with the solid suspended on one arm (which is graduated) and a counterpoise on the other arm. The counterpoise is adjustable from one notch to another on its arm, to balance the solid at a convenient distance along the other, but it must remain in the same notch throughout any one experiment. The specimen is moved along the graduated arm so as to bring the end of the beam opposite a fixed index mark and its position read from the graduations on the arm (reading a). The

specimen is then immersed in the displacement liquid and moved farther from the fulcrum until the beam is again opposite the index mark. Its new position gives reading b. The lengths a and b are inversely proportional to the weights in air and liquid. If the liquid is water

$$D_{tw}^{tm} = \frac{\dfrac{1}{a}}{\dfrac{1}{a}-\dfrac{1}{b}} = \frac{b}{b-a}.$$

Because large pieces of material are needed, the method is not often suitable for minerals. It is defeated by the problem of inhomogeneity.

The Jolly spring balance

Two small pans, one above the other, the lower one immersed in the liquid, are suspended from a weak spiral spring (Fig. 81a). The position of an index mark, in the form of a small bead on the pan-support, is read by a scale on the post supporting the spring, parallax being avoided by having the scale engraved on a mirror and taking the reading with the bead and its reflection superimposed. The scale is read with the balance unloaded (reading a), with the mineral on the upper pan (reading b) and with the mineral on the lower pan (reading c). The differences $b-a$ and $c-a$ are directly proportional to the weights in air and liquid. With water as the displacement liquid

$$D_{tw}^{tm} = \frac{b-a}{(b-a)-(c-a)} = \frac{b-a}{b-c}$$

The Jolly balance itself is not much used now-a-days, but the principle, using glass-fibre springs, is sometimes useful. Its limitations are the same as those of direct weighing on a balance.

THE PYKNOMETER OR SPECIFIC GRAVITY BOTTLE

This is a small flask (from 10 to 50 ml capacity) with a long stopper fitting an accurate ground-glass

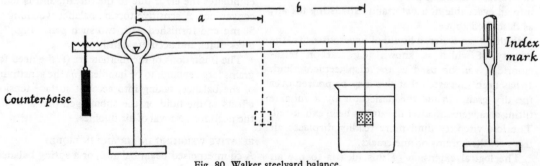

Fig. 80. Walker steelyard balance.

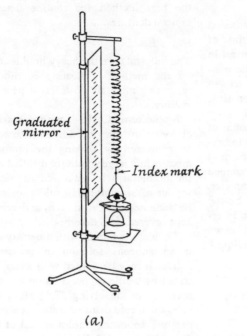

Fig. 81. (a) Jolly spring balance. (b) Pyknometer.

joint at the neck (Fig. 81b). The stopper has a capillary hole down the centre. The objective is that when the flask is filled to the brim and the stopper is inserted it should seat always at exactly the same level, and the liquid should rise through the capillary. When the surplus liquid has been wiped off, the volume of liquid should be the same on every occasion.

To determine the density of a liquid, the bottle is filled and weighed, and the weight divided by that of the bottle filled with water.

To determine the density of a solid weighings are made in the following order, to avoid transfer and possible loss of mineral powder:

(a) pyknometer, dry and empty,
(b) pyknometer + mineral (usually as crushed grains)
(c) pyknometer + mineral + liquid,
(d) pyknometer + liquid alone (this should be constant for a given liquid at a given temperature).

Then $b - a$ gives the weight of the mineral in air, and $c - d$ the weight of the mineral less the weight of an equal volume of water (or other liquid). Using water

$$D_{tw}^{tm} = \frac{b-a}{(b-a)-(c-d)}.$$

The flask is necessarily rather large and heavy. If it is made too small the following errors become important:

(i) variation in the seating of the stopper;

(ii) evaporation from the capillary during weighings. This may be counteracted with a chip of cover-glass on top of the stopper, the chip, of course, always being weighed along with the other parts of the pyknometer;

(iii) atmospheric condensation on the flask, or imperfect wiping. This means that a larger bottle is better, but a large bottle demands a relatively large amount of powder.

The influence of temperature is appreciable, and a constant temperature bath is desirable to bring the bottle and its contents to the same temperature at each measurement. The time taken to achieve this will allow evaporation from the capillary, unless precautions are taken.

The air must be extracted from amongst the crushed mineral grains when liquid is added to the bottle (stage (c) above), by placing the bottle in an evacuable desiccator as described earlier.

The less involved procedure of determining the density of a liquid by the pyknometer is useful in the method of floating equilibrium described below.

THE METHOD OF FLOATING EQUILIBRIUM

The central problems of homogeneity and exclusion of air are largely solved by the indirect method of placing a small number of grains of the mineral in a liquid of high density, and adjusting the concentration of the liquid until the grains remain suspended, showing no tendency to rise or sink. The density of the liquid is then determined.

This procedure demands only a small amount of the mineral, the grains can be inspected individually for purity, and a sufficient amount of liquid can be used to allow its density to be determined without serious error using a large pyknometer. Alternatively the Westphal balance (see below) may be used, or for more approximate work, interpolation between the densities of two grains of known density that respectively sink and float in the liquid.

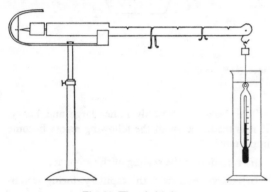

Fig. 82. Westphal balance.

The Westphal balance (Fig. 82), a modification of the steelyard balance, has a glass sinker combined with a thermometer hung at the end of the graduated arm, and a counterpoise trimmed so that a pointer attached to it is opposite an index mark when the sinker is immersed in water. When the sinker is hung in a liquid denser than water, the addition of calibrated riders to the graduated arm to bring the pointer back to the index mark enables the density of the liquid to be read directly.

A set of sinkers of varying densities can be made by sealing small pieces of steel inside short lengths of glass tubing and these are useful for approximate measurements. A set of mineral grains of known densities is also useful in adjusting liquids to roughly the density required for a particular determination.

Because of its accuracy, flexibility and convenience, floating equilibrium is here regarded as the best method for routine determinations of mineral densities.

Heavy liquids

The following are the heavy liquids commonly used for the method of floating equilibrium, and for separating minerals of different densities from one another.

Bromoform, $CHBr_3$, with a maximum density of 2·90 g/cm^3 may be diluted with acetone. It may be recovered by shaking the bromoform-acetone mixture in a large volume of distilled water, allowing the bromoform to sink, and drawing it off. The vapour of bromoform is mildly toxic; work should be done near an extract fan, and vessels should be kept covered. It is decomposed by light.

Thoulet's solution, with a density of 3·19 g/cm^3, is an aqueous solution of potassium mercuric iodide. It may be diluted with water and reconcentrated by gentle evaporation on a water-bath. It is prepared by dissolving 270 g HgI_2 and 230 g KI in 80 ml of cold distilled water, by stirring, and then gently evaporating until a crystal of fluorite floats. A small excess of KI will not matter, but HgI_2 should not be in excess. The solution is poisonous and tends to corrode the skin.

Methylene iodide (di-iodomethane), CH_2I_2, with a density of 3·325 g/cm^3, may be diluted with chloroform and recovered by gentle volitilization of the diluent. It is decomposed by exposure to light.

Clerici's solution, with a density of 4·4 g/cm^3, is an aqueous solution of thallium formate malonate. To prepare it, neutralize two equal parts of Tl_2CO_3 with equivalent parts of malonic and formic acids, separately, and then mix the two solutions and concentrate by gentle evaporation. Clerici's solution corrodes the skin, but, because of its high density and the fact that it can be diluted with water and readily reconcentrated, it is one of the most useful liquids. Directions for purifying it, or recovering the thallium are given by Kalervo Rankama (1936. *Bull. Comm. géol. de Finlande*, No. 115, 65–67).

Density of a mineral heavier than a liquid

This case will quite often arise. Make a small buoy a few mm across by enclosing a small piece of heavy metal in a ball of wax, so that its density is between 1 and 2 g/cm^3. Weigh this, and weigh some small grains of the mineral to be determined. Lightly press the mineral grains into the wax and determine the density of buoy + mineral by flotation. Remove the mineral grains and determine the density of the buoy alone, by flotation in the diluted liquid.

To find the density, Dm, of the mineral we have the following data:

Wb, the weight of the buoy,

Db, the density of the buoy,

$Vb = Wb/Db$, the volume of the buoy,

Wm, the weight of the mineral,

Dbm, the density of buoy and mineral combined.

Vm, the volume of the mineral, will equal Wm/Dm.

Then
$$Dbm = \frac{Wb + Wm}{Vb + Vm}$$
$$= \frac{Wb + Wm}{Wb/Db + Wm/Dm}.$$

Whence

$$Dm = (Dbm \cdot Wm) \Big/ \left(Wb + Wm - \frac{Dbm \cdot Wb}{Db} \right).$$

Instead of a wax buoy, a horseshoe loop of glass, the springiness of which allows it to grip the mineral grain, may be a more satisfactory and permanent float.

THE VOLUMENOMETER

The use of air as a displacement fluid is limited, because of the difficulty of measuring small volumes of air accurately. Its use is imposed upon us in two cases:

(i) for fine powders,

(ii) for substances soluble in, or reacted upon by displacement liquids.

It is also useful for porous materials if the pore spaces communicate with one another and with the surface. If the cavities are sealed, however, the material must be crushed to obtain the volume of the solid part.

The principle of the apparatus is shown in Fig. 83. The volume of air, v_1, in the test chamber is compressed with the manometer to volume v_2, by raising the level of the liquid in the chamber from the lower to the upper of two fixed marks. The head, h_1, required to do this is measured.

The liquid level is lowered to the bottom mark, and a weighed amount of the substance whose density is to be determined is placed in the specimen cup on the gauze support.

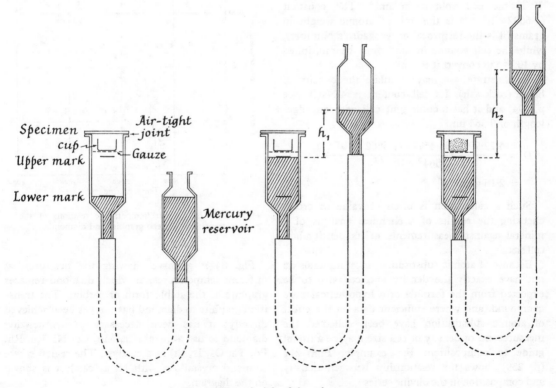

Fig. 83. The principle of the volumenometer.

The chamber is closed, the liquid brought again to the upper mark, and the head, h_2, required is noted. The volume, v_3, of the introduced substance is to be found.

At an atmospheric pressure of P mm mercury, with h_1, h_2 measured in mm of mercury also, $P_1 = P+h_1$, $P_2 = P+h_2$. Then at any given temperature

$$P_1 v_2 = k = P_2(v_2 - v_3)$$
$$P_1 v_2 = P_2 v_2 - P_2 v_3$$
$$P_2 v_3 = P_2 v_2 - P_1 v_2$$
$$v_3 = \frac{v_2(P_2 - P_1)}{P_2}.$$

Calculation of density

The density of a mineral is directly related to the volume of the unit cell of the structure and the atomic weights of the atoms in the cell. The relationship is

$$D = \frac{AW \times (1 \cdot 6602 \times 10^{-24})}{V \times 10^{-21}}$$

where D is the density in g/cm^3, AW is the sum of the atomic weights of the atoms in the unit cell, and V is the cell volume in nm^3. The constant $(1 \cdot 6602 \times 10^{-24})$ is the unit of atomic weight in grams (it is the reciprocal of Avogadro's Number), while the cell volume in nm^3 must be multiplied by 10^{-21} to convert it to cm^3.

To illustrate, we may calculate the density of halite (rock salt). Its cell content is 4 NaCl (see p. 194) and it has a cubic unit cell with an edge-length of 0·564 nm.

$$D = \frac{4(22 \cdot 997 + 35 \cdot 457) \times (1 \cdot 6602 \times 10^{-24})}{0 \cdot 564^3 \times 10^{-21}}$$

$$= 2 \cdot 16 \text{ g/cm}^3.$$

Such a calculation is often of value in cross-checking the results of a chemical analysis of a mineral against measurements of its density and cell size.

Because of atomic substitution, most minerals do not have exactly the density and cell size to be expected from the formula of a hypothetical pure compound, and where sufficient data on the effect of atomic substitution have been collected the measurement of density or cell size may be a useful guide to composition. For example, Fig. 211 (p. 239) shows the relationship between density and composition in the olivine series.

DENSITY AND POLYMORPHISM

Polymorphism has already been discussed in Chapter 1 (p. 22) and the effect of high confining pressures in producing polymorphs illustrated by the case of SiO$_2$. As one would expect, the atomic arrangement resulting from high pressure is one where the volume per atom is reduced, and the previously quoted example may be reinforced by giving here the figures for the two forms of carbon, graphite (low pressure) and diamond (high pressure).

System	Graphite Hexagonal	Diamond Cubic
Cell sides (nm)	a 0·2464	a 0·3567
	c 0·6736	
Cell vol. (nm^3)	0·0354	0·0454
Z (atoms per cell)	4	8
Vol. per C atom (nm^3)	$8 \cdot 8 \times 10^{-3}$	$5 \cdot 7 \times 10^{-3}$
D measured	2·09–2·23	3·516–3·525
calculated	2·253	3·514

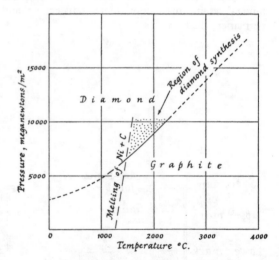

Fig. 84. The pressure/temperature relations of the transition between graphite and diamond.

Fig. 84 is a curve showing the pressures, at different temperatures, at which diamond replaces graphite as the stable form of carbon. The transition graphite to diamond has not yet been achieved directly: it has been necessary in synthesizing diamond to use a metal (Cr, Mn, Co, Ni, Ru, Rh, Pd, Ta, Os, Ir, Pt) as a catalyst. The region where diamond crystallizes with Ni as catalyst is shown on the diagram.

MECHANICAL PROPERTIES OF MINERALS

HARDNESS AND PROPERTIES OF COHESION

Under this heading is considered the resistance of minerals to various forms of mechanical assault. This resistance depends ultimately, of course, on the strength of the atomic bonding in the mineral; but it may exhibited in a variety of ways. Here we consider the quantitative measurement of hardness, and the qualitative properties of cleavage, brittleness, malleability, sectility, flexibility and elasticity.

HARDNESS

The hardness of a mineral is a characteristic property and a valuable guide in identification. Traditionally the hardness recorded by the mineralogist is the *scratch hardness*, measuring the capability of a sharp corner of one mineral to scratch a smooth surface of another. The basis of the test is a set of minerals selected by the Austrian mineralogist F. Mohs in 1824, numbered 1 to 10 in order of increasing hardness, each of which will scratch the one below it in the scale and will not scratch the one above. The list follows.

Mohs's Scale of Hardness

1. Talc	6. Orthoclase
2. Gypsum	7. Quartz
3. Calcite	8. Topaz
4. Fluorite	9. Corundum
5. Apatite	10. Diamond

Mohs was well aware that the steps on this scale were not equal, but he pointed out that this ought not to detract from its usefulness, and experience has confirmed his opinion.

To the list of testing materials may be added, for convenience, the thumb nail which will, for most people, scratch gypsum but not calcite, the point of a good quality pocket-knife which will barely scratch orthoclase, and common window glass which can be scratched by orthoclase and is readily scratched by quartz.

For making hardness tests a set of chips of the minerals of the scale should be kept available. It is possible to cement a chip of each, with 'araldite', into the end of a short length of metal tubing to make a set of hardness 'pencils' which are easy to handle. A *smooth* surface of the substance under test must be chosen, because many minerals are brittle and the edges of irregularities may crumble without a true scratch being made. On a smooth surface, when a scratch or indentation is made, the material, though brittle, appears to yield in part by plastic flow. The corner of the testing chip is pressed hard against the surface and a *small* rubbing movement made—less than a millimetre is ample. Care should be taken to avoid defacing a mineral specimen with an ugly scratch. When the material being tested is close in hardness to the standard point being used, the resulting mark should be lightly wiped and examined with a hand lens to make sure that a true scratch has been made.

Mineral hardness determined in this way may be quoted as, for example, $3\frac{1}{2}$, but not as a decimal fraction, because the accuracy of the determination does not warrant subdivision of the scale intervals into tenths.

Relationship of scratch hardness to indentation hardness

There is a standard method of measuring the hardness of metals by pressing a pyramidal diamond point into the surface under a known load and measuring the diameter of the resulting indentation (see p. 138). This method applies fairly well to brittle minerals, which seem to deform plastically under localized load, and may be used for comparison with the scratch hardness scale of Mohs.

The results of scratching tests on metals show that for a point to scratch a surface its hardness must be about 1·2 times that of the surface. This being so, as the hardness goes up, the intervals between the standards on a scale like that of Mohs will become progressively greater, each standard being ideally 1·2 times as hard as the last. When the diamond indentation hardness, IH, for the minerals of Mohs's scale is plotted against the Mohs number, we find that the intervals do become steadily larger, except that the interval between corundum and diamond is excessively large (Fig. 85). Although there is some spread of values of IH measured by different investigators, each standard on Mohs's scale is about 1·6 times as hard as the last, up to number 9, corundum. This indicates that Mohs selected his minerals with great care and skill to obtain not equal, but *just* intervals of scale, except that diamond by far surpasses in hardness the next mineral below it.

Hardness, like other physical properties, is affected by anisotropy of the mineral structure and will vary on different planes. This is true even of cubic minerals. Except for a few cases like kyanite, with $H = 4$–5 on $\{100\}$ parallel to x-axis, 6–7 on $\{100\}$ parallel to y-axis and 7 on $\{010\}$ the effect is not large enough to be reported. In the case of diamond, however, the difference in hardness of the faces is important since it permits the cutting of diamond by abrasion with diamond powder.

Hardness of different mineral groups

Some generalizations about the hardness of minerals are useful. Appendix III, p. 305, gives further data.

Native elements, with the notable exception of diamond, are generally soft. Platinum ($H = 4$–$4\frac{1}{2}$) and iron ($H = 4\frac{1}{2}$) are however, fairly hard and iridosmine ($H = 6$–7) is hard.

Compounds of the heavy metals, silver, copper, lead and mercury, are soft ($H < 4$).

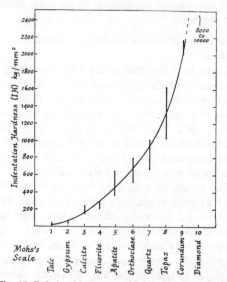

Fig. 85. Relationship between Mohs's Hardness Scale and indentation hardness, based on data of TABOR, D., 1954, p. 249.

Most sulphides and sulpho-salts are relatively soft (though the common iron disulphide, pyrite, has $H = 6$–$6\frac{1}{2}$).

Halides are soft.

Carbonates and sulphates are mostly soft.

Phosphates are of intermediate hardness, $H \simeq 5$.

Anhydrous silicates are mostly hard, $H = 5\frac{1}{2}$–8; hydrous silicates (micas, zeolites) are, however, softer.

Oxides tend to be hard: hydroxides are, by contrast, relatively soft.

CLEAVAGE

The tendency of many minerals to break smoothly along particular planes of atoms in the structure is a manifestation of weaker bond strength across those than in other directions in the structure.

Cleavage planes are always planes with a high density of atoms, and so are always parallel to a possible crystal face. The cleavage plane is identified by the appropriate Miller symbol, in the same way as a crystal face.

Cleavage is studied by observing the regular systems of cracks within transparent crystals like fluorite or calcite, or the smooth reflecting surfaces that result from breaking the grains, as in feldspar, pyroxene, galena, etc. It is a constant and reliable property, which may give valuable guidance to the symmetry of a mineral. The traces of cleavage planes are invaluable as reference lines in the optical study of anhedral grains (those without well developed faces) under the microscope.

According to the ease with which the mineral breaks along the regular planes, cleavage is described by the following terms:

eminent: extreme ease of breaking, so that it is difficult to prevent it, e.g. the cleavage of mica parallel to {001}; that of molybdenite along {0001}.

perfect: easily broken, e.g. fluorite along {111}; calcite along {10$\bar{1}$1}; barytes along {110}.

distinct: as in hypersthene along {110}.

imperfect: as in apatite along {0001}.

Other categories are *difficult* and *in traces*.

A *parting* is a tendency to break along discrete surfaces, as opposed to cleavage which occurs along any one of a set of planes separated by distances on the interatomic scale. Parting is related to lamellar growth in the crystal, often produced by exsolution of a separate crystalline phase along a certain crystallographic plane in the host crystal. It is shown, for example, in some augite, which has lamellae of pigeonite or hypersthene parallel to {001}.

A knowledge of the atomic structure of the mineral often explains the direction of cleavage. This is well illustrated by the micas, where the atomic arrangement is one of sheets of atoms parallel to {001}, the sheets being grouped into 'sandwiches' (Fig. 239). The slices of bread in each sandwich are linked sheets of Si^{4+}—O^{2-} tetrahedral groups with their apices pointing inwards. The sandwich filling is a layer of Al^{3+} or (Mg^{2+}, Fe^{2+}) octahedrally linked to the tetrahedral apices. Successive sandwiches are joined to the ones above and below in the pile by weak bonds to K^+ which is in 12-fold coordination. The single charge on the K^+ is thus shared between 12 neighbours, each bond is electrostatically weak, and the cleavage along planes of K^+ ions is easily understood.

The cleavage of pyroxenes and amphiboles is also directly related to their structure of chains of linked

Fig. 86. Conchoidal fracture in obsidian.

Si^{4+}—O^{2-} tetrahedral groups running parallel to z crystallographic axis. The cleavage takes place between the chains as shown in Figs. 227 and 233.

FRACTURE

When a mineral is broken in a direction other than a cleavage plane the nature of the fractured surface may be distinctive.

Conchoidal (shell-like) fracture is the most notable variety. The material breaks, under a blow, along concave surfaces marked with ridges roughly concentric about the point of impact, the whole surface being reminiscent of the impression of a mussel shell. It is the fracture shown by glass, and is well shown by the volcanic glass *obsidian* (which is a rock and not a mineral, by the way). Crypto-crystalline silica in the form of flint shows conchoidal fracture well, and early man exploited this property in making flint implements with sharp cutting edges produced by the intersection of conchoidal surfaces of fracture (Fig. 86). Amongst minerals, quartz and olivine show conchoidal fracture well.

Other less distinctive fractures are described as *even*, *uneven*, and *hackly*, the last being applied to a surface with small but sharp and jagged irregularities.

TENACITY

Under this heading are summed up the reactions of minerals to shock, crushing, cutting and bending actions.

The native metals, copper, silver and gold may be flattened under light blows of a hammer. They are described as *malleable*. Most minerals, however, are *brittle* and will crumble under light blows, or pressure. Some minerals fly to pieces more readily than others, anglesite ($PbSO_4$), for example, being very brittle. At the other extreme, a few minerals can be cut by a knife, though they still powder under a hammer. These are *sectile* minerals of which cerargyrite or horn silver (AgCl) is an example.

Cleavage flakes of mica are *flexible* and *elastic*. Flakes of molybdenite are *flexible*, but *inelastic*, and the same is true of scales of talc.

PROPERTIES INVOLVING EXCITATION OF CRYSTAL ENERGY

Under this heading are considered the properties of colour, fluorescence and phosphorescence; piezo-electricity and pyroelectricity; magnetism; and radioactivity. These are all changes in the energy

distribution within the atoms, and all except the last are called forth by external stimulation. The optical properties of minerals in transmitted and reflected light might perhaps be thought to belong here, too; but it is the modification of the light rather than the modification of the crystal field that is studied in optical mineralogy. In any case, the optical properties are given so much attention in this book that a separate chapter is obviously called for.

COLOUR, FLUORESCENCE AND PHOSPHORESCENCE

Colour in a mineral is, in most cases, due to absorption of certain wavelengths of light energy by the atoms making up the crystal. The remaining wavelengths of white light that are not absorbed give the sensation of colour to the eye.

Crystals may also absorb radiation of wavelengths beyond the visible range, and part of the energy thus gained may be radiated again, as light within the visible range. This phenomenon is called *luminescence*. If the luminescent emission goes on only as long as the original exciting radiation is falling upon the crystal, the effect is called *fluorescence* (from the mineral fluorite, CaF_2, which shows it well). When the emission of light continues after the input of exciting radiation has ceased, the crystal is said to be *phosphorescent*. In this last case, the crystal stores light-energy gained from the incident radiation, for later emission.

These properties are varied expressions of one fundamental mechanism, namely the absorption of radiant energy by electrons in the outer electron shells of the ions of the crystal lattice, or by electrons associated with defects in the lattice. As we have seen (Chapter 1) each electron shell of an atom is associated with a certain quantum of energy, an outer shell electron being more energetic than one in an inner shell. When the influence of the energy field of the crystal lattice as a whole is superimposed on this energy pattern of the individual atom, the picture that emerges is that of a series of *energy levels* or *bands* in which electrons lie, separated by forbidden zones (Fig. 87).

The electrons involved in colour production are those of the outer shells of the atoms, that is the valency electrons. Electrons of the deeper, X-ray, levels do not seem to be affected.

Absorption of energy by an electron may promote it from an inner (less energetic) to an outer band, and when the amount of energy absorbed is great enough, the electron may pass into a *conduction band* in which it becomes independent of the atomic nucleus and is free to move about through the crystal, under the influence of external electric fields. Absorption of radiation is associated with promotion of electrons to higher energy bands, and emission of radiation with their fall (partial or complete) back to their original level.

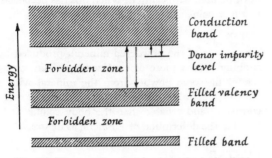

Fig. 87. Diagram of energy levels in the lattice field that are involved in colour and luminescence.

F CENTRES (COLOUR CENTRES)

An F centre (from German *Farbe*) is a lattice defect that absorbs visible light. Such defects may be due to:

(*i*) An excess of atoms of one element in the compound over that required by the chemical formula, for example an excess of Na atoms in crystal of NaCl.

(*ii*) The presence of foreign (impurity) ions in the lattice.

(*iii*) Colloidal particles in the lattice, due to coalescence of excess atoms like those mentioned under (*i*).

(*iv*) Mechanical deformation of the lattice.

In general, disturbance of the lattice regularity will produce vacant anion and cation positions. A vacant anion position (the absence of a negative charge) behaves electrostatically like a positive charge, and can capture an electron. The F centre is believed to be a positively charged vacancy with an electron moving about it.

When colour is produced in an initially colourless crystal by irradiating it with ultra-violet light or X-rays, it is supposed that some of the anions have lost an outer electron which has absorbed a quantity of energy that permits it to move into a conduction band. If the lattice were perfect it would fall back again when excitation ceased, but lattice defects

provide local energy levels between the excited state and the original or *ground state*, into which the electron may now move. When irradiation ceases the energy distribution in the crystal will have been changed and F centres that absorb light-energy will have been formed. The crystal will now be coloured.

Each anion that has lost an electron will have a vacancy in its outer electron energy level. Such anions are called *positive holes* (a term also much used in connection with transistors) and have power to capture electrons.

Experiments with alkali halides have shown that these crystals may be coloured by heating them in alkali metal vapour. This leads to an excess of alkali metal atoms in the lattice. (The colour produced depends only on the crystal, not on the vapour used.) When part of a crystal is coloured in this way, and an electric field is applied to it at high temperature the colour moves through the crystal to the anode, showing that the colour centres move in the same way as negatively-charged particles.

FLUORESCENCE

When certain minerals, such as fluorite (CaF_2), scheelite ($CaWO_4$), willemite (Zn_2SiO_4), uranium-bearing minerals, some diamond, and many others, are irradiated in darkness by ultraviolet light (which is invisible) they glow with their own emitted light. Fluorescence can also be induced by irradiation with cathode rays, or with X-rays, by heating or contact with flames (*thermoluminescence*), or by friction and crushing (*triboluminescence*).

The possession and strength of fluorescent properties varies in different specimens of the same mineral, and even from zone to zone in one crystal. It is often, perhaps always, related to the presence of small amounts of impurity elements in the crystal lattice. The impurity is called an *activator* and its concentration may vary widely. Transition metal ions, manganese in particular, have been found to be effective activators for silicates in amounts of about 1%, while copper and silver activate ZnS and other sulphides when present in amounts of only 1 part in 10 000.

In terms of our energy level model for the crystal lattice, electrons of the active centres (impurities) are lifted into a higher energy band by the incident radiation and then fall back into their unexcited (ground) state with the emission of visible light-energy. Some may be lifted into a conduction band, giving rise to *photoconductivity* during irradiation.

The frequency of the light emitted during fluorescence is almost always lower (and hence its wavelength is greater) than that of the exciting radiation. That is, the emitted radiation is of lower energy, some energy being lost in the form of heat during excitation of the electrons and their return to the ground state.

The use of an ultra-violet lamp at night or underground is a very effective way of finding minerals that show fluorescence. Wavelengths of 250 nm and 360 nm are particularly useful as these are strongly absorbed by silicates and sulphides respectively.

In the laboratory, fusion in borax of minerals containing uranium and other activators may give a fluorescent glass bead.

The fluorescence of some minerals is fairly constant; but, as might be expected from its origin in variable impurity content, it is a variable property, and specimens from different localities show different behaviour.

The fluorescence of artificial crystals is now highly important as a means of detecting and measuring ionizing radiations. The scintillations of light from thallium-activated potassium iodide, for example, are the basis for one of the common types of radiation counter.

PHOSPHORESCENCE

The period of light emission from an excited crystal after radiation has ceased is called the *after-glow*. It may be anything from 10^{-8} s to hours. This delayed emission is *phosphorescence*, and it is exhibited by many of the minerals that show fluorescence. Study of the decay-time of the emission leads to a model in which the excited electrons from atoms ionized by the radiation return to their ground state and recombine with the ionized centres by stages, being *trapped* on the way, at intermediate energy levels provided by impurity centres, positive holes, or other defects. These traps are energy 'pits' in which the electron may be held until a high jump in the statistical fluctuations of its thermal vibration allow it to escape. In phosphors with slow decay times, the release of stored light-energy can often be accelerated by treating them with infra-red radiation. Absorption of energy from this radiation presumably helps the trapped electrons to escape from the traps and return to the unexcited state, with the emission of light.

Phosphors with long decay times are useful for making radar and television screens and cathode-ray oscilloscopes.

COLOUR IN MINERAL IDENTIFICATION

Because of its variability and dependence upon impurities, colour is said to be a poor criterion of a mineral. Nevertheless, because it is so obvious, it inevitably influences mineral identification strongly. Some colorations are quite reliable, e.g. the green of malachite, the blue of azurite, the red of cinnabar, the yellow of sulphur. But in many cases colour must be used with caution. When this has been said, it still remains a valuable diagnostic feature.

STREAK

The colour of the powdered mineral, produced by rubbing it on a piece of unglazed porcelain, or by scratching some powder from the mineral with a knife or file, or by crushing a small portion, is more consistent and reliable than the body colour. It is listed in the description of a mineral as its *streak*.

LUSTRE

This is defined, rather vaguely, as the amount and quality of the reflection of light from the mineral surface. It is the amount of sparkle. Although the definition lacks precision, lustre is a very specific and useful property in mineral identification. Its value lies in the fact that it is determined by simple inspection, and yet is the product of two fundamental properties, each of which requires much more elaborate means if it is to be measured. These are (*i*) the nature of the atomic bonding in the crystal and (*ii*) the refractive and absorptive indices, together with the dispersion. A third influence, not fundamental, is the degree of roughness of the reflecting surface. This factor may modify the lustre of a mineral, but may also confer a characteristic quality upon it.

Terms describing the *amount of light reflected* are these:

splendent: brilliant reflection, giving sharp images of the light source (reflections of window bars clearly seen, for example);

shining: showing blurred images only;

glistening: a general reflection, but no image.

On the *quality of reflected light*, lustre is divided into two main kinds, the *metallic* and the *non-metallic*.

METALLIC LUSTRE

This is the lustre of opaque substances of high absorptive index which are good reflectors (see p. 134). It indicates metallic bonding, or a high degree of covalent bonding between the atoms, and is exhibited by the native metals, sulphides and sulpho-salts. The substances concerned nearly all have reflectivities of 20% and usually over 30%, with high refractive indices, though the latter are rarely directly measurable.

SUBMETALLIC LUSTRE

This is shown by some of the semi-opaque oxides (e.g. rutile, TiO_2; haematite, Fe_2O_3) with refractive indices of 2 to 3.

NON-METALLIC LUSTRE

Transparent and translucent substances possess this type of lustre. It is subdivided as follows.

Adamantine

A high degree of sparkle, associated with high translucency. Covalent bonding, as in diamond, and the presence of heavy metal atoms, as in cerussite, $PbCO_3$, or transition elements as in rutile, TiO_2, contribute to this lustre. Minerals possessing adamantine lustre have high refractive indices of from 1·9 to 2·6, and high dispersion.

Resinous

The lustre of resin is shown by sphalerite, ZnS, and other semi-transparent crystals of refractive index greater than 2.

Vitreous

The lustre of broken glass, or of quartz. A large number of translucent minerals with predominantly ionic bonding of elements of atomic number less than 27 (i.e. below Fe in the Periodic Table) display this lustre. In particular, many silicates are vitreous in lustre. Their refractive index range is from 1·5 to 2·0.

Pearly

This is shown by layer lattice silicates, like talc, $Mg_3Si_4O_{10} (OH)_2$, and chlorite, $(Al, Mg)_{5-6} (Si, Al)_4O_{10} (OH)_8$, with eminent basal cleavage. Dolomite, $Ca Mg (CO_3)_2$, and the cleavage faces of gypsum (variety selenite), $CaSO_4 . 2H_2O$, also reflect with this soft light.

Silky

This is shown by minerals of fibrous crystallization, and is more related to surface texture than to internal structure. Fibrous gypsum exemplifies it.

Greasy

Nepheline, $(K, Na)AlSiO_4$, has the lustre of grease, perhaps partly because of surface alteration.

These phenomena are due to reflection of light from exsolution lamellae or other inhomogeneities in the crystal. The variety of feldspar called moonstone is a good example. In this mineral there are regularly-orientated alternating lamellae of sodium-rich and potassium-rich feldspar due to exsolution in crystals that were homogeneous at higher temperatures. In fact, the crystals may be homogenized again by appropriate heating with the loss of the play of colours. The variation in refractive index between the lamellae causes internal reflection of light and a play of colours.

Schiller lustre is a similar effect in pyroxene, also as a result of exsolution.

The play of colours in opal, an amorphous mineraloid, is ascribed to a submicroscopic structure of minute spheres of colloidal silica (perhaps with differing refractive indices) whose interfaces give rise to internal reflections and spectral dispersion.

PIEZOELECTRICITY

It was shown in 1880, by the brothers Curie, that when some crystals are stressed along certain directions they acquire an electric field, one surface of the crystal becoming positively charged and the opposite surface negatively charged. This is called the *direct piezoelectric* effect*. The polarized charge per unit area is given by

$$P = d\sigma$$

where σ is the stress and d is a constant called the *piezoelectric modulus*. The disposition of the charges is reversed when the stress is changed from tension to compression.

Conversely, when an electric field is applied to a piezoelectric crystal its dimensions change slightly and this is known as the *converse piezoelectric effect*.

This property of crystals has been widely employed in technology, particularly in quartz resonators for frequency control, in crystal pick-ups for gramophones, in propagators of ultrasonic wave-trains, and in pressure-gauges.

The piezoelectric effect is a vector property, varying with the direction of applied stress. When some values of the modulus d in known directions have been measured experimentally, it is possible, by vector analysis, to calculate the magnitude of the charges produced by any given load acting in a specified direction. The reader is referred to the

* Pronounced like the words *pie ease oh electric*.

account given by J. F. Nye (1957. *Physical properties of crystals*, Oxford Univ. Press, Chap. 7) for the mathematical procedure.

By applying the vector matrix formula that Nye derives to each crystal class in turn, it is found that the *piezoelectric moduli become zero in all centrosymmetrical classes*. Also in class 432, although it has no centre of symmetry, the high degree of axial symmetry causes all the moduli to become zero. This leaves 20 symmetry classes in which the piezoelectric effect will be manifested.

The fact that a crystal shows the piezoelectric effect may therefore be helpful, in conjunction with other observations, in assigning it to its symmetry class.

A QUALITATIVE TEST

The measurement of the charges produced by stressing crystals involves the use of very perfect crystal plates, free from twinning and impurities, and accurately cut, whose surfaces are coated with a conducting metal, and great care must be taken in insulating the circuit.

A qualitative test for the effect, useful in investigating the symmetry, is however readily made with the Giebe-Schiebe apparatus. Either grains or plates of the mineral may be used. The mineral is placed between two metal plates forming a condenser, which is connected to a variable condenser so that the frequency of the circuit may be changed, and the output from the circuit is amplified and fed to a loud-speaker. As the frequency is varied it will pass through the resonant frequency of the piezoelectric crystal. The resulting variations in current give a series of sharp sounds, or, with several fragments, a train of sounds through the loud-speaker. A circuit for such an apparatus is given by W. A. Wooster (1938. *A textbook on crystal physics*, Cambridge, p. 208).

ORIGIN OF THE PIEZOELECTRIC FIELD

The charges produced in piezoelectric crystals originate in the disturbance and unbalancing of the electrostatic bonding forces between the atoms in the crystal. We still await a full explanation of the energy relations, but in the case of low-quartz it is held that pressure along any of the diad axes distorts the network of atomic linkage shown in Fig. 17(c). An estimate of the electric effect produced by distortion of the dipoles on the assumption that the linkage is fully ionic is of the right order of magnitude to account for the observed surface charges.

PYROELECTRICITY

It has been known for centuries in India and Ceylon that tourmaline, when heated in the embers of a fire, first attracted ashes and then repelled them. This phenomenon, due to electrical charges on the surface, is called *pyroelectricity*. It occurs in crystals of the same classes as those that show piezoelectricity— that is those without a centre of symmetry—and it must be closely related to piezoelectricity. As crystals when heated and cooled expand and contract, they are in a state of strain, and the separation of pyroelectric and piezoelectric effects is thus a matter of difficulty. We must, therefore, be content here to note the probable close link between the two phenomena.

Of particular interest is the easy demonstration of the electrical property of crystals by heating tourmaline. A clean, dry crystal of tourmaline is heated to about 200°C in· an oven, then ·passed a few times through the flame of an alcohol burner to dissipate the surface charges, and placed on a glass plate to cool. As it cools, opposite charges will separate on its surface. A bell-jar filled with the smoke of burnt magnesium ribbon is placed over the crystal on the plate and filaments of white magnesium oxide will grow out along the lines of force around the crystal. Alternatively, a mixture of powdered red lead and sulphur may be sifted down on the crystal, when the red lead goes to the negatively charged surfaces and the sulphur to the positive areas. Again, if the crystal is cooled in liquid air or liquid nitrogen, and then suspended in the room, the moisture in the air will condense as threads of tiny ice crystals extending along the lines of force. Cady, in a notable indictment of old-fashioned museum methods, illustrates the pattern of dust that had gathered by electrostatic precipitation around tourmaline crystals lying unmoved for many years on the shelf of a mineralogical cabinet!

MAGNETIC PROPERTIES OF MINERALS

The behaviour in a magnetic field of all crystalline materials may be classified under the headings diamagnetic, paramagnetic, ferromagnetic, antiferromagnetic or ferrimagnetic.

The magnetic properties reside in the electrons of the atoms or ions. On the principles of wavemechanics, the electron, moving in a closed path about the nucleus, can be considered as a current behaving like a wave, and this moving current generates a magnetic field. When the crystal is placed in an external non-uniform magnetic field, there will be a force tending to align the magnetic fields of the atoms to produce a magnetic moment for the whole crystal. The magnetic susceptibility, χ, is the ratio of the resulting magnetic moment, M, to the strength of the external field, H.

$$\chi = M/H.$$

Diamagnetic substances have a small negative value of χ, and are slightly repelled by the field.

Paramagnetic substances have a small positive value of χ, and are weakly attracted by the field.

Neither diamagnetic nor paramagnetic substances retain any magnetic moment in the absence of a field.

Ferromagnetic substances possess a magnetic moment even in the absence of an applied field.

The other two categories will be mentioned later.

PARAMAGNETISM AND DIAMAGNETISM

Physical experiments and theory show that, whereas paramagnetism is associated with the spins of the electrons, diamagnetism is related to their distribution in space. Diamagnetism is a property possessed by all atoms; but when the atom contains an odd number of electrons, or has incomplete electron shells (as in the transition elements) imbalance of the electron spins causes the paramagnetic effect to overshadow the diamagnetic part of the total magnetic susceptibility. Paramagnetism is also found in metals where there is a cloud of free conduction electrons.

These statements about atoms apply in a general way only, to crystals, because the internal crystal field as a whole modifies the magnetic effects. The electronic energy levels in a crystal are described as being split, and the total magnetic susceptibility depends on the distribution of the electrons in the different levels. In complex compounds, therefore, prediction of magnetic properties is not yet possible. Amongst minerals we can say, at least, that iron-bearing structures are paramagnetic. There are, however, paramagnetic minerals without iron.

The differences in magnetic susceptibility · are sufficient to be of very great help in separating pure fractions, or concentrations, of minerals from crushed rocks or ores, by the use of a high intensity magnetic field. Electromagnetic separators are much used for this work in the laboratory and in industry.

FERROMAGNETISM

Ferromagnetic substances are strongly attracted by even a weak magnetic field, and are permanently magnetized by it. They can, however, also exist in an unmagnetized condition. In these substances, at ordinary temperatures, the electronic magnetic moments are permanently in alignment, as a result of interaction between neighbour atoms. To explain how ferromagnetics can exist in an unmagnetized state, a model is proposed in which the crystal is subdivided into small volumes, called *domains*, within which the moments are aligned; but, in the unmagnetized condition, the alignments of the different domains are not parallel. They are randomly orientated.

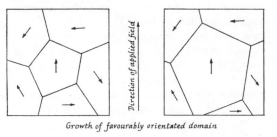

Growth of favourably orientated domain

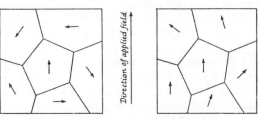

Realignment of domain moments

Fig. 88. Alternative ideas to explain the alignment of electronic magnetic moments in the magnetization of a ferromagnetic crystal.

When an external field is applied, we suppose that either the domains whose moments are nearly parallel to the field grow at the expense of the others, or that the directions of alignment in the domains change until they are all parallel with the field. The two ideas are shown in Fig. 88. When the domains have been induced into parallelism the material is magnetized; it will possess polarity, and if it is allowed to swing, it will set itself parallel to the earth's field.

ANTIFERROMAGNETISM AND FERRIMAGNETISM

The details of the alignment of electron spins in certain crystals introduce two new concepts. The interaction of adjacent atoms may be such as to align the spins in parallel but opposed directions, called *antiparallel spins*. This gives rise to *antiferromagnetism*. The two sets of moments cancel one another, and there is no permanent magnetic moment.

Further than this, we have cases of antiparallel alignment in which the components in opposed directions are not equal, so that there is a resultant permanent moment. This is *ferrimagnetism*.

Ferrimagnetism is shown by the important mineral magnetite $(Fe^{2+}Fe_2^{3+}O_4)$ which is a member of the *spinel group*. Members of this group have a chemical formula of the type $M^{2+}M_2^{3+}O_4$. Their structure is a cubic close-packing of O^{2-} ions between which there are 64 tetrahedral interstices, each surrounded by four O^{2-}, and 32 octahedral interstices each surrounded by six O^{2-} ions. Of these, 8 tetrahedral sites (the A sites) and 16 octahedral sites (the B sites) are occupied by metal ions, in such a way that chains of octahedral groups run parallel to the cube face-diagonals, the chains being cross-linked by tetrahedral groups (Fig. 185).

The ions M^{2+} and M^{3+} may be distributed in various ways. One extreme is the *normal spinel* with the A sites occupied by divalent ions M^{2+}, and the B sites by trivalent ions M^{3+} (e.g. Mg Al_2O_4). The other extreme is an *inverse spinel*, in which the A sites contain half the trivalent ions and the B sites are filled by the rest of the trivalent ions plus the divalent ions (e.g. $Fe^{3+}[Mg\ Fe^{3+}]O_4$). Artificial spinels called ferrites can be made, in which M^{3+} is Fe^{3+} and M^{2+} may be Mn, Co, Cu, Mg, Zn, or Cd.

The magnetic properties of magnetite and other ferric iron spinels is explained as follows. The A and B sites form two interpenetrating sub-lattices within the main lattice. There may be interactions of types A—A, B—B and A—B between adjacent metal ions when the sites are occupied by mixed ions. But an adequate, though simplified, explanation of the magnetism is given if we assume that A—B interactions are much the strongest, and that the interaction causes antiparallel (antiferromagnetic) arrangement of the A and B spins. In an inverse Fe^{3+} spinel, in this case, the Fe^{3+} A spins cancel the Fe^{3+} B spins and the resulting magnetization is due to the moments of the M^{2+} ions in B positions.

Measurement of the magnetization of magnetite supports the view that its moment arises only from the Fe^{2+} ions and implies that its structure is that of an inverse spinel whose formula might be written $Fe^{3+}[Fe^{2+}\ Fe^{3+}]O_4$. Fig. 89 shows how the magnetization of inverse Fe^{3+} spinels in general depends upon the properties of the M^{2+} ion.

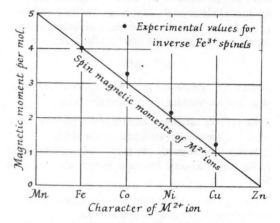

Fig. 89. Diagram showing the dependence of the magnetism of inverse spinels on the nature of the divalent metal ion. (After WILLIAMS, D. E. G., 1966.)

LOADSTONE

The attraction of certain varieties of magnetite[*] for iron and its capacity to impart magnetization to iron was known in classical times. Legend has it that the shepherd, Magnes, discovered this property when the iron ferrule of his staff adhered to the rock.

This variety, called loadstone (lodestone), assumed great practical importance in the advance of navigation. It was recognized at an early date that loadstone possessed polarity, and that if an elongated piece of it were allowed to swing it would align itself north and south. The use of this property as a direction-indicator—in fact as a magnetic compass—is recorded from about A.D. 1200. The loadstone was used to magnetize a needle, which was more conveniently pivoted, but the mineral itself, mounted and shod with iron tips, was carried by voyagers to remagnetize the needle or make a new compass, as required, until the end of the eighteenth century. Well-known localities for loadstone were Calabria, in Spain, and Monte Calamita in Elba.

It is probable that lightning striking the magnetite rock exposed at the surface induces the condition of polarity. It is a common experience that

---

[*] Perhaps including maghemite, γ-Fe_2O_3, a cubic close-packing of O^{2-} with only $21\frac{1}{3}$ out of 24 metal ion sites occupied, which is also ferromagnetic.

on mountain summits in terrain built of rocks containing very moderate amounts of magnetite, which elsewhere are non-magnetic, the rock may produce strong deflection of the compass-needle. This is attributed to lightning striking the higher points of ground.

RADIOACTIVITY

The most prominent naturally-occurring radioactive elements are uranium and thorium. These atoms decay spontaneously by emitting first an α-particle (identical with the nucleus of a helium atom) from the nucleus. They then lose a β^--particle (an electron), and by successive emission of further α and β particles they pass through a series of unstable daughter products to yield, finally, a stable isotope of lead. At the same time, the disintegration yields energy in the form of γ-rays, which are an electromagnetic radiation like X-rays, but of shorter wavelength.

The radiation produced by U and Th minerals can be detected by a portable Geiger-Müller counter or a scintillation counter (see p. 88) and these instruments are a valuable prospecting tool for U and Th minerals.

One consequence of radioactive decay is that the radiation produced damages the crystal structure of the radioactive mineral. Such damaged crystals are said to be *metamict*. The damage may produce a volume change that cracks the surrounding crystals (e.g. allanite), the optical properties may be affected, and the diffracting power of the lattice for X-rays may diminish until it is *X-ray amorphous*. The diffracting properties may sometimes be restored by heating the crystal, which allows the damaged lattice to re-organize itself. Small crystals of zircon, containing thorium, embedded in biotite, may by their α-radiation produce an intensely pleochroic halo round the zircon grain. The diameter of the halo is related to the nature of the radioactive isotope present, the energy and hence penetrating power of α-particles from a particular type of nucleus being constant; while the intensity is related to the length of time since crystallization, when the bombardment started.

Radioactive decay proceeds at a constant rate, regardless of temperature, pressure, or the state of chemical combination of the atoms. This rate is known, so that, by determining the ratio of the radiogenic lead to uranium (or uranium + thorium) in minerals containing these elements, we can measure the time since the mineral crystallized. This, and the similar methods based on the decay of potassium to argon and rubidium to strontium, are the basis of the absolute geological time-scale.

A final important result of radioactivity is its contribution to the internal heat of the earth. A. Holmes[*] states that the whole heat loss from the earth could be supplied by 36 g of uranium to every 10^9 tons of rock. The average crust rock contains fifty times as much, though clearly the interior materials must contain much less. To this must be added the heat from the radio-isotopes of the more abundant potassium, which is also concentrated in the crust. The conclusion seems to be that we have here an ample diffused source of heat for geological processes like eruptive activity. How it becomes concentrated at particular places is another question.

[*] *Principles of physical geology*, 1944, p. 409. Edinburgh, Nelson & Sons.

Further Reading

DENSITY

HOLMES, A. 1921. *Petrographic methods and calculations*. London, Murby & Co.
JOHANNSEN, A. 1918. *Manual of petrographic methods*. New York, McGraw-Hill Co.
MASON, B. 1944. The determination of the density of solids. *Geol. Fören. Förhandl., Stockholm*, 66, 27–51.

COLOUR, FLUORESCENCE AND PHOSPHORESCENCE

GLEASON, S. 1960. *Ultraviolet guide to minerals*. Princeton, N.J., Van Nostrand Co.
PRZIBRAM, K. (trans. J. E. Caffyn). 1956. *Irradiation colours and luminescence*. London, Pergamon Press Ltd.

PIEZOELECTRICITY

CADY, W. G. 1946. *Piezoelectricity*. New York, McGraw-Hill Co.
NYE, J. F. 1957. *Physical properties of crystals*. Oxford, Clarendon Press.

MAGNETIC PROPERTIES

BROMEHEAD, C. E. N. 1948. *Ships' loadstones*. Min. Mag. 28, 429–37.
WILLIAMS, D. E. G. 1966. *The magnetic properties of matter*. London, Longmans Green & Co.

RADIOACTIVITY

FAUL, H. 1966. *Ages of rocks, planets and stars*, New York, McGraw-Hill Co.
HOLMES, A. 1937. *The age of the earth*. 2nd ed., London, Nelson & Sons.

4

THE OPTICAL PROPERTIES OF CRYSTALS

RADIATION

We are all familiar with the experience that when something is burned its atoms dissociate in some degree and recombine in new compounds, some of which are gases; and at the same time a certain amount of light and heat is produced. The light and heat differ from the other products in that they are not corporeal—they are not bodies that can be collected and weighed, like the gases, the smoke and the ash, but are *radiant energy*, or simply *radiation*.

Radiation of this kind can travel through empty space, as witness the fact that light and heat reach us from the sun, and radio-waves from stars in immensely remote parts of the Universe. It is described, by equations derived by James Clerk Maxwell in 1865,* as an electromagnetic disturbance generated in pulses, with an *electric energy vector* at right-angles to the direction of travel, and an associated *magnetic energy vector* also at right-angles to the direction of propagation and at the same time normal to the electric vector. One of the consequences of the Clerk Maxwell formulation is that the velocity of all such disturbances in a vacuum would be the same.

Electromagnetic radiation in a vacuum is, then, all of the same kind. But when the energy it represents encounters matter—that is, atomic particles of the chemical elements—it interacts with it, and by its behaviour in this interaction it can be sorted out into different kinds. This division of electromagnetic radiation into different kinds is arbitrary and purely a matter of convenience, for each kind grades continuously into the variety

* *Phil. Trans. R. Soc.* **155**, 459–512.
See also PHEMISTER, T. C. 1954. *Am. Miner.*, **39**, p. 172.

on either side of it, and we call the entire range of types the electromagnetic spectrum (Fig. 90).

When we come to describe the interaction of electromagnetic radiation with matter the most satisfactory way of doing so is by using the analogy of wave-motion. The oscillating energy field pulses forward like the train of waves moving outwards across a pool when a stone is dropped into the water. As may be seen by watching a cork floating on the pool, the water is not moved away from the centre during the passage of the wave-train. The water particles simply move up and down in response to the disturbance until, in this case, the initial energy is exhausted by working against gravity and internal friction.

The wave analogy will be used in this chapter to describe the behaviour of light, and later to describe the interaction of X-rays with crystals. Some definitions are needed, therefore, for use in the discussion.

Fig. 91 is a cross-section of a wave-form with a scale of distance travelled, d, and a scale of time, t, showing what is meant by the wavelength, λ, the period, T, and the amplitude, a, of the wave.

The *velocity* of the wave is the distance travelled in unit time

$$V = d/t.$$

The *frequency*, v, is the number of complete oscillations made by a particle* transmitting the wave-motion in unit time. One swing takes place in the period, T, so

$$v = 1/T.$$

* Note that the idea of a particle moving to and fro normal to the direction of travel is only a notion derived from the analogy with a water wave with particles of water rising and falling. There are no particles in electromagnetic radiation, but the analogy with water particles is helpful.

94

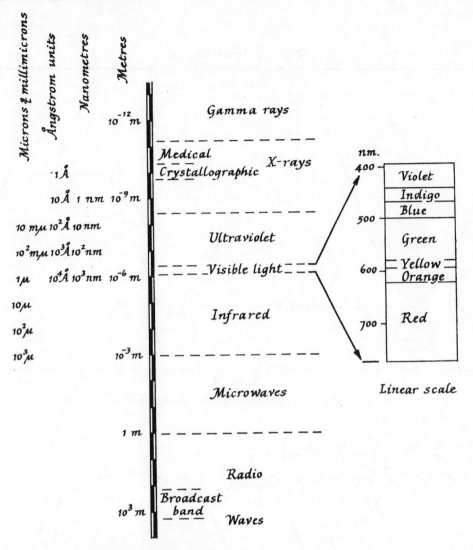

Fig. 90. Wavelengths of the electromagnetic spectrum.

The frequency is the fundamental property of an electromagnetic disturbance that remains unchanged during its interaction with matter. It is related to λ and V by the equation.

$$v\lambda = V.$$

In a vacuum where the velocity of all electromagnetic radiation is the same, the frequency and wavelength vary together. This is true for the passage through any other uniform medium also. Since the velocity can change from medium to medium, it follows from the statement $v\lambda = V$ that there is

a related wavelength change when it does so. In two media, 1 and 2,

$$V_1/V_2 = \lambda_1/\lambda_2.$$

In the case of light, the sensation of colour which we receive on the retina of the eye is, of course, due to radiation passing through the fluid of the eye. Therefore a given frequency of light will always have a velocity and wavelength appropriate to that medium of transmission, and so *we may associate colour with wavelength*, though basically it is dependent upon the frequency. During the passage

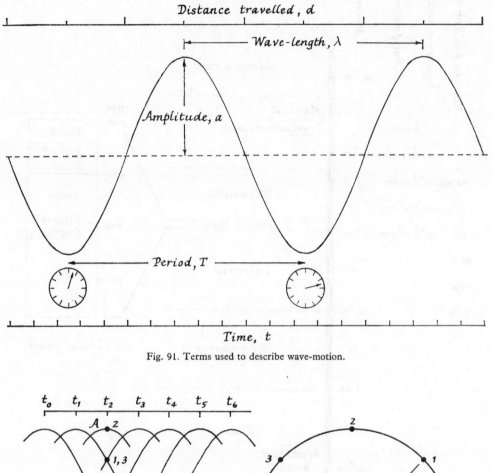

Fig. 91. Terms used to describe wave-motion.

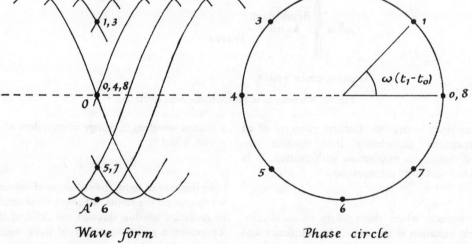

Fig. 92. Phase circle in relation to wave-form.

of light through a coloured substance certain frequencies are more strongly absorbed than others, so that the portion of the spectrum emerging from the substance contains a different frequency distribution from that of white light and the emerging light appears coloured.

DESCRIPTION OF WAVE-MOTION

Consider the motion of an individual particle participating in the transmission of the wave-form. As the wave passes, it occupies positions 0 to 8 in Fig. 92. Its velocity of movement to and from the zero position on the wave-path through $OAOA'O$ can be described in terms of the movement of a particle around a circle at constant velocity, when the circle is viewed edge-on. Its velocity will be at a maximum when passing through O and will diminish towards A and A' to zero at these points where its direction is reversed. The radius of this circle, which we will call the *phase-circle*, represents the amplitude of the wave-form.

If the velocity of the particle round the circle is ω radians per second, then the angle swept out in a period of time t which is a fraction of the wave-period T, will be ωt and is called the *phase-angle*. The particle makes a complete oscillation (or vibration) in the wave-period T, and the radius of the phase-circle has then swept out an angle of 2π radians.

FREQUENCY, AMPLITUDE, INTENSITY AND ENERGY

Radiation carries energy. The energy per unit volume multiplied by the velocity gives the energy-flow per second across a unit area, and this energy-flow is called the *intensity* of the radiation. There is a relation between the frequency (and wavelength) of a wave-motion and the energy of the radiation, the energy increasing in proportion to the square of the frequency. The higher the frequency (and hence the shorter the wavelength in a given medium) the greater the energy-flow or intensity.

For radiation of any given frequency, the intensity varies as the square of the amplitude. In other words, the energy represented by the displacement of a particle propagating the wave is proportional to the square of the distance through which it is displaced.

COMBINATION OF WAVE-MOTIONS

If two waves travelling in the same direction are propagated by the vibration of particles moving in the same plane, their effect upon the vibration of an individual particle is algebraically additive, counting upswings as positive and downswings negative. Fig. 93a shows two waves of the same frequency that are vibrating *in phase*, that is with crests in step, combined in this way. They reinforce one another to give greater amplitude, hence greater intensity. In Fig. 93b the two waves are one-half wavelength out of step or, on our phase-circle diagram they are $180°$ (π radians) out of phase. They *interfere* with one another destructively, and since they have the same amplitude, no resultant motion of the particle occurs and the wave is *extinguished*. In Fig. 93c the two waves are out of step by an arbitrary amount. The resultant wave has an amplitude different from that of either component vibration.

The combinations of wave-motions are represented graphically in the figure by the phase-circle diagrams, where amplitudes and phase differences are drawn to scale by radii of appropriate lengths separated by the appropriate phase-angle difference. The amplitude and phase of the resultant is given by vector addition using the well-known parallelogram rule that the resultant of two movements is given by graphing their individual effects on the particle as though they took place successively.

THE BEHAVIOUR OF LIGHT

Applying these concepts to the visible part of the electromagnetic spectrum, with wavelengths from about 390 nm to 770 nm, the first point to note is that the velocity of the radiation is reduced when it is transmitted through matter. Light is, indeed, slowed up during its passage through air compared with its velocity of 3×10^5 km/s in a vacuum, but for practical purposes we can use the velocity in air as the standard.

The underlying reason for this slowing up is that the electric vector of the radiation distorts, or polarizes, the electronic charge clouds of the atoms in its path, imposing on them vibrations that will vary in amplitude, depending upon both the electronic structure and the state of chemical combination of the atom concerned. This interaction reduces the velocity of propagation of the light, and the measure of this retardation is the *refractive index* of the substance traversed. At the same time, the energy of the light is absorbed to a greater or less degree, so that its intensity is reduced. This is the *absorption* of light by the substance, and as absorption increases the substance becomes more opaque.

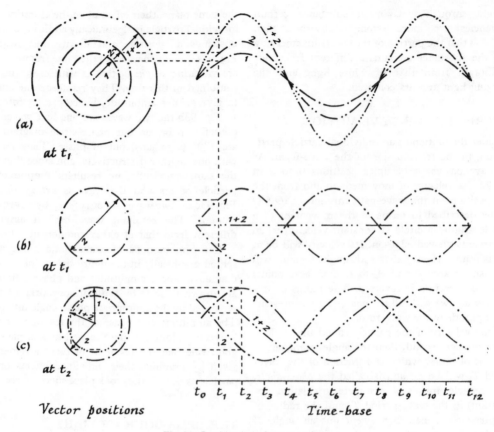

Fig. 93. Combination of wave-motions.

REFRACTIVE INDEX

The refractive index (n) of a substance is defined as the ratio of the velocity of light in air to the velocity of light in the substance,

$$n = V_a/V_s$$

ISOTROPIC AND ANISOTROPIC SUBSTANCES

Liquids and glasses (the latter being virtually supercooled liquids) do not have a regular atomic structure, but consist of atoms or molecules in random arrangement. Because of this, all directions through them are statistically the same, and in accordance with this they possess only one value of the refractive index (for a given wavelength of light), which is the same whatever the direction of travel of light through the substance.

Substances crystallizing in the cubic system also have only one value of the refractive index (for a given wavelength), whatever the direction of travel of light through the crystal. They have a regular atomic arrangement, and the symmetry of this arrangement is very high, so that the atomic groupings are closely similar along any chosen direction, and light travelling in any direction is retarded to the same degree.

Substances of these classes, having only one value of the refractive index, are described as *optically isotropic* ('equal-turning' in Greek).

Crystals of less symmetrical systems, because they possess a regular orientation of atomic groups on a lattice with different repeat patterns in different directions, affect the light differently according to its direction of travel through the atomic array. They have a range of values of the refractive index related to the direction of travel of the light and are called *optically anisotropic*.

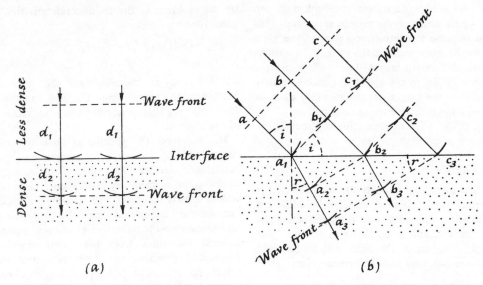

Fig. 94. Light-rays entering an isotropic medium (*a*) at normal incidence and (*b*) with oblique incidence.

REFRACTION BY ISOTROPIC SUBSTANCES: SNELL'S LAW

When a beam of light passes from a less dense isotropic medium into a denser one, travelling *normal to the interface* between them, it is slowed up, but its direction of travel is unchanged, as in Fig. 94a where d_1 and d_2 are the distances travelled in unit time in the two media.

When the light beam crosses the interface *obliquely*, however, the slowing up of the rays causes the beam to be bent or *refracted* at the interface. In Fig. 94b the arcs described about *a*, *b*, and *c* represent the distance travelled in unit time in air. The common tangent to the arcs represents the advancing *wave-front* of the light. When ray *a* crosses the interface into the denser substance (here glass) it advances at reduced velocity, reaching in unit time a point on the arc drawn about a_1. Ray *b* still travelling in air has meanwhile advanced to b_2. In the next unit of time, *a* advances to a_3, *b* to b_3 and *c* to c_3. The common tangent to these arcs represents the new wave-front which now advances at an angle to its direction in air.

When light passes obliquely from a rarer to a denser medium the *wave-normal* (at right-angles to the wave-front) is refracted *towards the normal to the interface* between the two media. Conversely, on passage from a denser to a rarer medium it is refracted away from the normal to the interface.

In Fig. 94b,

$$c_1c_3/a_1a_3 = V_a/V_s$$
$$c_1c_3/a_1c_3 = \sin i; \, a_1a_3/a_1c_3 = \sin r$$
$$\sin i/\sin r = c_1c_3/a_1c_3 . a_1c_3/a_1a_3 = c_1c_3/a_1a_3$$
$$= V_a/V_s = n$$

This relationship, known as *Snell's Law*, may be stated thus:

The refractive index, *n*, is given by the ratio of the sine of the angle of incidence in air, *i*, to the sine of the angle of refraction, *r*, both angles being measured from the normal to the interface.

DISPERSION

When sunlight passes through a glass prism (or any piece of glass with two non-parallel surfaces, like

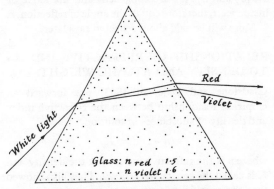

Fig. 95. Dispersion of light by a prism.

the bevel at the edge of a mirror) it is broken up into the separate colours of the visible spectrum. This happens because the refractive index of the glass differs for different wavelengths of light, vibrations of shorter wavelength (say 440 nm, which gives a sensation of violet to the eye) being more strongly refracted than the longer wavelengths (say those of about 760 nm which give the sensation of red). This is shown in Fig. 95. The separation of the different wavelengths is called *dispersion*.

REFLECTION

When light travelling through one medium strikes an interface with another medium of different optical density, a proportion of the light enters the second medium, and is refracted in the way just described; but part of the light will be reflected from the interface back into the first medium.

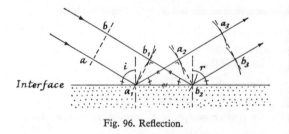

Fig. 96. Reflection.

In Fig. 96, ab and a_1b_1 are successive positions of the wave-front after unit time. After ray a meets the interface and is reflected, ray b travels a distance b_1b_2 to meet the interface in its turn. During this interval ray a, moving in the same medium and at the same velocity, will have travelled the same distance along its new path a_1a_2, normal to the wave-front which is tangent to the arc at a_2 and passes through b_2. The two right-angled triangles $a_1b_1b_2$ and $a_1a_2b_2$ are congruent and the angle of incidence, i, therefore equals the angle of reflection, r.

More will be said about reflected rays later.

RELATIONSHIP OF REFRACTIVE INDEX TO DENSITY AND ATOMIC WEIGHT

Gladstone and Dale, in 1863, put forward an empirical relationship between the refractive index and density of substances in solution,

$$(n-1)/d = K,$$

where n is the refractive index, d is the density and K is called the *specific refractivity*. They also showed that the specific refractivity of a solution is given

by the addition of the specific refractivities of its constituents:

$$((n_1-1)/d_1)w_1 + ((n_2-1)/d_2)w_2$$
$$= ((n-1)/d)(w_1+w_2)$$

where w_1, w_2 are the weights of the constituents. Moreover, the change from the liquid to the solid state only very slightly affects the specific refractivity of a substance.

If, therefore, the specific refractivities of the constituents of a mixture or a compound can be determined (either from glasses of appropriate composition or from crystals of pure compounds) and its density measured, the refractive index of the mixture or compound may be approximately calculated. Specific refractivities of certain standard mineral molecules have been determined, and reasonable agreement found between observed and calculated refractive indices of complex natural glasses.[*]

A more complex formula for the relationship between refractive index and density was put forward, independently of one another, by Lorentz and Lorenz in 1880. It is

$$(n^2-1)/(n^2+2) \times 1/d = K$$

For silicate glasses and feldspars, this and the simpler Gladstone and Dale relationship hold about equally well.

Although we are dealing in this section with isotropic substances, it is convenient, while on the subject of the relation between refractive index and composition, to anticipate a little and look at the properties of some crystalline compounds (including some that are anisotropic and hence have more than one principal value of the refractive index) in this connection.

It is sometimes said that, in general, crystals containing heavy elements have high refractive indices. This may be true if we take extreme cases, but there are many examples which do not bear out this generalization. The case of diamond (cubic system and hence isotropic) comes to mind immediately, for although composed of the light element carbon (atomic weight 12) it has the high refractive index of 2.417. Looking further, the examples of Table 14, where groups of minerals with similar and simple structures are arranged in order of increasing refractive index, show that it would be quite wrong to assume any simple relationship between the atomic weight of the element

---

[*] TILLEY, C. E. 1922. Density, refractivity and composition relations of some natural glasses. *Min. Mag.*, **19**, 275–94.

Table 14. Refractive indices of some related groups of minerals.

Mineral	Formula	Refractive Indices			Cation	
		α	β	γ	At. No.	At. Wt.
Strontianite	$Sr\,CO_3$	1·520	1·667	1·668	38	87·63
Witherite	$Ba\,CO_3$	1·529	1·676	1·677	56	137·36
Aragonite	$Ca\,CO_3$	1·530	1·680	1·685	20	40·08
Cerussite	$Pb\,CO_3$	1·804	2·076	2·078	82	207·21
		ω		ε		
Calcite	$Ca\,CO_3$	1·658		1·486	20	40·08
Dolomite	$Ca\,Mg\,(CO_3)_2$	1·682		1·502	$\begin{cases} 20 \\ 12 \end{cases}$	40·08 24·32
Magnesite	$Mg\,CO_3$	1·717		1·515	12	24·32
Rhodochrosite	$Mn\,CO_3$	1·820		1·600	25	54·93
Smithsonite	$Zn\,CO_3$	1·849		1·621	30	65·38
Siderite	$Fe\,CO_3$	1·873		1·633	26	55·84
		n				
Sylvine	$K\,Cl$	1·490			19	39·09
Halite	$Na\,Cl$	1·544			11	22·99

present and the refractive index, even within one class of compounds.

Whatever relationship exists must be dependent upon the character and strength of the atomic bonds and only secondarily upon the inner electronic configuration of the cation, while its nuclear mass may exercise very little influence.

DETERMINATION OF REFRACTIVE INDEX

Before going on to the behaviour towards light of anisotropic substances, we shall describe the methods of determining refractive index. These are the same for isotropic and anisotropic substances, except that in the latter the result will vary with the direction of vibration of the light in the substance.

MEASUREMENT OF THE CRITICAL ANGLE

When light passes from a denser to a rarer medium, it is refracted away from the normal to the interface. In Fig. 97a, as the angle of incidence, i, increases, the angle of refraction, r, of the successive rays 1, 2 becomes greater, until a ray such as 3 is refracted into parallelism with the interface. Any ray, such as 4, with a greater angle of incidence than this, will be unable to cross the interface and will be totally internally reflected. The value of the angle i at which angle r becomes 90° is fixed by the relative indices of the two media, and is called the *Critical Angle*.

Reversing the direction of a ray such as 3, so that it strikes the interface with *grazing incidence* we

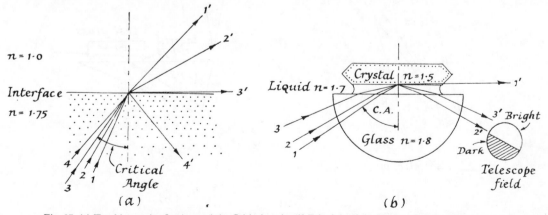

Fig. 97. (*a*) Total internal reflection and the Critical angle. (*b*) Principle of the Herbert Smith refractometer.

may write

$$n_{\text{glass}} = V_{\text{air}}/V_{\text{glass}}$$

or, by Snell's Law $= \sin i/\sin r$

with grazing incidence $= \sin 90°/\sin \text{C.A.}_{\text{air/glass}}$

$$= 1/\sin \text{C.A.}_{\text{air/glass}}$$

APPLICATION IN A REFRACTOMETER

Herbert Smith refractometer

The measurement of the critical angle by total internal reflection is applied in the Herbert Smith refractometer, and others of similar type (Fig. 97b). Here a flat surface of a crystal is placed on the flat surface of a hemi-cylinder of high-index glass with a liquid of higher index than the crystal between them to exclude air. Light enters one quadrant of the hemi-cylinder, and that part of the light which suffers total internal reflection at the lower surface of the crystal produces a bright field in a telescope trained on the other quadrant. The boundary between bright and dark fields gives the critical angle between glass and crystal. No refraction takes place at the hemi-cylinder surface, as all rays cross this at normal incidence. As the liquid forms a thin film with parallel surfaces its effect is balanced at entry and exit of the rays and can be ignored. Although the effective critical angle is that between liquid and crystal, this is corrected to the critical angle between glass and crystal by the deviation of the light at the glass/liquid interface.

The refractive index of the glass being known, the position of the boundary between bright and dark fields can be calibrated directly in refractive index values for the crystals tested.

$$n_{\text{crystal}} = V_{\text{air}}/V_{\text{glass}} \cdot V_{\text{glass}}/V_{\text{crystal}}$$

$$= 1/\sin \text{C.A.}_{\text{air/glass}} \cdot \sin \text{C.A.}_{\text{crystal/glass}}$$

$$= n_{\text{glass}} \cdot \sin \text{C.A.}_{\text{crystal/glass}}$$

The instrument is particularly useful for gemstones.

Abbé Refractometer

The Abbé refractometer, used primarily for liquids, employs grazing incidence of light passing from the liquid into a prism of high-index glass. Fig. 98a shows the light path. The pair of prisms, with a film of liquid between them, is rotated until the boundary between light and dark fields intersects the crosswires in a fixed telescope. The instrument is calibrated against a solid of known index, again using an arrangement with grazing incidence, as in Fig. 98b. It can be used in this mode for direct determination of the refractive index of suitably cut crystal blocks, but its principal use is as an adjunct to the immersion method for crystals (see below).

METHOD OF MINIMUM DEVIATION OF LIGHT BY A PRISM

This may be used for crystals with a suitable angle between a pair of faces, minerals or glasses cut into prism shape, or liquids in hollow glass prisms.

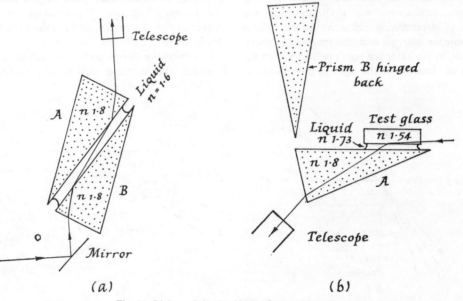

(a)   (b)

Fig. 98. Light path in the Abbé refractometer.

It depends upon the fact that when the angle at which light leaves a prism is the same as the angle of incidence on the other face, the beam suffers the minimum deviation (Fig. 99). To show that this is so, let us assume i does not equal i' when deviation is a minimum. Then, because light rays are reversible, there must be two angles of incidence at which deviation is a minimum. But experimentally we find only one. Therefore the light path must be symmetrically disposed and $i = i'$, $r = r'$.[*]

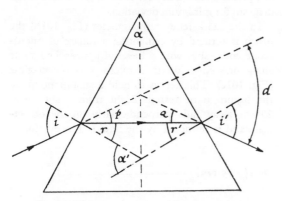

Fig. 99. Minimum deviation of light by a prism.

To determine the refractive index of the prism material two quantities are measured, α the prism angle, and d the angle of minimum deviation of the beam of light. Then

$$\alpha = \alpha' = r + r' = 2r$$
$$r = \alpha/2 \qquad (1)$$
$$i = p + r = q + r'$$
$$2i = p + q + 2r$$
$$d = p + q$$
$$2i = d + 2r = d + \alpha$$
$$i = \tfrac{1}{2}(d + \alpha) \qquad (2)$$
$$\sin i/\sin r = \sin \tfrac{1}{2}(\alpha + d)/\sin \tfrac{1}{2}\alpha.$$

The measurement is conveniently made on a single-circle goniometer. With the telescope at about 90° to the collimator, the prism is set up with the intersection of the two faces vertical by centring the reflected signal from the collimator on the telescope cross-wires with the adjustable arcs, just as in setting a zone for crystallographic measure-

ment (p. 42). The angle between the prism faces is read by rotating the prism on the central pillar which is clamped to the horizontal circle. The required angle α is the supplement of the one read.

The telescope is now unclamped and the central pillar released from the horizontal circle. The telescope is moved round (carrying the verniers with it) to be aligned with the direct beam from the collimator, and the verniers are read. The prism is turned so that its angle points to the observer's left, and the telescope moved to the right to pick up the deviated beam. With a white light source, this will appear as a band showing the colours of the spectrum, red to the left and violet to the right, because of the dispersion of the prism. (With a prism cut from an anisotropic substance two deviated beams, in general, will be seen—see below.) The central pillar with the prism is now turned (leaving the horizontal circle undisturbed) so that the signal moves to the left, and it is followed with the telescope. A point will be reached at which, no matter which way the prism is turned, the signal will not move further to the left, but changes direction and moves to the right. It is set at its most left-ward position, the position of minimum deviation of the beam, and the telescope is moved so that the cross-wires are set on the colour-band for the wave-length at which n is required (often the yellow line of the spectrum). The verniers are then read, and the difference between this and the last reading gives d, the angle of minimum deviation used in the calculation.

Since r must equal $\tfrac{1}{2}\alpha$ (Fig. 99), whilst as n goes up i increases, the size of the prism angle places an upper limit to the refractive index that can be measured, and for high indices α must be small. In practice a prism angle of 15° is satisfactory for refractive indices of about 2·0, and the angle may be increased as the index to be measured decreases, 30° being a useful angle at indices around 1·5. Although a small value of α reduces somewhat the accuracy attainable, the method is very useful for translucent substances of high index, where the immersion method is attended with difficulties.

The method of minimum deviation is applied to liquids by making a glass prism with one corner truncated, and cementing bevelled glass slips to its faces (Fig. 100) to make a cell. For refined work, water may be circulated through the block to control the temperature of the cell. With a specially large cell it is possible to establish a small column of heavy liquid, such as Thoulet's solution (p. 80),

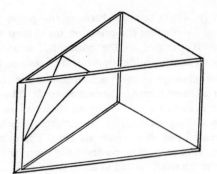

Fig. 100. A hollow prism to measure minimum deviation by a liquid.

A mask is used to restrict the incident beam to the level at which the grain is floating (JONES, J. M., 1961, *J. Scient. Instrum.*, **38**, 303–304).

The Leitz-Jelley Refractometer

This simple and effective instrument uses the deviation of light through a prism by the Method of Perpendicular Incidence, to determine the refractive index of small amounts of liquid. The geometry of the light path incident perpendicularly on one face of a prism is shown in Fig. 101a. It is one half of the diagram for minimum deviation.

In the Leitz-Jelley refractometer (Fig. 101b) the prism is formed by cementing a square of microscope cover-slip, with a bevel of appropriate angle along one edge, on to a microscope object-glass (Fig. 101c). This serves to hold a small amount of the liquid to be tested. Light from a slit about 25 cm away falls perpendicularly on the object-glass and is refracted by the liquid. The eye forms

of graded density (light at the top and denser at the bottom) and, making use of the relation between refractive index and density of the solution, to determine the density of mineral grains in floating equilibrium in the solution at a particular level.

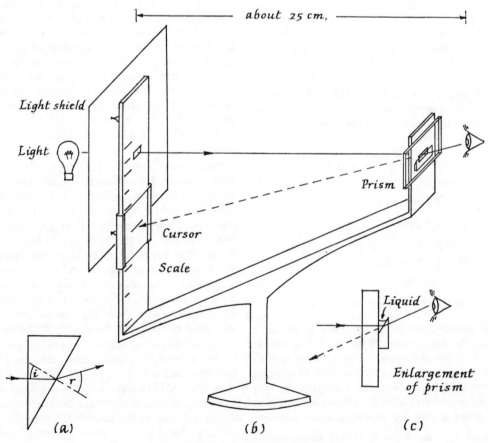

Fig. 101. The Leitz-Jelley refractometer.

a virtual image of the refracted spectrum* at the
nearest distance of distinct vision on a scale which
may be calibrated in refractive index. A filter
eliminating the yellow of the spectrum may be
placed behind the source slit and the reading taken
at the dark band. The instrument is useful for
refractive indices up to 1·9.

THE IMMERSION METHOD

For translucent minerals of refractive index from
1·4 to 1·9 this is the most generally suitable and
convenient method.

The mineral, broken into grains generally less
(often much less) than 0·5 mm across, is placed in a
drop of liquid of known refractive index on a microscope slide, covered with a cover-slip and examined

* The spectral colours here are mainly due to dispersion in
the glass. The case is thus different from those of white light
through a prism of a mineral solid, and of monochromatic
light varied step-wise in wavelength through a liquid in a
hollow prism at minimum deviation, both of which yield
information on dispersion.

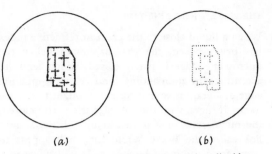

(a)   (b)

Fig. 102. Crystal outlines in immersion liquids.
(a) Refractive index of grain considerably higher or
lower than that of liquid. (b) Indices of grain and liquid
nearly equal.

under an intermediate power objective of a petrographic microscope in plane-polarized light (see
p. 110). If the grain has a refractive index much
different from that of the liquid it will appear with
a heavy dark outline. In a liquid nearer to it in
refractive index its outline becomes more difficult
to see, and a colourless transparent substance will
almost completely disappear (Fig. 102).

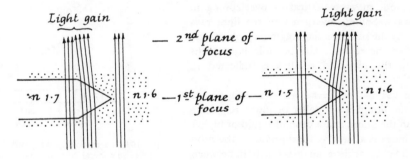

(a) Lens effect

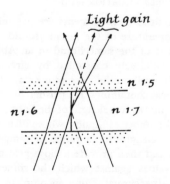

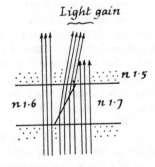

(b) Vertical grain junction,
convergent light rays

(c) Oblique grain junction,
parallel light rays

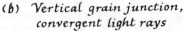

Fig. 103. Explanations of the Becke white line.

THE BECKE WHITE LINE TEST

When a liquid close to the grain in refractive index has been chosen, the iris diaphragm below the microscope stage is partly closed, which tends to accentuate the grain outlines, and a suitable tapering edge of a grain is sought for and brought to precise focus. If, now, with the substage iris partly closed, the microscope tube is racked up a little, a light line called the Becke White Line will appear to move from the grain boundary. It will move into the medium of *higher* refractive index when the microscope tube is *raised*.

The simplest explanation of this effect is that the tapering edge of the grain acts as a lens, converging the light when the grain index is higher than that of the liquid and diverging it when the grain index is lower, as shown in Fig. 103a. Experience shows that in mounts of grains in oil those with gradually tapering edges give the best white lines, and it is wise to be selective. The white line is also seen, however, in thin sections of rocks, at boundaries between grains of different refractive index, and here junctions may be vertical or overlapping in various ways. Some convergence of the light rays produced by the substage diaphragm may contribute to the effect here, and some possible light paths leading to the observed result are indicated in Fig. 103b, c.

By the use of the Becke line we may decide whether the liquid is of higher or lower refractive index than the grain, and a drop of liquid of higher or lower index is added, as appropriate, to the drop on the slide, to achieve an exact match between liquid and grain.

In practice an exact match will not be achieved in white light, because the dispersion of the liquid, usually greater than that of the grain, causes the white line to break up into coloured fringes. The variation in refractive indices of liquid and grain for different wavelengths of light is shown schematically in Fig. 104. When the refractive index of liquid and grain are exactly matched for, say, yellow light the grain index will exceed that of the liquid for orange light, and an orange-red line will move into the mineral, while a pale blue line will move into the liquid on raising the tube. When this effect is seen the refractive index of the liquid may be accepted as that of the grain for the spectral wavelength thus bracketed, or monochromatic light from a sodium-vapour lamp may be used to make the final adjustment to the liquid to cause the disappearance of the single light line then seen.

For accurate work, when the match is exact a small amount of the liquid from around the grains is transferred either to an Abbé, or to a Leitz-Jelley refractometer and its refractive index measured. The transfer is most easily done with the help of a short length of 3 mm glass tube drawn out into a capillary at one end. The liquid is easily picked up in the capillary and expressed on to the refractometer with a little pressure of the finger over the large end of the tube.

For rough work, an estimate of the refractive index after matching may be made by interpolating between the known values of index for the liquids used in securing the match.

Careful measurements, made with the help of a refractometer as described above, give a reproducibility of ± 0.003 for minerals with $n = 1.6$ to 1.7 and better than this in the range 1.5 to 1.6.

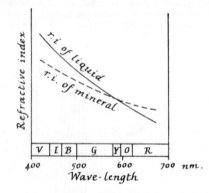

Fig. 104. Schematic dispersion curves for an immersion liquid and mineral. Spectral colours indicated along the abscissa.

THE DOUBLE VARIATION METHOD

More refined measurements are possible, in which the temperature of the liquid around the mineral, and some of the same liquid in an Abbé refractometer, is closely controlled by circulating water through a cell on the microscope stage and through the block containing the prisms of the refractometer, and the wavelength of the light is varied also. A small number of liquids, whose dispersion curves with wavelength for a range of temperatures are known, can then provide a close grid of refractive index values against which the mineral may be tested. Inaccuracy from incomplete mixing of liquids on a microscope slide is also avoided. For most purposes, however, the simpler method given before is quite adequate.

IMMERSION LIQUIDS

For the simple immersion technique the following liquids, each of which is miscible with the ones on either side of it in the list, can be used to make up a set, graded in steps of about 0·01, through the range of the common transparent minerals. The liquids, standardized on the refractometer and kept sealed in dark bottles away from the light, remain stable and reasonably constant in refractive index for years.

Paraffin (Kerosene)	1·40
Clove Oil	1·54
α-monochlornaphthalene	1·64
α-monobromnaphthalene	1·66
methylene iodide	1·73
(= di-iodomethane)	
methylene iodide saturated with sulphur	to 1·79

THE SHADOW TEST (SCHROEDER VAN DER KOLK METHOD)

Less sensitive than the Becke method, this provides a quick test whether grains are higher or lower in index than the liquid in which they are immersed. It is carried out under the microscope with a lower-power objective. The light coming from the microscope mirror is partly cut off by insertion of a card from one side, below the stage. First, while viewing the grains through the microscope, bring the image of the edge of the card to focus in the same plane as the grains, by moving the substage condenser up or down. Then *raise* the condenser until the image of the card becomes blurred. The reason for this is that the shadow effect will be reversed if the condenser focus is below the plane of the mineral grains.

It is a wise precaution to test an unfamiliar optical system first with grains of a known mineral, such as quartz.

Having adjusted the condenser, partly interrupt the light with the card; if the mineral differs in index from the liquid one side of the grain will appear brightly illuminated and the other dark. With the condenser adjusted as above, when the shadow in the grain is on the same side as the shadow in the microscope field, the mineral has a higher index than the liquid. When the shadow in the grain is on the opposite side from the shadow in the field the grain has a lower index than the liquid (Fig. 105).

This test is useful in inspecting a sample of mineral grains for purity, as it readily shows up foreign grains having an index different from the majority in a relatively large field of view.

ANISOTROPIC SUBSTANCES

Crystalline substances of lower symmetry than cubic, and certain kinds of fibres with molecules orientated in a preferred direction, are *anisotropic* towards light. In these substances the refractive index for light of a given wavelength is different for different vibration-directions of the light ray, and the substances exhibit *double refraction* of light or *birefringence*.

THE CALCITE RHOMB EXPERIMENT

Double refraction is most strikingly shown by calcite, in its highly transparent variety, Iceland Spar. Calcite cleaves readily into rhombs along

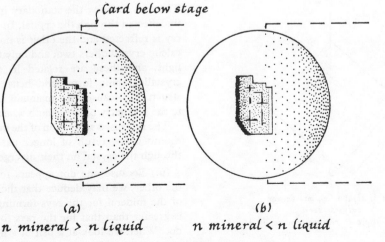

Card below stage

(a)
n mineral > n liquid

(b)
n mineral < n liquid

Fig. 105. Shadow test for refractive index.

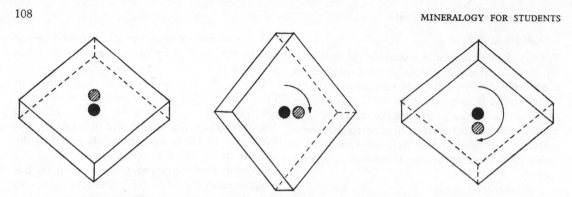

Fig. 106. Rotation of a calcite rhomb above a dot. Shaded dot appears less intense and a little lower.

planes of the form {10$\bar{1}$1} and if such a cleavage rhomb of Iceland Spar is placed over a dot on a piece of paper, two images of the dot are seen when it is viewed through the mineral (Fig. 106). One image of the dot will appear to lie a little higher above the surface of the paper than the other, and the line joining the two images will be parallel to the diagonal of the upper surface of the rhomb that joins the two obtuse angles. When the rhomb is rotated on the paper, one image (the apparently higher one) remains stationary while the other (the lower-seeming one) moves round the first so as to remain between the stationary dot and the obtuse angle of the rhomb that overhangs the paper.

If rhombs of different sizes are available, it will be seen that the degree of separation of the dots is greater the thicker the rhomb through which they are observed.

If the rhomb is now tilted up on to the obtuse

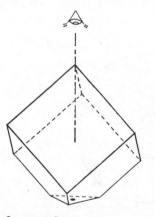

Line of sight when one image
only is seen
(Solid angle of rhomb ground off)

Fig. 107. Viewing the dot along the optic axis of calcite.

corner under the overhanging part of the cleavage block, while the eye continues to look vertically down on the dots (Fig. 107), the apparently lower dot will move nearer to the higher one until the two images fuse into one. This fusion takes place when the line of sight is parallel to the line equally inclined to the three {10$\bar{1}$1} planes that meet at the obtuse solid angle of the rhomb. This can be done roughly with an ordinary rhomb, but it is better if the obtuse solid angle is symmetrically truncated by grinding, so that the light rays cross the interface perpendicularly.

From these observations the following conclusions can be drawn.

(i) Because two images of the dot are seen when the rhomb lies on its face, the light is coming from the paper through the crystal by two paths.

(ii) Because one image of the dot moves round the other, the rays forming the moving image are inclined at a fixed angle in a fixed direction (parallel to the diagonal joining the obtuse angles of the face). The rays forming the stationary image are coming undeflected through the crystal. In other words, one ray is refracted and the other is not. Moreover, the calcite crystal offers two, and only two, paths to the light, and these are related in direction to the crystallographic form (and hence to the internal structure) because they maintain the same relation to the geometry of the rhomb when it is rotated.

The increased separation of the images in a thicker rhomb is the result of longer passage of the rays through the crystal on their divergent paths.

(iii) Because one dot appears to lie higher than the other, we may deduce that the refractive index of the mineral for the rays forming the higher dot is greater than that for the rays forming the lower dot. When we look at a fish in a fish tank, it looks nearer to the glass than it really is because the eye

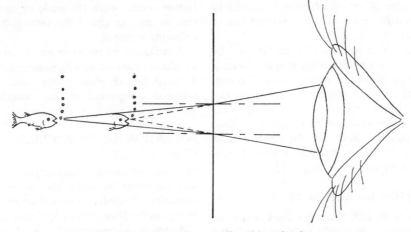

Fig. 108. Apparent distance affected by refraction.

forms an image of it at the convergence of the light rays in air, after refraction between water and air, not at the point on the fish from which the rays originated (Fig. 108). The effect is greater the greater the refractive index difference between the two media.

(*iv*) Because the two images fuse when the rays travel parallel to a line equally inclined to planes {10$\bar{1}$1}, that is parallel to crystallographic z, no double refraction occurs parallel to this direction, and the light all follows one path.

ORDINARY AND EXTRAORDINARY RAY

The light in the calcite rhomb experiment is being reflected from the paper through the rhomb, and as the rhomb is lying on the paper and we are looking straight down, the light reaching the eye is incident normal to the rhomb face in its passage from below. In studying the passage of light into an isotropic substance from air, we saw that light at normal incidence is not refracted, but merely slowed up. It is light incident obliquely that is refracted.

But in the calcite, light entering normal to the interface is broken up into two rays, one of which is refracted and the other not. The refracted ray is called the *extraordinary ray*, or *E-ray*, and the unrefracted ray, which obeys the ordinary rule for isotropic substances, is the *ordinary ray* or *O-ray*.

RAY DIRECTION, WAVE-NORMAL DIRECTION AND VIBRATION DIRECTION

After refraction in an anisotropic substance, the direction of propagation of the light energy—the ray direction—may no longer be the same as the normal to the wave-front of the advancing waves that make up the beam of light—the wave-normal direction. The wave-normal direction is the one with which the optical-instrument user is concerned, and the wave-normal velocity is the one to which Snell's Law applies.

More important still to the mineralogist is the fact that wave-normal velocity depends, not on the wave-normal direction or the ray direction, but upon the *vibration direction* of the light. Waves travelling in different directions in crystals have the same speed if their vibration directions are crystallographically the same. The student is therefore strongly counselled to think about the behaviour of light in crystals in terms of vibration direction, and to return to this concept when directions of travel threaten to become confusing. The reason for the importance of vibration direction becomes clear if we think of the electric displacement[*] of the light interacting with the electronic charge clouds of the atoms, as it sweeps through the crystal. The direction of its vibration governs the nature and succession of atoms that it interacts with in its passage.

POLARIZATION OF LIGHT

When light passes through an anisotropic crystal it becomes polarized, that is the light making up the E-ray vibrates in one plane only, and the O-ray also vibrates in one plane, which is at right angles to the plane of the E-ray. In order to study the

[*] There is a distinction between the electric displacement and the electric vector which we need not pursue here.

optical properties of crystals further, we need to understand and make use of the characteristics of *polarized light.*

Light radiating from a source in air can be represented as a wave motion oscillating or vibrating in all directions normal to its direction of travel. There are three ways in which a beam of light can be produced which vibrates in one plane only, that is can be converted into *plane-polarized light,* or *polarized light* for short. All these methods of producing polarized light diminish the total light flux by eliminating part of the original beam.

POLARIZATION BY REFLECTION

We noted above (p. 100) that when light strikes an interface between media of different refractive index part crosses the interface and is refracted in the new medium, and part is reflected. Light reflected in this way is partly polarized, so that most of it vibrates in a plane normal to the plane containing the the incident ray and the normal to the reflecting surface. (The plane containing the incident ray and the normal to the interface is called the *plane of incidence.*) The refracted ray that crosses the interface is polarized so as to vibrate at right angles to the vibration of the reflected ray (Fig. 109). The polarization of the reflected ray is not complete, but it is at a maximum when the angle between the reflected and refracted rays is 90°. This relationship is called Brewster's Law. The appropriate angle of incidence for maximum polarization depends on the relative refractive indices of the two media. In Fig. 109b, when $\angle BOC = 90°$, r is the complement of i. Hence, $\sin i/\sin r = \sin i/\cos i = \tan i$, so that if the less dense medium is air, maximum polar-

ization results when the angle of incidence is such that its tangent equals the refractive index of the reflecting material.

Polarization by reflection can be used to produce a beam of polarized light by passing the light through a succession of glass plates, and rejecting the reflected component at each interface. But the method is not much used in practice.

POLARIZATION BY DOUBLE REFRACTION

In older polarizing microscopes, polarized light was produced by using the property of double refraction of calcite, described above, in the device known as the Nicol Prism, or simply the *nicol.*

A prismatic cleavage piece of calcite is cut diagonally through the centre, and the pieces cemented together again with Canada balsam (Fig. 110). The ends of the prism are ground and polished and the diagonal cut made at angles such that, of the two beams produced by double refraction when light enters the end of the prism one, the ordinary ray ($n = 1·658$) meets the Canada balsam film ($n = 1·54$) at an angle greater than the critical angle (see p. 101) so that it is totally reflected into the blackened wall of the prism holder. The extraordinary ray ($n = 1·516$) is able to pass, with refraction, through the balsam film and emerge from the prism. This ray is completely polarized, as can be confirmed by a simple experiment described below (p. 111).

POLARIZATION BY ABSORPTION

Some natural crystals show a marked difference in their absorption of light vibrating in different

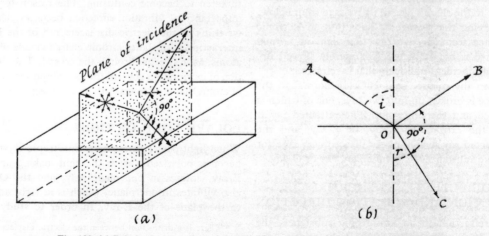

Fig. 109. (a) Polarization by reflection. (b) The Brewster angle of maximum polarization.

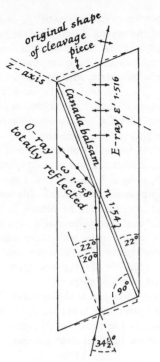

original shape
of cleavage
piece

z-axis

Canada balsam

E-ray ε, 1.516

O-ray
totally reflected

ω 1.658

n 1.549

22°   22°
20°
90°
34½°

Fig. 110. Nicol's Prism.

imposes new vibration directions before the light reaches the second plate.

In 1928 E. H. Land, using this principle and exploiting the properties of newly-discovered plastic films, invented the polarizing sheet which, in a more developed form, is now almost universally used to produce polarized light in microscopes. In the early form of *polaroid*, as this sheet is called, myriads of tiny needle-shaped crystals of an organic iodide that shows strong dichroism were deposited in plastic film, which was then greatly stretched so that all the needles were drawn into parallelism. Light vibrating parallel to their length (in this case) is so strongly absorbed that hardly any passes the film. Light vibrating normal to the length of the crystals passes freely, to provide plane-polarized light. Later types of polaroid make use of iodine compounds, or other dye compounds, of the stretched and aligned molecules of the plastic itself, to achieve the same result.

CONFIRMATION OF POLARIZATION BY CALCITE

Now that we have discussed polarized light, we can carry our calcite rhomb experiment a stage further. Wearing a pair of polaroid sun-glasses, look again through the calcite rhomb lying on a dot on a piece of paper, and turn the paper, with the rhomb on it while your head remains still. First one and then the other image will disappear. This shows that the light forming the two images is polarized, and when its plane of vibration parallels the absorbing direction of the polaroid the image disappears. It is easy to confirm that the E-ray and the O-ray are polarized in planes mutually at right angles.

THE OPTICAL INDICATRIX

That the two rays transmitted by calcite have different refractive indices may easily be proved by examining small cleavage rhombs immersed in refractive index oils, under a polarizing microscope. First one diagonal, and then the other, is set parallel to the plane of vibration of the polarizing screen below the microscope stage. Two different values of the index will be found.

In anisotropic substances the refractive index is a vector property, its value varying with the direction of vibration of the light ray. It is found, as might be expected, that this vector property is intimately connected with the crystallographic symmetry. Indeed this was foreshadowed by the fact that cubic crystals are optically isotropic. The optical indicatrix

directions in the crystal. In tourmaline (formula $Na(Mg, Fe)_3Al_6(BO_3)_3(Si_6O_{18})(OH)_4$, a member of the trigonal system), for example, the rays vibrating parallel to the length of the crystal are much less strongly absorbed than those vibrating transverse to the length. This property is called *dichroism* because the crystal shows different colours for the two vibration directions. In tourmaline it is explained by the atomic structure, which has Si_6O_{18} rings and BO_3 triangular groups both with their planes normal to crystallographic z, the triad axis. When the electric vector of light vibrates normal to z-axis and parallel to the plane of these groups it interacts strongly with them, and visible light is strongly absorbed. When the electric vector vibrates parallel to z most wavelengths are freely transmitted.

Mineralogists have used this property for many years, in tourmaline tongs employed to observe double refraction on a coarse scale. Two plates of tourmaline are cut parallel to z-axis and the z-axes of the two are in crossed orientation in the tips of the tongs. Light that has been able to pass one plate is thus vibrating in the direction of strong absorption in the other, and so no light gets through the pair unless a doubly refracting crystal in between is a device to help us to see the relationship between refractive index and crystal symmetry.

To construct the indicatrix we use the values of refractive index, *n*, in different directions in the crystal as radii, drawn to scale, to define the surface of an ellipsoid. In a cubic crystal this surface will become a sphere, since there is only one value of *n*. This correlates with the morphological symmetry, in which the crystallographic axes (the principal symmetry axes) are equal and interchangeable.

THE UNIAXIAL INDICATRIX

In crystals of the tetragonal, hexagonal and trigonal systems, measurement shows two principal values of *n*, one for vibrations parallel to crystallographic *z* (the tetrad, hexad, or triad axis) and the other for all directions of vibration normal to this. The indicatrix becomes, thus, an ellipsoid of revolution, with a single circular section and a semi-axis normal to this which may be greater or smaller than the radius of the circular section (Fig. 111). If it is *greater* the ellipsoid is prolate, and the *crystal is defined as optically positive*; if *less* the ellipsoid is oblate and the crystal is defined as *optically negative*.

The direction normal to the circular section (and

coinciding with crystallographic *z*) is called the *optic axis* of the crystal. Light incident normally on a surface of a crystal that is at right angles to the optic axis, will be vibrating in the plane of the circular section of the indicatrix, wherein *n* is the same in all directions, so that it will not be doubly refracted, but will behave as though the crystal were isotropic. This is true only for light whose wave-normal is parallel to this unique direction, the optic axis. Light incident normally* on a surface of the crystal in any other orientation will be doubly refracted.

Consider, now, light incident on a surface that is not at right angles to the optic axis. If a section of the ellipsoid, passing through its centre, is cut normal to the direction of the incident light, this section will be an ellipse whose major and minor semi-axes give the refractive indices of the two rays into which the light is divided by double refraction. All such central sections of the ellipsoid will have one semi-axis the same, and equal to the radius of the circular section. The refractive index for light vibrating in this direction is the same for all sections, and the same as that for light incident parallel to the optic

---

* In this discussion we shall consider normal incidence only. This is the condition obtaining, to a close approximation, in most situations in optical mineralogy.

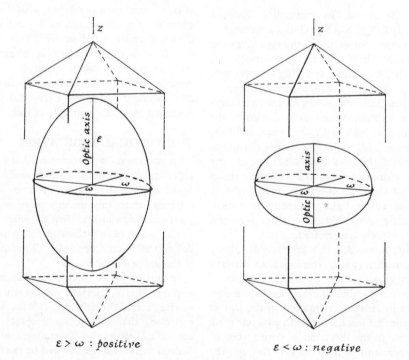

$\varepsilon > \omega$: *positive* $\varepsilon < \omega$: *negative*

Fig. 111. The uniaxial indicatrix.

axis. It represents the refractive index of the ordinary ray, n_o or ω. The length of the other semi-axis of the elliptical section will vary with the direction of the slice, between the extremes given by (*i*) the radius of the circular section, and (*ii*) the length of the ellipsoid axis parallel to crystallographic *z*. This variable semi-axis of the elliptical section represents the refractive index of the extraordinary ray, polarized in the plane of the incident ray and the optic axis (= crystallographic *z* in uniaxial minerals). The extreme value of the refractive index of the extraordinary ray is found in sections parallel to *z*, and is given the symbol n_e or ε. Intermediate values for the E-ray in sections between this and the circular section are symbolized by n_e' or ε'.

SUMMARY OF THE OPTICAL PROPERTIES OF UNIAXIAL MINERALS

From this we see that every section of a translucent uniaxial mineral:

(*i*) offers two vibration directions to light passing through it;

(*ii*) these two vibration directions are at right angles to one another;

(*iii*) one vibration direction (the O-ray) has refractive index ω; the other (E-ray) has a refractive index dependent on the direction of the section, but lying between limits ω and ε;

(*iv*) the E-ray vibrates in the plane of the incident ray and the optic axis; the O-ray vibrates normal to this plane;

(*v*) *except* that a section normal to the optic axis has one value of the refractive index only, which is ω, and superficially behaves like an isotropic crystal;

(*vi*) if $\varepsilon > \omega$ the mineral is optically positive, if $\varepsilon < \omega$ it is negative. *Epsilon Greater than Omega* = positive—the EGO IS POSITIVE.

THE BIAXIAL INDICATRIX

Crystals belonging to the orthorhombic, monoclinic and triclinic systems differ from the optically uniaxial group in that no two of their crystallographic axes can be considered equivalent for descriptive purposes. This, of course, reflects a corresponding lack of equivalence of any two atomic structural directions. The directional variation in refractive index is now expressed by a triaxial ellipsoid surface (Fig. 112), with three unequal semi-axes *X*, *Y*, and *Z*, the refractive indices along which are symbolized by α, the least, β, the intermediate,

and γ, the greatest. (Note that β is *not* the mean of α and γ, but an independent intermediate value.)

It is a property of a triaxial ellipsoid that it has two circular sections sharing the *Y* (β) axis as a diameter, which lie symmetrically about the *Z* (γ) axis, and at an angle to *Z* that depends upon whether β is nearer in length to α or to γ. The biaxial indicatrix has, then, two such circular sections, and light incident normal to either of them will find one value of the refractive index, namely β, in all directions normal to its path. It will not be doubly refracted and will behave as though the crystal were isotropic—that is, it will behave like light incident along the optic axis of a uniaxial mineral. So minerals of the orthorhombic, monoclinic and triclinic systems have two optic axes instead of one; they are *biaxial minerals* and their optical properties are expressed by a *biaxial indicatrix* (Fig. 112).

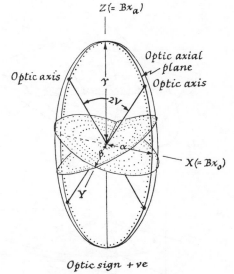

Optic sign +ve

Fig. 112. The biaxial indicatrix.

The plane in which the two optic axes lie (the α–γ plane of the ellipsoid) is called the *optic axial plane*. The angle between the optic axes is called the *optic axial angle*, symbolized by $2V$, and is a function of the relative lengths of α, β and γ. If $2V$ measured over γ is acute, the mineral is said to be optically positive, and γ is called the *acute bisectrix*, symbolized Bx_a. If $2V$ measured over γ is obtuse the mineral is negative, γ is then the *obtuse bisectrix* (Bx_o) and α becomes Bx_a. A point of indeterminancy of optic sign occurs when $2V = 90°$, but this causes no trouble and is itself a distinctive property that can be used in mineral determination.

The indicatrix axis $Y(\beta)$ normal to the optic axial plane is called the *optic normal*.

Light incident normally on a randomly cut plate of a biaxial mineral will have its vibration directions, and the refractive indices for those vibrations, given by the semi-axes of an ellipse formed by a slice through the centre of the indicatrix in the same orientation as the mineral plate, just as in the uniaxial case. But the rule governing the vibration directions of the two rays is now different. It was enunciated by Biot and proved by Fresnel.

THE BIOT–FRESNEL RULE

There will be two planes, intersecting in the line of the incident light (which is normal to the mineral slice), each also containing one of the optic axes of the crystal. The vibration directions in the mineral slice will bisect the angles between the traces of these two planes on the mineral slice. In effect this rule is simply a definition of the directions of the major and minor semi-axes of any elliptical section through the centre of a triaxial ellipsoid.

To visualize this, think of two circles (the circular sections of the indicatrix), intersecting in a common diameter ($Y = \beta$ of Fig. 113a), and inclined at any angle to one another. Imagine another plane (the plane of the mineral slice) cutting these two, and passing through their common centre (it is drawn as a slab in Fig. 113a). Its lines of intersection with the circles will be two straight lines, diameters of the two circles, passing through the common centre (AB, CD of Fig. 113a, b). The four points

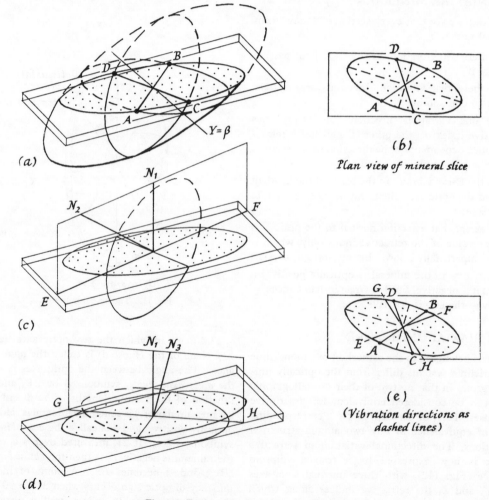

(b)

Plan view of mineral slice

(e)

(Vibration directions as dashed lines)

Fig. 113. Explanation of the Biot–Fresnel rule.

where these lines of intersection meet the circumferences of the circles must be points on the elliptical section (stippled in the figure) cut through the indicatrix by the mineral slice. We have, then, two pairs of equal radii of the elliptical section. But equal radii of an ellipse are equally inclined to its semi-axes. Hence the bisectors of the two pairs of angles between the pairs of radii are the semi-axes of the elliptical section (Fig. 113b).

Now, thinking of one of the circles and the mineral slice cutting it (Fig. 113c), we see that the plane containing the normal to the circle (i.e. the optic axis), N_2, and the normal to the mineral slice, N_1, will stand at right angles to the diameter in which circle and slice intersect. This will be true, of course, for both circles (Fig. 113d, N_1, N_3). Hence the Biot–Fresnel Rule, as stated above, using the bisectors of the angles between the traces on the section plane of planes containing the normal to the section and each optic axis in turn, also gives the directions of the semi-axes of the elliptical section, which are the vibration directions of the slice (Fig. 113e).

CALCULATION OF $2V$

The optic axial angle may be calculated from the refractive indices of a biaxial mineral.

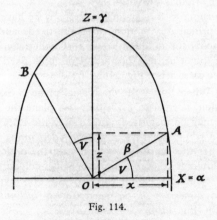

Fig. 114.

The $\alpha\gamma$ section of the indicatrix is an ellipse, the equation of which is

$$\frac{x^2}{\alpha^2} + \frac{z^2}{\gamma^2} = 1. \tag{1}$$

In Fig. 114 let OA be the trace of the circular section of radius β. OB is the corresponding optic axis.

From the figure (in which V is measured from γ)

$$z = \beta \sin V_\gamma \tag{2}$$

and

$$\beta^2 = x^2 + z^2. \tag{3}$$

Therefore

$$x^2 = \beta^2 - z^2. \tag{4}$$

Substituting (4) in (1)

$$\frac{\beta^2 - z^2}{\alpha^2} + \frac{z^2}{\gamma^2} = 1$$

$$\frac{z^2}{\gamma^2} = 1 - \frac{\beta^2 - z^2}{\alpha^2}$$

$$= \frac{\alpha^2 - \beta^2 + z^2}{\alpha^2}$$

$$\frac{z^2}{\gamma^2} - \frac{z^2}{\alpha^2} = \frac{\alpha^2 - \beta^2}{\alpha^2}$$

$$z^2 \left(\frac{1}{\gamma^2} - \frac{1}{\alpha^2} \right) = \frac{\alpha^2 - \beta^2}{\alpha^2}$$

$$z^2 = \frac{\alpha^2 - \beta^2}{\alpha^2} \times \frac{\gamma^2 \alpha^2}{\alpha^2 - \gamma^2} = \frac{\gamma^2(\alpha^2 - \beta^2)}{(\alpha^2 - \gamma^2)}.$$

From (2)

$$\sin^2 V_\gamma = \frac{\gamma^2(\beta^2 - \alpha^2)}{\beta^2(\gamma^2 - \alpha^2)}$$

Calculation of $2V$ from the refractive indices in this way is not usually very accurate, because small errors in the refractive indices produce a relatively large change in the calculated angle. Of more frequent use is a rough estimate from the approximate relation

$$\tan^2 V_\gamma \simeq (\beta - \alpha)/(\gamma - \beta).$$

RELATION BETWEEN APPARENT AND TRUE OPTIC AXIAL ANGLE

The apparent optic axial angle in air, symbolized by $2E$, will be greater than the true angle, $2V$, in the mineral, because of the refraction of the light passing from the mineral into air. In Fig. 115, where OA, OB represent the optic axes in the mineral and AC, BD the same directions refracted into air, $\sin E/\sin V$ equals n, the refractive index of the mineral. Substituting β, the intermediate value of the refractive index in a biaxial mineral, for n, we have the relation

$$\sin E = \beta \sin V.$$

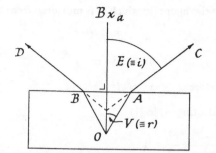

Fig. 115. Relation between true and apparent optic axial angle.

THE RELATION OF THE INDICATRIX TO CRYSTAL SYMMETRY

In uniaxial minerals, as we have seen, the optic axis coincides with the principal symmetry axis (tetrad, hexad or triad) of the crystal, and as the indicatrix in these minerals is an ellipsoid of revolution, its position is thereby fixed (Fig. 111).

In the orthorhombic system we find that the three principal indicatrix axes always coincide with the three diad axes shown by crystals of this system. The particular vibration direction α, β or γ that coincides with crystallographic x, y and z varies from one mineral to another (and must be stated in the description of the mineral's properties) but the two sets of directions always correspond.

In the monoclinic system one indicatrix axis always coincides with the single diad (= crystallographic y) characteristic of this system. Apart from this, the indicatrix may lie in any position relative to the crystallographic axes. In describing the properties of a monoclinic mineral we must state which principal vibration direction corresponds with y and also the *angle* between one of the other principal vibration directions and crystallographic z (or x).

In the triclinic system there is no restriction on the position of the indicatrix in relation to the crystallographic axes, and in describing these crystals it may be necessary to quote angles referring two principal vibration directions each to the three crystallographic axes. In practice it is easier to specify the vibration directions in terms of extinction angles (p. 125) on prominent pinacoids (cleavage planes and twin-planes in particular). These pinacoids will often define the plane of two of the crystallographic axes.

In optical mineralogy the determination of the relation of the indicatrix to physical directions in the crystal (that is to crystallographic planes) and through them to the crystallographic axes, is one of the chief aims. When this data has been recorded we can use the orientation of the optical directions to identify the mineral. Small changes in the optical orientation may be related to changes in composition brought about by atomic substitution in the mineral lattice.

THE EFFECT OF DISPERSION
(see also p. 133)

Because the refractive indices of a medium differ for different wavelengths of light, the indicatrix, uniaxial or biaxial, will be of slightly different shape for different colours of light. Since the uniaxial indicatrix is fixed with respect to the crystallographic axes, the semi-axes of the ellipsoids for different colours of light will be parallel and we shall have dispersion of the refractive indices only.

In the orthorhombic system the semi-axes of the indicatrix are again fixed, but because the positions of the circular sections, and hence of the optic axes, depend upon the relative lengths of α, β, and γ, the optic axial angle may vary for different colours of light. This phenomenon is called *dispersion of the optic axes*, or *rhombic dispersion* (Fig. 116a). In extreme cases, of which the mineral brookite (TiO_2) is an example, the optic axial angle may reduce to zero with changing wavelength of light, and then open out again in a plane at right angles to the original one (Fig. 189).

Turning to the monoclinic system, we have only one principal vibration direction fixed, and the ellipsoids for different colours of light may, as it were, pivot in their position about this fixed axis. Three cases arise (Fig. 116b–f).

(*i*) When $\beta = y$ the optic axial plane is fixed in position, normal to β. Within this plane the optic axes may be dispersed, and so may the bisectrices α and γ. The total effect will vary depending upon whether the degree of separation of the optic axes for different colours is great or small compared to the separation of the bisectrices. This type is called *dispersion of the bisectrices* or *inclined dispersion* (Fig. 116b, c).

(*ii*) When the obtuse bisectrix, Bx_o, is parallel to y the optic axial planes for different colours of light may be dispersed. Within the optic axial plane for any colour the optic axial angle may be different in size from its value in the nearby optic axial plane for another colour. The acute bisectrix will, however, always lie in a plane normal to y and to Bx_o. Because the line joining the optic axes for any colour is parallel to that joining the optic axes

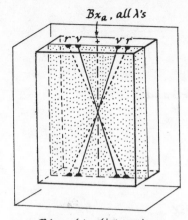

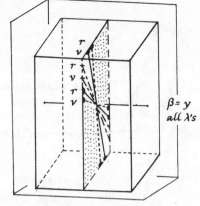

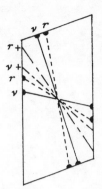

(a) *Rhombic dispersion*
Optic axial plane stippled

(b) *Inclined dispersion*
Optic axial plane stippled

(c) *View of (b) on*
010

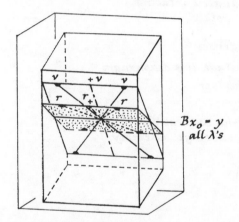

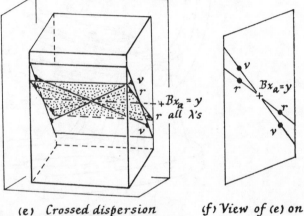

(d) *Horizontal dispersion*
Optic plane for red stippled

(e) *Crossed dispersion*
Optic plane for red stippled

(f) *View of (e) on*
010

Fig. 116. Dispersion of the indicatrix. Red light is denoted by r, violet by v.

for another colour this type is called *horizontal dispersion (of the optic axial plane)* (Fig. 116d).

(*iii*) When the acute bisectrix is parallel to y, dispersion of the optic axial plane and of the optic axes may occur as in the previous case. The lines joining the pairs of optic axes for different colours of light through Bx_a will now cross one another, intersecting in Bx_a, and the arrangement is called *crossed dispersion (of the optic axial plane)* (Fig. 116e, f).

THE EXAMINATION OF MINERALS IN TRANSMITTED LIGHT

We shall now turn to the practical measurements used to determine the relationship of the indicatrix to the crystallographic directions in minerals.

THE PETROLOGICAL MICROSCOPE
(Fig. 117).

This consists essentially of a rotating stage on which the mineral under examination is placed, with a polarizing screen called the *polarizer* below the stage, and another removable one, called the *analyser*, above the mineral. Important accessories not always found on other microscopes are cross-hairs at right angles in the plane of focus of the eyepiece arranged so that they lie N–S and E–W in the field of view, an iris diaphragm below the stage to limit the incident beam of light, a slot in the microscope tube at 45° to the cross-hairs for the insertion of accessory plates, and an arrangement of condensing lenses below the stage that produces converging light rays in the plane of the mineral under study when required.

E

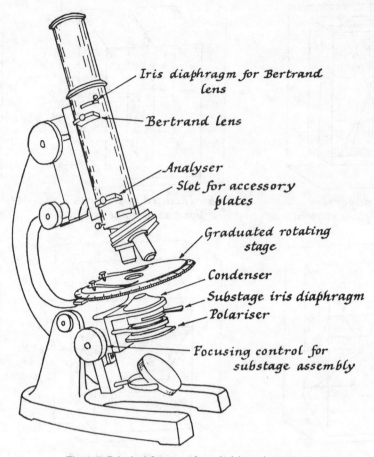

Fig. 117. Principal features of a polarising microscrope.

Many refinements and variations in arrangement are available.

PREPARATION OF MATERIAL

In order to examine minerals by transmitted light one must, in general, use small crystals or fragments of crystals that will allow the passage of light, or else prepare a thin section. Grains or small crystals are usually immersed in a drop of oil of known refractive index, to prevent too great refraction of light at their edges, and the drop is covered with a cover-slip. A thin section is made by grinding a flat surface on the mineral or mineral aggregate (usually a rock), and cementing it to a slip of glass with Canada balsam or a synthetic resin of known and stable refractive index. By long tradition a resin

with $n = 1.54$ is used. The mineral or rock is then ground down with successively finer grades of carborundum powder to a suitable thickness (0.03 mm is standard) and a cover-slip cemented on it.

OBSERVATIONS IN PLANE POLARIZED LIGHT

The polarizer below the microscope stage is set with its plane of polarization (that is the vibration direction of light passing through it) parallel to one of the eyepiece cross-hairs, usually the east–west cross-hair, using the ordinary map-orientation convention. This position will usually be indicated by graduations and a catch on the rotating holder of the polar. If there is any doubt about the direction of vibration of the polarizer, the pleochroism of biotite, which shows its strongest absorption of light when the vibration direction is parallel to its cleavage, may be used to determine the vibration direction of the light. The vibration direction of the analyser will be normal to that of the polarizer.

ISOTROPIC MINERALS

(*i*) Insert the analyser above the mineral, and rotate the microscope stage. If the grain remains dark through the complete rotation, and if this is true for a number of grains lying in random orientation, the mineral is isotropic. Like the glass of the object slip supporting it, it has no effect on the polarized light from below, which is extinguished by the analyser as though no grain were present. If the grain permits the passage of light at some positions of rotation it is anisotropic. In this case it will become dark four times in a complete rotation, at intervals of 90°.

(*ii*) With the analyser removed from the tube, the next property to be recorded is the *colour* of the mineral in transmitted light, which will often differ from that in reflected light.

(*iii*) Refractive index may next be measured. In grain mounts this is done against oils, with high accuracy. Carefully done, it is virtually a finger-print for identification of the mineral (see p. 306). In thin section refractive index is estimated by the Becke white line test, either against grains of a known mineral in contact with the grain under study, or against the cement ($n = 1·54$) used in making the thin section.

The fact that the cement coats all free surfaces of the grain and penetrates cracks in it, makes it an immersion medium for the grain. The difference in refractive index between grain and cement confers an appearance of greater or less *relief* on a grain in thin section. If there is little difference in index the grain appears flat and featureless, without well-marked outlines. If the difference is great (say 0·05 or more) the grain outlines become marked, cracks become more conspicuous, and as the difference becomes greater the surface roughness of the grain left by the grinding of the thin section becomes noticeable. The effect is much enhanced and made more sensitive by partly closing the iris diaphragm below the microscope stage. The experienced microscopist pays great attention to adjusting this iris to produce the critical contrast for the mineral he is examining.

Relief may denote an index above or below that

of the cement. The direction of the difference is determined by the Becke test. An isotropic mineral, having only one value of the refractive index, will respond to these tests in the same way, whatever its orientation with respect to the vibration direction of the polarized light coming from below.

(*iv*) Colour and relief enable us to see the *form* of the mineral grain, which must be recorded. Remembering that we are seeing only one view of a grain, or the shape of a random slice through it, it is clear that as many grains as possible must be inspected to get an idea of the crystallographic forms present and the habit of the mineral.

Isotropic minerals will be cubic, and are likely to be equidimensional (*equant*) in shape, unless they are completely irregular (*anhedral*).

(*v*) Cleavage, appearing as regular sets of parallel cracks, is to be noted, and the angle between different sets estimated. Where two sets of cracks are normal to the thin section, indicated by the fact that they do not appear to move sideways on changing the focus of the microscope, an exact measurement of the angle between them is possible, using the rotating stage to bring cracks of each set in turn parallel to the same cross-wire.

(*vi*) Inclusions of foreign materials, or cavities in the crystal, may be regularly arranged on crystallographic planes and are then useful indicators of crystallography and growth conditions.

(*vii*) The mineralogist will be concerned not only with the identification of a mineral, but also with its relationship to the associated minerals in terms of sequence of formation, relative rapidity of growth, related orientation, chemical interaction or co-precipitation, later alteration, and similar questions. Under *textural relationships* is summed up a variety of observations too numerous and particular to catalogue here, that will be recorded along with the data used in identifying the mineral species.

ANISOTROPIC MINERALS

The observations to be made are the same as those for isotropic minerals, with the following additions.

(*i*) Colour may now vary with the direction of vibration of light in the mineral. This is the phenomenon of *dichroism* in uniaxial minerals, already mentioned on p. 111. In these, there are two contrasting absorption directions. In biaxial minerals the effect is called *pleochroism* and may involve three distinctive absorption directions.

Dichroism and pleochroism are described by

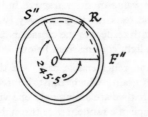

*Phase circle gives amplitude, OR, of
resultant vibration for a phase
difference of $2\pi/5.5 + \pi = 245.5°$*

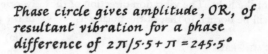

*Components of the two vibrations are
composed into the plane of vibration
of the analyser, AA. Phase difference
then causes interference of the vibrations
with extinction of some wave-lengths.*

*The two beams of light, vibrating in planes
at right angles, do not interfere with one
another; but one is retarded relative to
the other. Retardation shown in figure is
about $\frac{1}{5.5}\lambda$*

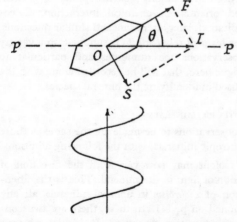

*Polarised beam, OI, resolved into
the two permitted vibration directions.
OF (fast), OS (slow), of the crystal plate.
PP = plane of polariser.*

*Plane polarised light
from polariser*

Fig. 118. Passage of light through the microscope with crossed polars.

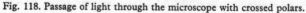

noting the colour of the light transmitted when each principal indicatrix axis is parallel to the plane of the polarizer. Along with this is recorded the relative strength of absorption in these directions. For example we may have such descriptions as

in a hornblende	or in a biotite
α or X = pale yellow	α or X = yellow
β or Y = yellow-green	β or Y = dark brown
γ or Z = green	γ or Z = dark brown
$X < Y < Z$	$X < Y = Z$

(*ii*) The refractive index will now (in general) have two values, those of the two permitted vibration directions for the particular plane of section examined. When one or other of the two permitted vibration directions is parallel to the plane of the polarizer all the light will pass through vibrating in that permitted direction. When neither permitted direction parallels the plane of the polarizer, the light will be doubly refracted and part will vibrate in each permitted direction. To make accurate refractive index comparisons we must rotate the grain so that the light uses first one, and then the other permitted vibration direction, and make observations on the refractive index of each. To place the grain in these two desired positions, we insert the analyser, turn the grain until it is completely dark (*in extinction*) and then remove the analyser to test the refractive index. The grain is then turned through exactly 90° to observe the other vibration direction.

General observations of relief are still of value with anisotropic minerals, and they are made in the same way as for isotropic crystals. In a few cases the two values of the index are so different that the whole relief appearance changes noticeably on rapid rotation of the stage, an appearance called (not very aptly) *twinkling*. Calcite, $\varepsilon = 1\cdot486$, $\omega = 1\cdot658$, shows this best, one value being far higher than the refractive index of the mounting cement ($n = 1\cdot54$) and the other much nearer this, and below it.

OBSERVATIONS BETWEEN CROSSED POLARS

When the analyser is placed in the optical train of the petrological microscope, above the rotating stage, with its plane of polarization at right angles to that of the polarizer below the stage, the polars are said to be *crossed*. If there is nothing on the microscope stage, no light will reach the eye, because light passing the lower polar vibrates in the plane of absorption of the upper one.

ISOTROPIC MINERALS

As we have already seen, an isotropic substance does not doubly refract the light reaching it from below, which passes through it unchanged (except for a drop in velocity during its passage) and is eliminated in the analyser. Isotropic substances are dark between crossed polars and remain so however they may be turned.

ANISOTROPIC MINERALS

Interference colour

In general, an anisotropic substance placed between crossed polars permits light to be transmitted through the analyser to the eye. This may be explained with the help of Fig. 118.

Plane polarized light from the polarizer is broken up, on entering an anisotropic crystal, into two beams vibrating in the permitted vibration directions of the crystal, OF denoting the fast and OS the slow vibration direction in Fig. 118. After passing through the crystal, the two beams are vibrating in planes mutually at right angles, and hence do not interfere with one another; but one wave train is retarded relative to the other, because the two trains have travelled through the crystal at different velocities as a result of the difference in refractive index between the two permitted vibration directions.

On reaching the analyser, components OF', OS' of the two beams are transmitted vibrating in the same plane AA, which is the plane of transmission of the analyser. From the construction, it will be seen that, in whatever pair of quadrants OF and OS are drawn, OF' and OS' will be equal.

$$OF' = OF \sin \theta = OI \cos \theta \sin \theta$$
$$OS' = OS \cos \theta = OI \sin \theta \cos \theta.$$

Their amplitude will depend upon the value of θ, the angle between the permitted vibration directions of the crystal and the plane of the polarizer, and it is easy to see that it will be greatest when $\theta = 45°$.

Considering first the case of an incident beam of monochromatic light, if the phase difference imposed by the crystal plate should be zero, OF' and OS' will be in opposite directions and will cancel one another, causing extinction of the light. If the phase difference is $2\pi, 4\pi, \ldots$, etc. (i.e. an integral number of wavelengths) OF' and OS' will again be opposed and extinction will occur.* For all other phase differences, light will be transmitted, the amplitude and hence the intensity of which can be found from a phase circle diagram.

Consider now the passage of white light through the system. The relative retardation produced by a given thickness of a particular crystal is a fixed quantity, which will equal an integral number of wavelengths for (ideally) only one wavelength of the white light spectrum. This wavelength is extinguished as a result of the relative retardation, but for the other wavelengths the phase difference is not an integral number of wavelengths, and they are transmitted by the analyser to give a colour equal to white light minus the extinguished wavelength. This colour, reaching the eye, is the *interference colour*.

The interference colour is governed by three factors:

(*i*) the difference between the greatest and least refractive index of the crystal. This is the *birefringence*, or strength of the double refraction and is measured as $\varepsilon - \omega$ (positive uniaxials), $\omega - \varepsilon$ (negative uniaxials) or $\gamma - \alpha$ (biaxials).

(*ii*) the direction in which the section is cut. For all sections except those parallel to the optic axis in uniaxials, or parallel to the optic axial plane in biaxials, the difference between the major and minor semi-axes of the elliptical indicatrix section will be less than the maximum given under (*i*). The difference between the semi-axes of the elliptical section is the *partial birefringence*. It is symbolized by $\varepsilon' - \omega$, $\omega - \varepsilon'$, or $\gamma' - \alpha'$, where the dashed symbols represent values tending towards ε, α, etc.

(*iii*) the thickness of the section. As we saw above, during the passage of light through the crystal,

---

* An interesting point arises here. In the case of two waves vibrating in the same plane, destructive interference will occur when the phase difference is $1/2\lambda$, $3/2\lambda$, $5/2\lambda$, . . . , etc. (i.e. π, 3π, 5π, . . . , radians) as in Fig. 93(b). But when the two waves are vibrating in planes at right angles, and are composed into one plane in the analyser, destructive interference takes place when they are an integral number of wavelengths out of phase. In Fig. 118, if we take *OF*, *OS* as upswings (positive) of the waves, and let the mineral plate impose $1/2\lambda$ retardation on the slow ray (which might be expected to produce extinction) on reaching the analyser *OS* would lie in the opposite quadrant from the one it occupies in the figure, because it would be making a downswing (negative) of the wave. The projection *OS'* on to *AA* would then be in the same direction as *OF'*, and the two would not extinguish but reinforce one another. The effect of the analyser then, is to impose an additional $1/2\lambda$ retardation to the slow ray. The phase angle between the two rays is increased by π, and the requirement for extinction is changed to a phase difference of a whole number of wavelengths.

vibrating as two beams polarized at right angles, the light of one beam is retarded relative to the other because its vibrations take place in a direction of higher refractive index. The longer the beams travel at different velocities, the farther behind will the slow one fall.

This is best shown by placing a plate of quartz, cut parallel to the optic axis and ground into a wedge, on the microscope stage at 45° to the cross-hairs. (Such a wedge is a usual accessory to the microscope.) In this position the maximum amount of light is resolved into each of the vibration directions of the wedge. Use a sodium vapour lamp, or place a yellow filter in front of a white lamp, to get approximately monochromatic light for the microscope, which should have its polars crossed.

When the quartz wedge is moved forward, feather-edge first, into the field of view, a succession of broader yellow bands, separated by evenly-spaced, narrower dark bands will be seen running across the length of the wedge. The explanation of this effect is that the yellow light, passing through the quartz in two beams, is composed in the analyser into light all of the same wavelength (since nothing has been done to change this), the amplitude and hence the intensity of which will vary along the length of the wedge, according to the amount by which the two beams are out of phase. Where the difference in phase is an integral number of wavelengths, the two beams interfere destructively, and the dark bands are produced. As the wedge thickens the first dark band represents a relative retardation of λ, the second 2λ, the third 3λ, and so on. The even spacing of the dark bands along the wedge reflects the fact that successive equal increments of thickness impose an extra 1λ retardation of the slow ray behind the fast. That is to say, that the extra thickness of quartz between one dark band and the next has caused the slow ray to fall a further 589 nm behind the fast ray. The brightest part of successive yellow bands occurs at a phase difference of $\lambda/2$, $3\lambda/2$, $5\lambda/2$, etc. where the two components transmitted by the analyser reinforce one another.

Using white light, instead of monochromatic light, the tip of the quartz wedge passes from grey to white and through yellow and orange to red, after which we see successive repetitions of Newton's scale of colours, violet, indigo, blue, green, yellow, orange, red. Four repetitions are usually distinct before the colours, which have become progressively paler, blend into a pale pinkish hue called the high-order white. Divided at the red bands these repetitions

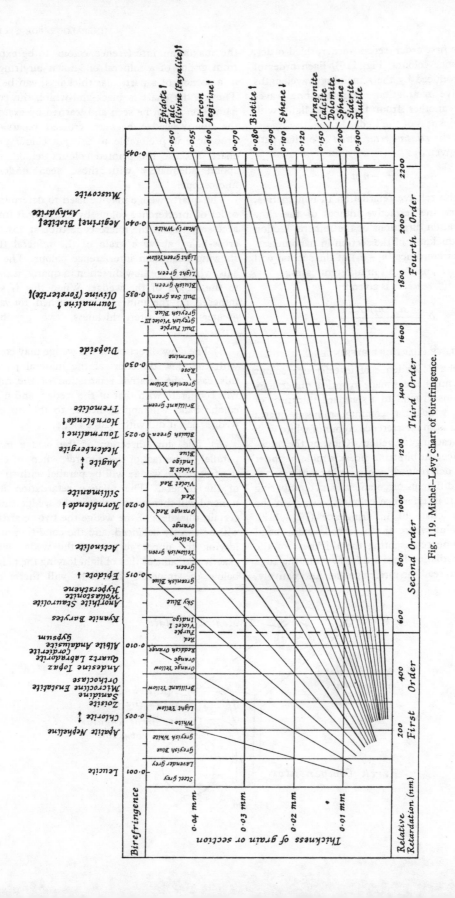

Fig. 119. Michel–Lévy chart of birefringence.

are called the first-order, second-order, third-order, etc. interference colours (Fig. 119). Each order of colours is produced by the remaining wavelengths as the relative retardation eliminates one wavelength after another from the white light with increasing thickness of the wedge.

In general the *relative retardation* or optical path difference is given by

$$\Delta = (n_s - n_f)t$$

where Δ is the relative retardation in nanometres, n_s and n_f are the refractive indices of the slow and fast vibration directions of the mineral section and t is the thickness of the section in nanometres. The partial birefingence $(n_s - n_f)$ of the section will be $\varepsilon' - \omega$, $\omega - \varepsilon'$, or $\gamma' - \alpha'$, in different cases.

The *phase difference*, δ, is given by

$$\delta = \frac{(n_s - n_f)t \times 2\pi}{\lambda}$$

THE MICHEL–LEVY CHART OF BIREFRINGENCE

Because of the dependence of interference colour on thickness, much use is made in optical mineralogy of thin sections ground to a standard thickness of 0·03 mm, at which the common mineral quartz shows a maximum interference colour of white with the very faintest tinge of yellow. Knowing the thickness of the slice, the birefringence of a mineral under study can be determined by examining a number of sections in different orientations and comparing the maximum interference colour shown with the Michel–Lévy chart of birefringence (Fig. 119). From the point where the observed colour band crosses the line representing the thickness of the section, the radiating lines are followed to the edge of the chart to read the birefringence. Alternatively,

the maximum interference colour to be expected from grains of a mineral of known birefringence, in a section of a particular thickness, can be read. Though the chart is black-and-white, the colours named on it can be seen and learned by examining the quartz wedge between crossed polars. This procedure is preferable to using a colour-printed chart, because the printed colours seldom correspond adequately with those seen under the microscope.

The quartz wedge may be used to determine the order of interference colour by inserting it into the accessory slot in the microscope tube at 45° to the cross-hairs, above a grain of the mineral that is showing maximum interference colour. The optic axis, which is the slow direction in quartz, is parallel to the length of the wedge. When this is superimposed on the mineral, and pushed forward to greater and greater thickness, two possibilities arise.

(*i*) The slow direction of the wedge may coincide with the slow direction of the mineral plate. In this case, the relative retardation of the mineral will cooperate with that of the wedge, and a given colour band will appear nearer to the tip of the wedge than it did with wedge alone.

(*ii*) The slow direction of the wedge may be parallel to the fast of the mineral, when, of course, the fast of the wedge will be parallel with the slow of the mineral. The relative retardation in the mineral will then oppose that in the wedge, and at a certain thickness of the wedge the two retardations will be exactly balanced, and the condition of light leaving the upper surface of the wedge will be exactly the same as that of light leaving the polarizer below the mineral. The light will therefore be

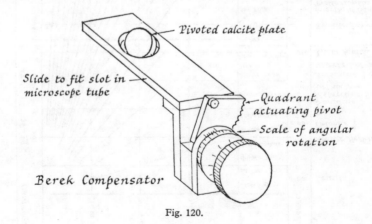

Fig. 120.

eliminated in the analyser. The resulting *dark grey band* across the mineral is called the *compensation band*, because the birefringent effect of the wedge has compensated for the birefringence of the mineral. The position in which it occurs in the colour-scale of the wedge gives the relative retardation in the mineral; this coupled with the thickness of the mineral gives its birefringence.

If on the first trial the possibility (*i*) is realized, the mineral section is turned through 90° to find the position of compensation.

THE BEREK COMPENSATOR

In this device a disc of calcite, cut normal to the optic axis, is placed in the microscope tube in a holder that allows the disc to be turned about an axis in its own plane (Fig. 120). This axis of rotation is arranged at 45° to the planes of the polars. With a mineral section showing maximum interference colours on the stage, the calcite disc is rotated. Its relative retardation rises rapidly as the effective plane of section of the calcite departs from the circular section of its indicatrix, and the effective thickness of the disc at the same time increases. When this increasing relative retardation compensates that of the mineral under examination, the angle of rotation of the disc is read. A calibration table can be made which gives the relative retardation directly from the angle measured. As in all compensation procedures, opposition between mineral vibration directions and those of the compensator must be found by trial.

MEASUREMENT OF GRAIN THICKNESS

Any accurate determination of birefringence by observation of interference colour, or by compensation, requires that the orientation of the grain and its thickness are both accurately known. If enough grains are examined, a suitable orientation can usually be found. The thickness measurement can be more difficult.

Grinding of the two surfaces of the material is generally necessary, to provide grains with a suitable area of uniform thickness. Most microscopes have graduations on the fine focusing adjustment, with the intention that, by focusing first on the lower surface of the grain, and then on the upper, the thickness can be read. In practice this is difficult to do accurately because the depth of focus of the system (desirable on other grounds) is too great, and because of lost motion in the rack and pinion of the microscope.

If pieces of a known mineral in favourable orientation are present, they provide the best means of determining the thickness of the preparation. If suitable minerals are not present, grains of, say, quartz can be added around the unknown mineral before the grinding, to act as a thickness-control.

The estimation of birefingence from interference colour in thin section is extremely useful, and is constantly employed in optical mineralogy. Nevertheless, it must be admitted that it is qualitative rather than quantitative, unless special precautions are taken. Often, the separation of grains of the unknown mineral and direct measurement of its principal refractive indices by immersion in oils may prove to be the most satisfactory way of determining birefringence accurately.

POSITION OF EXTINCTION

The description just given, of the production of an interference colour, applies to an anisotropic section in a general orientation, when neither of its permitted vibration directions coincides with the plane of polarization of the polarizer. As the mineral section is rotated between crossed polars, first one and then the other of its vibration directions will be brought parallel to the plane of the polarizer. When this happens, the light from the polarizer will pass through the mineral vibrating wholly in the permitted direction that is parallel to the plane of the polarizer, and will be eliminated in the analyser. In these positions the mineral will appear dark, and is said to be *in extinction*. Extinction occurs four times in complete rotation of the mineral.

The *position of extinction* permits the location of the vibration directions of the mineral section under study; when extinction takes place they are parallel to the cross-hairs of the microscope eyepiece, which are set parallel to the planes of polarizer and analyser. The position of the vibration directions can be recorded by measuring angles between them and physical directions, such as straight edges of the grain (representing the traces of crystal faces) or cleavage cracks (also marking the traces of sir ￼e crystallographic planes) or just to the length of the crystal. The angle is measured by bringing the physical direction and the vibration direction successively parallel to the same cross-hair, and reading the graduations of the rotating stage.

Straight extinction occurs when the vibration direction is parallel to the physical direction, or to the length of the crystal.

Inclined (or oblique) extinction denotes an angle between the physical direction, or length of the crystal, and the vibration direction, and its value must be quoted in description.

Symmetrical extinction is the term used for cases where the vibration directions bisect the angles between two sets of cleavages, as in augite and hornblende cut normal to z-axis.*

In studying extinction angles, observations must be made on grains in as many different orientations as possible to build up a picture of the relationship of the indicatrix to the crystallographic axes.

FAST AND SLOW DIRECTION AND SIGN OF THE ELONGATION

The two vibration directions of an anisotropic section transmit light vibrating at different velocities. The vibration direction with the lower refractive index retards the light less than that with the higher index. In working out indicatrix relations from a series of sections in different orientations it is often useful to identify the *fast* (low r.i.) and *slow* (high r.i.) *vibration directions*. This is done by turning the section from extinction through 45°. Through the accessory slot in the microscope tube at 45° to the cross-hairs is inserted a plate of a known mineral, quartz or gypsum, with a known vibration direction along its length, that is cut to a thickness that yields a first order red interference colour. This plate, known as a *sensitive tint*, will have its vibration directions marked on it, slow (Z) along the length for quartz, fast (X) along for gypsum, and is a standard accessory with petrological microscopes.

Taking a quartz sensitive tint, for example, when this is superimposed on the mineral section two possibilities exist.

(*i*) If the slow vibration direction of the sensitive tint is parallel to the slow direction of the mineral, then the fast will be parallel to the fast, and the relative retardation in the test mineral will be added to that of the sensitive tint, so that its interference colour will rise to a blue, or other higher colour appropriate to the increased relative retardation.

(*ii*) If the vibration directions are opposed in the test grain and the sensitive tint, the effect will be to diminish the total relative retardation, and the colour of the plate will fall to a yellow appropriate to this decrease.

Alternatively, the quartz wedge may be used for this determination, and is more convenient with highly birefringent test grains. In this case the

*It must be noted that symmetrical extinction angles used in feldspar identification refer to vibration directions in twinned crystals, symmetrical to the trace of the twin-plane.

position of compensation (see above, p. 125) is sought; when it is found the retardations of grain and wedge are opposed. If the position of compensation should prove difficult to determine, it may help either to remove the microscope eyepiece, or to insert the Bertrand lens (if there is one), during the test, when the compensation position will often show as a well-defined dark band across the sequence of colours seen as the wedge is advanced.

The *sign of the elongation* is taken as positive when the slow vibration direction is parallel to, or close to the direction of elongation of a sample of grains of the mineral being examined; it is negative if the crystals have the fast direction along the length. The terms *length-slow* and *length-fast*, being self-explanatory, are better than the sign convention. When the indicatrix is orientated with $Y (\beta)$ parallel to the length of the crystals, some individuals will be length-fast, others length-slow, an observation which is also of diagnostic value.

OBSERVATIONS IN CONVERGENT LIGHT

We have been concerned, so far, with observations in approximately parallel beams of light, all following the same path through the atomic array of the crystal. This is the *orthoscopic* method. We may obtain more exact information about the orientation and optical character of anisotropic crystals by observing them in a beam of light whose rays are strongly converging at the plane of the crystal. This is the *conoscopic* method.

In the conoscopic method, what is observed is not an image of the mineral, but the pattern of interference effects it produces. In order to see this pattern the optical system is modified in the following way.

(*i*) The condensing lens system below the stage must produce a converging beam. On older microscopes this was done by inserting an auxiliary lens; but on modern instruments the light produced when the sub-stage iris diaphragm is fully open suffices, so that the extra lens is not needed.

(*ii*) In order to gather as wide a cone of rays as possible diverging above the mineral, a high power objective of large numerical aperture must be used.

(*iii*) The polars are crossed.

(*iv*) It is required to view the back of the objective lens. This may be done by removing the eyepiece of the microscope and looking down the tube. The interference figure is more easily seen if the eyepiece is replaced by a cap with a small hole drilled at its centre to align the eye on the axis of the instrument. Alternatively, the eyepiece is left in, and an extra lens, the Bertrand lens, is placed in the optical path, between the analyser and the eyepiece, by means of a slide or a lever arrangement. Simple petrological microscopes, however, may not be provided with a Bertrand lens.

CHOICE OF CRYSTALS

The method depends on birefringence, and applies only to anisotropic crystals. The only useful sections to examine are those cut approximately normal to an optic axis of uniaxial or biaxial minerals, those approximately normal to a bisectrix in biaxials, and sometimes those normal to Y (β) in biaxials.

In thin section, grains cut normal to an optic axis

are recognized by the fact that they remain nearly dark throughout rotation of the microscope stage with crossed polars. Those normal to a bisectrix have relatively low interference colour, but may be harder to recognize and must be found by trial. Those normal to Y will have the maximum interference colour for the particular mineral species. The same criteria apply to grain mounts, but they may be more difficult to use because of variation in thickness of grains, and more trials may be needed to find the required orientation.

THE UNIAXIAL INTERFERENCE FIGURE

A uniaxial grain of high birefringence with its optic axis directed up the microscope tube, viewed conoscopically in white light, produces a pattern consisting of a dark cross with its arms parallel to

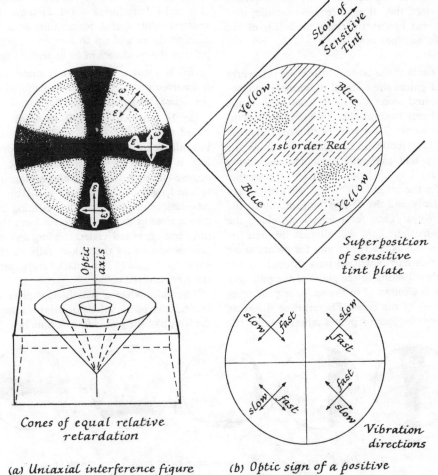

(a) Uniaxial interference figure

(b) Optic sign of a positive mineral $\varepsilon > \omega$

Fig. 121. The uniaxial interference figure.

the cross-hairs, and a series of concentric coloured rings each of which is a sequence of the colours of the Newton's scale (Fig. 121). This pattern is explained as follows.

The light rays, converging on the mineral from below, diverge upwards through the mineral forming a series of cones of wider and wider angle going outwards from the central ray which is incident normally upon the mineral (Fig. 121a). The central ray, being parallel to the optic axis, is not doubly refracted; but the diverging rays do not travel along the optic axis and hence are doubly refracted. Moreover, each successive cone of rays outward from the centre has a longer distance of travel through the mineral, and greater departure from parallelism to the optic axis. Hence, as we proceed from the centre of the field to the edge, both $(n_s - n_f)$ and t of the relative retardation equation $\Delta = (n_s - n_f)t$ increase. Relative retardation thus increases radially outwards, so that concentric rings made up of the colours of the spectrum are produced between crossed polars.

The dark area at the centre of the cross is the area where light enters the mineral at normal incidence and is directed along the optic axis, so that the mineral behaves isotropically. To explain the arms of the cross we recall the rule for uniaxial minerals, that the E-ray vibrates in the plane containing the incident ray and the optic axis, and the O-ray vibrates normal to this plane. From this we see that the E-ray for the cones of light rays will everywhere vibrate radially and the O-rays tangentially to the concentric rings. Thus for points lying along the E–W cross-hair the light will all pass through the crystal vibrating as the E-ray, and for points along the N–S cross-hair it will all pass through as the O-ray. Light vibrating in the plane of the polarizer in this way is eliminated in the analyser producing the dark arms of the cross. The arms of the cross are therefore the loci of points where one of the vibration directions of the mineral is parallel to the plane of the polarizer (Fig. 121a).

USE OF THE UNIAXIAL INTERFERENCE FIGURE

The formation of this figure yields the following information.

(*i*) The mineral belongs to the tetragonal, hexagonal or trigonal system.

(*ii*) The strength of the double refraction is indicated by the sharpness of the cross and the number of coloured rings produced by a section of standard thickness (0·03 mm). Apatite ($\omega - \varepsilon = 0·003$) and nepheline ($\omega - \varepsilon = 0·004$) give a broad poorly-defined cross with paler quadrants. Quartz ($\varepsilon - \omega = 0·009$) gives a definite black cross with grey-white quadrants, but no coloured rings. Pigeonite ($\gamma - \alpha = 0·03$) shows yellow in the quadrants, zircon ($\varepsilon - \omega = 0·06$) shows one orange-red ring and calcite ($\omega - \varepsilon = 0·172$) shows two sets of spectral colours, each terminating with a red, on a centred figure and up to four (the outermost very pale) in the quadrants of an uncentred figure.

(*iii*) If a quartz sensitive tint plate (length slow) is inserted at 45° to the cross-hairs while viewing the figure two possibilities arise:

(*a*) In positive minerals ($\varepsilon > \omega$) the relative retardation of mineral and sensitive tint plate will be additive in the pair of opposite quadrants whose join is parallel to the length of the plate, and opposed in the pair of quadrants arranged transverse to the length of the plate (Fig. 121b). Where the retardations are additive, the resulting interference colour rises in the spectral scale from the sensitive tint (first-order red-violet) to blue, and in the other pair of quadrants the colour falls in the spectral scale to yellow. The black cross assumes the sensitive tint red colour.

The rule, when a length-slow plate is used, is that the join of the yellow quadrants forms a + with the length of the plate for +ve minerals.

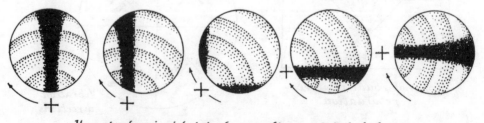

*Uncentred uniaxial interference figure rotated clockwise.
+ marks successive positions of optic axis*

Fig. 122. Uncentred uniaxial interference figure.

(b) In negative minerals the effect is reversed, and the join of the yellow quadrants is parallel to the length of the plate. That is, it forms a − (minus) in relation to the length of the plate in −ve minerals.

The quartz wedge may be used in a similar way, the coloured rings moving inwards in the quadrants in which retardations are additive, and outwards in those in which retardations are opposed.

Fig. 122 shows the appearance given by an imperfectly centred uniaxial interference figure on rotation of the stage; that is by a figure in which the optic axis is just outside the field of view of the microscope. The arms of the cross move across the field successively, *remaining parallel to the cross-hairs* during their passage. Such a figure is quite useful, and can yield nearly all the information given by a centred figure. It is distinguished from an uncentred biaxial figure by the fact that in the latter the black brushes sweep across the field with constantly changing orientation. There remains, however, the possibility of being misled by an uncentred biaxial figure of small optic axial angle, which may behave very like an uncentred uniaxial figure.

THE BIAXIAL INTERFERENCE FIGURE

The acute bisectrix figure

In a section cut normal to the acute bisectrix of a biaxial mineral with a moderate optic axial angle—say less than 45°—both optic axes emerge within the field of view of common types of high power microscope objective.

Fig. 123 shows the appearance when such a section is viewed conoscopically (i) with the optic axial plane at 45° to the cross-hairs, and (ii) with the optic axial plane parallel to the E–W cross-hair.

In the 45° position the figure shows two curved dark zones, the *isogyres* (or brushes), each narrowest at the point of emergence of an optic axis. The isogyres are always convex towards the acute bisectrix, and their degree of curvature increases as the optic axial angle becomes smaller. Coloured rings may be seen around the optic axes in highly birefringent minerals or in thick sections. These are the *lemniscate rings*, and the outer ones fuse to produce an hour-glass shape embracing both optic axes. The rings are not usually seen in standard sections of the common rock-forming minerals, but may often be seen in grain mounts. Flakes of muscovite are very convenient for demonstrating this figure, as they lie on the microscope slide with the acute bisectrix normal to the surface and may readily be cleaved to the required thickness.

On rotation to the *parallel position* the isogyres move and fuse into a cross, one arm of which, parallel to the optic axial plane, is more sharply defined than the other, and narrows at the optic axes.

The explanation of the biaxial interference figure follows from the Biot-Fresnel Rule (p. 114) that the vibration directions for any ray passing through a biaxial mineral bisect the angles between planes containing the ray and each optic axis in turn. If the traces of such planes are drawn on the diagram of the biaxial interference figure and the angles between these traces are bisected (Fig. 124) it can be seen, to a first approximation,* that the isogyres are the loci of points where the vibration directions

* Approximate because the planes whose traces form the angle bisected are in fact inclined to the section plane. The stereographic construction (Fig. 125) gives true angular relationships.

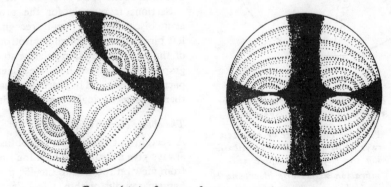

Biaxial interference figure normal to Bx_a

(a) *45° position* (b) *Parallel position*

Fig. 123. Acute bisectrix interference figure of a biaxial crystal.

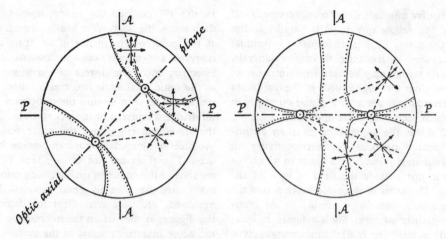

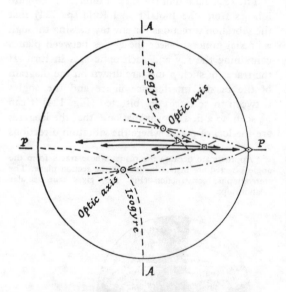

Vibration directions for the Bx_a figure from
an approximate application of the Biot-Fresnel rule

PP, AA : planes of polariser & analyser

Fig. 124.

Stereogram of Biot-Fresnel construction for
rays emerging along the isogyre

 Points of emergence of selected
rays, with great circles contain-
ing the ray & each optic axis

⟵———⟶ *Vibration directions bisecting the*
angles between pairs of great circles

PP and AA   Planes of polariser & analyser

Fig. 125.

are parallel to the cross-hairs, and their movement as the optic axial plane is rotated is explained. The lemniscate rings are the loci of points of equal relative retardation of the doubly refracted part of the light, expressed as curves of the same inter-ference colour, as in the case of the uniaxial figure.

In cases where the optic axial angle is large, the emergence of the optic axes will be beyond the field of view of the microscope; so that the isogyres will pass out of the field of view as the optic axial plane is turned to the 45° position. The line joining the pair of quadrants in which they disappear gives the direction of the optic axial plane in such cases.

Sections favourable for the production of the acute bisectrix figure may be recognized by showing low birefringence, though not isotropy. A knowledge of the optic orientation relative to other crystallo-graphic features, such as cleavages, may help in particular cases. They are less easily recognized than sections normal to an optic axis.

The obtuse bisectrix figure

A section cut normal to the obtuse bisectrix will always give a figure in which the isogyres disappear from view on rotation to the 45° position. This will happen rapidly, within a few degrees of rotation, when the optic axial angle (measured over the acute bisectrix) is small. When it is large, there may be difficulty in deciding whether the section is normal

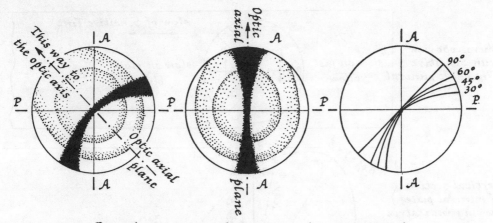

Biaxial interference figure normal to an optic axis
(a) 45° position   (b) Parallel position   (c) Curvature of isogyre
(PP, AA : planes of polariser & analyser)   for different optic angles

Fig. 126. The optic axis interference figure of a biaxial crystal.

to the acute or the obtuse bisectrix. In such cases it is wise to seek an optic axis figure for optical sign determination.

The optic axis figure

When one of the optic axes points directly up the microscope tube, a single isogyre is seen (with coloured rings in thick sections, Fig. 126). This isogyre will pivot about the centre of the field as the section is rotated, having maximum curvature when the optic axial plane is in 45° position, and straightening into parallelism with each cross-hair in turn as the optic axial plane becomes parallel to them. The trace of the optic axial plane crosses the isogyre like an arrow across a bow, at the position of maximum isogyre curvature, and the arrow-head in this analogy is pointed towards the acute bisectrix, which always lies on the convex side of the isogyre.

The degree of curvature of the isogyre in the 45° position is dependent upon the size of the optic axial angle. When this is small the isogyre is sharply curved; as the optic angle approaches 90° the isogyre becomes straight (Fig. 126c).

The optic normal figure (flash figure)

A section at right angles to the optic normal Y (β) yields an interference figure in the form of a broad dark cross, almost filling the field of view, that breaks up into two parts which vanish rapidly from the field on a small rotation of the stage. This behaviour gives it the name of a flash figure.

Use of the biaxial interference figures

(*i*) Biaxial minerals belong to the orthorhombic monoclinic or triclinic systems.

(*ii*) The strength of the double refraction is indicated by the sharpness of the figure and the appearance of coloured rings, as with uniaxial figures.

(*iii*) When an acute bisectrix figure, or an optic axis figure, is set in the 45° position, and a quartz sensitive tint (length-slow) is superimposed with its length parallel to the optic axial plane, one of two possibilities will arise.

A blue colour will appear in the concavity of the isogyre or isogyres, and a yellow on the convex side. In this case the mineral is positive ($Bx_a = Z$ or γ). (The isogyres themselves assume the sensitive tint.) Alternatively, yellow will appear in the concavity and blue on the convex side of the isogyre. The mineral is in this case negative ($Bx_a = X$ or α).

The test may be applied even in the case of an acute bisectrix figure in which the isogyres have moved out of the field of view on rotation to the 45° position, the colour between them (i.e. on their convex sides) being diagnostic.

Fig. 127 shows schematically how the colours arise in the case of a positive mineral cut normal to an optic axis, by consideration of a section of the indicatrix parallel to the optic axial plane.

Fig. 128 shows the appearance given by an uncentred biaxial interference figure on rotation of the stage. It is distinguished from an uncentred uni-

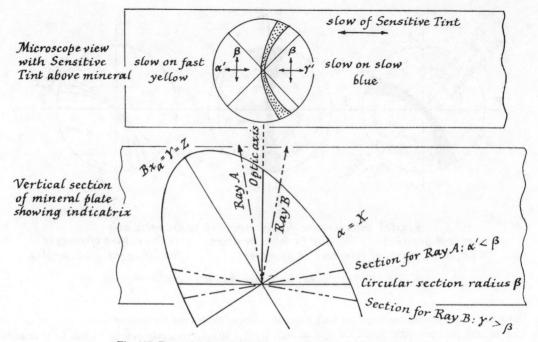

Fig. 127. Determination of sign from an optic axis interference figure.

axial figure by the fact that the isogyre moves into and out of the field of view with a definite curvature, and swings as though pivoted on a point just outside the field, whereas the arms of the uniaxial cross remain parallel to the cross-hairs in their passage across the field.

(*iv*) Obtuse bisectrix figures may be used for the determination of sign, on the same principles as acute bisectrix figures. Yellow between the isogyres on inserting a length-slow sensitive tint will now indicate a negative mineral. Obtuse bisectrix figures are, however, uncertain in use because of

difficulty in recognizing them and in estimating how well they are centred, and they should in general be avoided if possible.

(*v*) Flash figures are useful, in conjunction with the observation of maximum interference colour under orthoscopic conditions, to check that the section is normal to β before recording the extinction angle in certain monoclinic minerals, such as some of the pyroxenes and amphiboles. It is the extinction angle normal to β, i.e. measured in the optic axial plane and crystallographic symmetry plane, that is diagnostic in classifying these minerals.

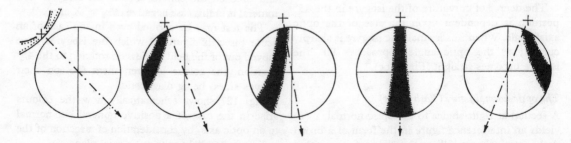

Uncentred biaxial interference figure rotated clockwise
+ marks position of optic axis: ⁻·⁻·→ indicates optic axial plane & direction of Bx_a

Fig. 128. Uncentred biaxial interference figure.

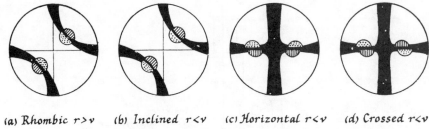

(a) Rhombic r>v   (b) Inclined r<v   (c) Horizontal r<v   (d) Crossed r<v

blue fringe (emergence of optic axis for red light)

red fringe (emergence of optic axis for blue light)

Dispersion effects in interference figures
Compare fig. 116

Fig. 129. Effects of dispersion.

DISPERSION AND THE INTERFERENCE FIGURE

The effect of dispersion on the indicatrix was described on pp. 116–7. This effect is best exhibited by the interference figure. The appearances seen are indicated in Fig. 129, which should be compared with the diagrams in Fig. 116. The coloured fringes are easily understood if it is realized that the position of the isogyres will be different for different wavelengths of light. The interference figures produced in white light will show a blue colour where the red light is in extinction at the isogyre, and a red colour where the blue wavelengths are in extinction. The isogyre is therefore a dark zone, representing extinction of the intermediate wavelengths of the spectrum, bordered by a red fringe on one side and a blue one on the other, when dispersion is sufficiently strong to give visibly different positions of the isogyre for these extremes of the spectral wavelength range. When the optic axial angle is greater for red than for violet light the dispersion is described as $r > v$ and vice versa.

MICROSCOPIC STUDY OF OPAQUE MINERALS

Minerals which absorb light so strongly that they are opaque even in thin sections, can only be examined microscopically in reflected light. In the study of a thin section, some information on opaque minerals can be gained simply by lifting the microscope lamp and shining it down obliquely on the top of the thin section. Under these conditions three common opaque constituents may be differentiated: magnetite appears black, ilmenite may sometimes be recognized by its alteration to yellowish-white

leucoxene (a mixture of sphene and titanium oxide minerals), and pyrite appears brass-yellow.

For any detailed study, however, opaque minerals (ore minerals, as they are often called) or metals are embedded in a resin block, and polished on rotating laps with a paste of fine diamond dust or alumina. The polished surface is then examined through a microscope fitted with a special attachment that illumines the surface with vertically incident light (Fig. 130). The use of reflected light is at present a developing study, and improvements in equipment are taking place rapidly. In order to exploit all the theoretically available optical properties related to crystal structure, small differences in light intensity must be measured. Many refinements are necessary to do this, the equipment is correspondingly costly, and its use becomes a task for the specialist.

The principal observations used at the present time for opaque mineral identification can, however, be made with a fairly simple adaptation of the microscope, and these simple methods only will be described here.

Identification is based upon:

(i) Colour and colour change on rotation,

(ii) Reflectance (reflectivity) and bireflectance,

(iii) Relative polishing hardness (and, with more elaborate equipment, indentation hardness),

(iv) Isotropic or anisotropic character between crossed polars,

(v) Presence of internal reflections,

(vi) Microchemical tests.

After a general statement of optical principles, these methods will be treated in turn.

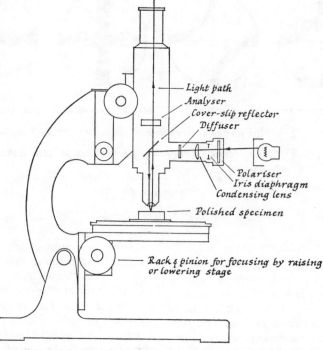

Fig. 130. Principal features of a polarizing microscope adapted for reflected light.

THE BEHAVIOUR OF REFLECTED LIGHT

GOOD AND BAD REFLECTORS

In physical optics, the main distinction between reflecting materials is that between *dielectrics*, or insulating substances, and electrical conductors, chiefly metals.

Dielectrics, which are generally transparent, have homopolar or ionic bonding, involving specific transfer or sharing of electrons between neighbour atoms. Conductors have metallic bonding with a generalized 'cloud' of bonding electrons which are free to move within the structure (Chapter 1). When light enters a conducting substance, much of the energy of its electric vector is dissipated as a conduction current imparted to the mobile electrons. This transfer of energy is additional to the displacement current that accounts for the energy loss in substances without a free electron cloud; and in metals, and good reflectors in general, it becomes dominant. Whereas the optical behaviour of a dielectric can be discussed in terms of its refractive index, that of a strongly light-absorbing substance must be discussed in relation to a value called the *complex refractive index*. This contains a term called

the *absorption coefficient*, k, derived from Lambert's Law

$$I_x = I_o e^{-kx}$$

where I_o is the initial intensity, and I_x the intensity after passing through a thickness x of the substance.

REFLECTANCE DEFINED

The ratio of the intensity of the reflected light from a surface to the intensity of the incident light is called the *reflectance*. For transparent substances (dielectrics), when the light is incident normal to the surface, the reflectance, r, is given by

$$r = \frac{(n-n_o)^2}{(n+n_o)^2}$$

where n is the refractive index of the substance, and n_o is that of the medium in which it is immersed. For observation in air this becomes

$$r = \frac{(n-1)^2}{(n+1)^2}.$$

For an absorbing substance these equations are modified to

$$r = \frac{(n-n_o)^2+k^2}{(n+n_o)^2+k^2} \quad \text{and} \quad r = \frac{(n-1)^2+k^2}{(n+1)^2+k^2}$$

respectively, where k is the coefficient of absorption.

Fig. 131 shows the curves of the last equation above, for a range of refractive index and three values of k, including $k = 0$ which corresponds to the case of a transparent substance. It will be seen that the higher the refractive index of a transparent substance the higher will be its reflectance. In absorbing materials, either a very small n or a large k can give high reflectance. Really good reflectors, like silver, have both low n and high k.

REFLECTION OF INCLINED LIGHT

In a reflected light microscope, convergence of the light by the objective means that the light is incident normal to the specimen only at the centre of the field of view, and towards the margins the rays are inclined. The higher the power of the objective the more marked is this effect. We must therefore consider the reflection of a beam of plane-polarized

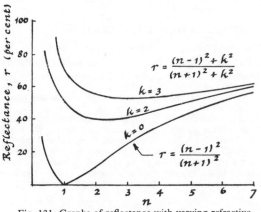

Fig. 131. Graphs of reflectance with varying refractive index and coefficient of absorption.

light inclined to the reflecting surface. We shall concern ourselves only with polarized light, because the light coming to the specimen is already partly polarized in being reflected down from the vertical illuminator and observations are usually made with a polar in front of the illuminator.

There are three quantities involved, the amplitudes of the electric vectors of the incident light, E, of the reflected light, R, and of the transmitted light entering the substance, E'. In an absorbing substance we need not be concerned with E': we are interested in the ratio of reflected to incident light, R/E.

In order to deal with inclined rays, their light may be resolved into two components, one vibrating parallel to the plane of incidence (i.e. the plane

containing the ray and the normal to the surface), called the p light; and one vibrating normal to this plane, called the s light (German senkrecht = perpendicular). These are shown in Fig. 132. The basic equations of reflection, due to Fresnel, state that

$$\frac{R_s}{E_s} = -\frac{\sin(\phi - \phi')}{\sin(\phi + \phi')}; \quad \frac{R_p}{E_p} = \frac{\tan(\phi - \phi')}{\tan(\phi + \phi')}$$

where ϕ is the angle of incidence, and ϕ' the angle of refraction.[*] The reflectances, which are the ratios

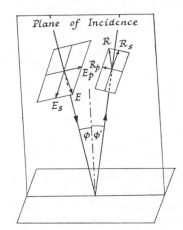

Resolution of incident (E) and reflected (R) polarised rays into components parallel and normal to plane of incidence.

Fig. 132. The components of reflected inclined light.

of the intensities of incident and reflected light, are proportional to the squares of the amplitudes:

$$r_s = \frac{(R_s)^2}{(E_s)^2} \quad \text{and} \quad r_p = \frac{(R_p)^2}{(E_p)^2}.$$

Fig 133 shows that, as the light departs from the vertical, these two intensities diverge, and then become equal again when the light falls on the surface at glancing incidence.[†] When the light is inclined, then, the reflected beam is made up of two unequal components. These two components are also out of phase, by an amount dependent upon the orientation of the incident ray, though the exact

[*] When the light is incident normal to the surface, the distinction between p light and s light disappears, because the plane of incidence becomes indeterminate, so that we have the simpler relationships given above (p. 134) for normal incidence.

[†] In dielectrics, as distinct from metals, the value of R_p falls to zero at the polarizing angle, giving the relations discussed in connection with Fig. 109, p. 110.

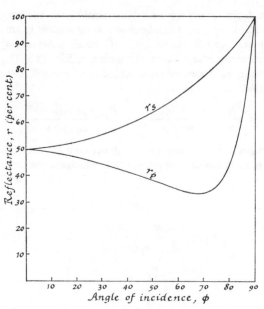

Fig. 133. Reflectances of the two components, *s* and *p*, of an inclined beam on reflection.

round the circle of diameter *x* represents the oscillations of a plane wave vibrating parallel to *X*, while a point moving round the circle of diameter *y* represents the oscillations of a wave vibrating parallel to *Y*. The waves being out of phase, let x_1 and y_1 be the oscillatory positions of the two at a given instant. If we project the displacement of each on to the trace of its own wave-plane to give the lengths OX and OY, the resultant of the two is the diagonal, $O(x_1y_1)$, of a rectangle of sides OX, OY.

In Fig. 134b this construction is applied to a series of pairs of points with a constant phase difference of $\pi/4$, numbered to show which ones correspond in time, and the successive resultants ($1', 2', 3', \ldots$) trace out the elliptical vibration. A whole series of ellipses (the Lissajous figures) may be produced as the phase difference is varied. In special cases (when the phase difference $= 0, \pi, 2\pi$) the ellipse becomes a straight line.*

THE EFFECTS OF ELLIPTICAL POLARIZATION

The elliptically polarized light produced by reflection of an inclined beam cannot be completely eliminated in the analyser of the microscope. At the

---

nature of this dependence need not concern us here.

The recombination of two unequal simple harmonic motions at right angles, that are out of phase, is in general an *elliptical vibration*, in which the amplitude varies with the vibration direction. In Fig. 134a two simple harmonic motions are represented by the movements of points round the two phase-circles (as on p. 97). A point moving

* In writing of the recombination of plane polarized beams in the analyser in transmitted light microscopy (p. 121), we used a different description, where the beams were recombined in accordance with the parallelogram of forces. We could have introduced elliptical polarization then, and said that the elliptically polarized beam from the recombination of the rays had a component parallel to the plane of transmission of the analyser which was transmitted to the eye. The two descriptions are equivalent.

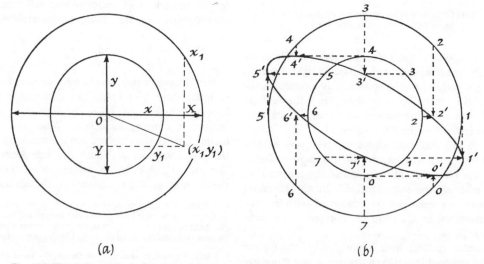

(a)   (b)

Fig. 134. Elliptical polarization resulting from the combination of two rays of unequal amplitude, that are out of phase.

centre of the field of view, the incident beam is truly perpendicular, whilst along the cross-hairs one component of the reflected light (either that parallel to or that normal to the plane of incidence) is parallel to the plane of the polarizer, so that in these regions of the field the light is not resolved into two parts and hence it is completely eliminated in the analyser.

In the outer parts of the field of view between the cross-hairs, however, extinction is not complete, even in isotropic crystals, because of the inclination of the incident beam and consequent elliptical polarization. When an isotropic crystal is put under the microscope, with a high power objective and the polars crossed, and the eyepiece is removed, a black cross will be seen showing the regions of complete extinction. The perfection of this cross serves as a check that the polars are exactly crossed.

OBSERVATIONS IN REFLECTED LIGHT

IN PLANE-POLARIZED LIGHT

1. Colour

The colours observed are, for the most part, pale shades with greys of various tones predominating. Colour contrasts are more easily recognized than the colours themselves, and comparison of an unknown with a known mineral is helpful. In a few minerals (e.g. covellite) the colour is more striking and of immediate use in recognition.

2. Bireflectance

Isotropic minerals remain unchanged in colour and degree of reflection when the stage is rotated. Anisotropic minerals will vary in reflectance, and may vary in colour, on rotation. Comparison between adjacent grains of the same mineral cut in different directions is often valuable in revealing these changes. The contrast in degree of reflection (brightness) is enhanced by using an oil immersion objective. From the relation

$$r = \frac{(n-n_o)^2 + k^2}{(n+n_o)^2 + k^2}$$

it can be seen that increase in n_o, the refractive index of the immersion medium, lowers the reflectance but increases the contrast between similar complex refractive indices.

3. Reflectance

Reflectance may be estimated visually by comparison with known minerals; but an instrumental method is required for any precise use of reflectance. Two systems have been employed for this.

One system employs a prism to split the incident light beam, one part being reflected from the mineral to the eyepiece, and the other travelling through a by-pass to the eyepiece. The 'direct' beam coming through the by-pass can then be matched with the reflected beam by dimming it with the aid of two polars, one of which can be rotated with respect to the other.

The other approach is to measure the light reflected from the specimen by means of a selenium photoelectric cell, or a photomultiplier, that replaces the eyepiece of the microscope. The current generated by either of these devices is amplified and fed to a sensitive galvanometer. The stability of the light source is very important in this method, and special precautions are taken to govern this, by placing a storage battery as a buffer in the power supply to the lamp.* In this method of direct measurement, calibration is made against a standard reflecting substance. The problem of providing reliable standards, capable of being polished to a uniform and reproducible condition of reflectance, at present imposes limits to the refinement of the method and to obtaining strict comparability between measurements made in different laboratories.

In both methods the result is expressed as the percentage of the incident light that is reflected. In refined measurements this value is determined for a series of different wavelengths of light.

In anisotropic minerals, the reflectance will vary with the vibration direction of the light.

4. Hardness

Relative polishing hardness. In polishing minerals for study in reflected light, variations in the polishing technique allow a controlled production of relief, whereby harder minerals stand up fractionally above the level of softer crystals. At the boundary between a softer and a harder grain there will be a sloping step or ramp. When this is focused under a medium power objective, with the aperture diaphragm in the illuminator partly closed, and the microscope tube raised (or the stage lowered), a light line called the Kalb line moves from the grain boundary towards the softer grain. This test enables the polishing hardness of an unknown mineral to be compared with that of a known mineral.

* Arrangements for this method are described by Bowie and Taylor, 1958. *Mining Mag.*, **99**, 265–77; 337–45.

Indentation hardness. This employs a balanced beam, pivoted near one end and carrying a pyramidal diamond point at about its mid-point, and a known load (10 to 100 grams) at the other end. The whole is so arranged that it can be swung into position in place of the microscope objective, with the diamond point exactly above the centre of the field of view. The point is brought to rest upon the surface of the specimen by racking down with the fine-focusing adjustment of the microscope. After about 15 seconds the point is raised and removed, and the objective brought back into position. The mean diagonal width of the pyramidal indentation made by the diamond point is measured with a micrometer eyepiece, and the indentation hardness is expressed as

Load/Area (kg/mm^2).

The value of this is given by

$$IH = 2P . \sin \theta / D^2$$

where IH is indentation hardness, P is the load in kg, θ is half the angle between opposite faces of the pyramidal point, and D the mean diagonal of the indentation in mm.

Indentations are small, and can be made in grains 10 μm across. The principle of the method is simple; but the device for making the measurements is a delicately constructed instrument requiring careful handling.

OBSERVATIONS WITH CROSSED POLARS

5. *Bireflectance*

In spite of the effect of elliptical polarization, isotropic minerals remain nearly dark on rotation between crossed polars, whilst anisotropic ones display bireflectance, in a manner analogous to birefringence in transparent minerals, with four positions of extinction 90° apart, alternating with positions of distinct illumination, during a full rotation of the microscope stage. As with transmitted light, the direction in which the crystal is cut governs the strength of the anisotropic effect. The position of extinction in relation to cleavages or crystal outlines can be used to help in identification, as in the case of transmitted light.

When a high power objective, giving strongly convergent light, is used and the eyepiece of the microscope is removed, sections of uniaxial minerals cut normal to crystallographic z-axis show a black cross that remains unchanged during rotation, like that given by isotropic minerals. Other sections show a cross that breaks up into two curved isogyres on rotation, the cross reforming four times

in a complete rotation. The amount of separation of the isogyres varies both with the orientation and with the mineral species.

The figures given by minerals of the orthorhombic, monoclinic and triclinic systems are the same as those for uniaxial minerals. Minerals belonging to these systems have four *axes of circular polarization*. Each pair of these is symmetrically disposed either side of an optic axis, so that the line joining them is normal to the optic axis plane. A section cut normal to an axis of circular polarization will show a polarization figure like that of a uniaxial mineral cut normal to crystallographic z.

6. *Dispersion of the reflectance*

This is the variation in reflectance with change in the wavelength of the light, and its effect may be seen in favourable circumstances as colour fringes in the study of the polarization figure. Determination of the dispersion is carried out by direct measurement of the reflectance in light of different wavelengths.

7. *Rotation of the plane of vibration of the incident light*

Anisotropic minerals produce a rotation, generally small, of the vibration ellipse of elliptically polarized light on reflection. This parameter has been used, especially by Cameron and his co-workers, as a determinative feature of opaque minerals.

Of these last two properties, dispersion and rotation, we may say that their use involves very accurate measurements of small differences in light intensity, and therefore demands sophisticated and carefully tested optical equipment. The results also vary somewhat with the type of equipment used. We shall not, therefore, go into any further details of them here.

8. *Internal reflections*

Some of the less strongly absorbing opaque minerals (e.g. chromite) allow penetration of a perceptible amount of the incident light, and characteristically show patches of coloured light, usually red, yellow or orange, along lines and in flecks, due to the internal reflection along cracks and imperfections in the crystal. The occurrence of these reflections is of help in identification of the mineral.

CHEMICAL TESTS FOR ELEMENTS PRESENT

Though not part of the optical study, these are often used in conjunction with reflected light microscopy, and it is perhaps worth mentioning them here. The procedure is to scrape off, or drill out with a dental drill, a little powder of the mineral under study, take it into solution, or leach it, and apply chemical tests to the solution, observing the results under the microscope. Except in scale, the methods are similar to qualitative chemical tests given in the descriptions of individual minerals in Part II, and the tests need not be repeated here.

If X-ray apparatus is available, it will in general be far better to take an X-ray powder photograph of the mineral powder obtained, in order to identify it. The chemical tests may, however, sometimes be a useful adjunct to this procedure.

Further Reading

CAMERON, E. N. 1961. *Ore microscopy*. New York, J. Wiley & Son.

GAY, P. 1967. *An introduction to crystal optics*. London, Longmans Green & Co.

HALLIMOND, A. F. 1953. *Manual of the polarizing microscope*. York, Cooke, Troughton & Simms Ltd.

JOHANNSEN, A. 1918. *Manual of petrographic methods*. New York, McGraw-Hill Co.

SHURCLIFFE, W. A., and BALLARD, S. S. 1964. *Polarized light*. Princeton, N.J., Van Nostrand Co.

5

X-RAYS AND CRYSTAL STRUCTURE

The interaction of light with crystals is dealt with in Chapter 4, where a general statement is given of the nature of electromagnetic radiation. We now turn to the behaviour of X-rays on passing through crystals.

The demonstration by Friedrich, Knipping and von Laue, in 1912, that X-rays were diffracted by passage through a crystal, began one of the most rapid and far-reaching bursts of progress in 20th century science. Almost at once, W. L. Bragg showed* how the diffracted beams could be treated as reflections from planes of atoms in the crystal structure, the angle of diffraction for a given wavelength depending upon the distance between the planes, and by the early 1930s, despite the intervention of the First World War, the structure of most important minerals had been worked out.

The principal result of X-ray study has been to establish that the structure of a mineral is more fundamental than its chemical composition. In place of the idea of molecules combined in fixed proportions by weight as the basis of mineral classification, we have the idea of a particular *structure* with a selection of atoms, suitably qualified as to size and charge, occupying the available positions (or sites) in it. The factors governing such atomic substitution are considered in Chapter 1.

CHARACTERISTICS OF X-RAYS

X-rays have wavelengths from 0·02 to 0·2 nm (see Fig. 90, p. 95). The shorter wavelengths are known as the *harder* and more penetrative X-rays; the longer wavelengths are *softer* and more easily absorbed. For mineral studies wavelengths around 0·1 nm are used.

* The diffraction of short electromagnetic waves by a crystal. *Proc. Camb. phil. Soc.*, **17**, 43–57 (1912).

X-rays are produced by a stream of electrons bombarding an anode, or *target*, of some pure metal. The popular model of X-ray tube is permanently evacuated; the electrons are supplied from a heated tungsten filament, forming the cathode, and are hurled against the target by a potential difference of 20 to 100 kV between cathode and anode (Fig. 135). The X-rays are emitted through thin foil windows of light metals (Al or Be) that do not absorb X-rays strongly.

THE X-RAY SPECTRUM

When the electron hits the target its energy is expended partly as heat and partly in the radiation of X-rays. The wavelength of the X-rays depends upon the energy expended in their production. If all the energy of an electron is converted into one quantum of X-ray energy, the wavelength of the ray produced is given by

$$eV = hc/\lambda$$

where e is the electronic charge, V the voltage, h Planck's constant, and c the velocity of light. Putting numerical values for e, h and c and expressing the voltage in kilovolts, we have

$$\lambda_{(nm)} = 1 \cdot 24/kV.$$

This gives the minimum wavelength of X-rays that can be generated by a given potential difference between the cathode and anode.

In practice most electrons will be stopped by several collisions, each involving part of their energies and producing X-rays of longer wavelengths than the minimum given by the equation above. The electron bombardment will therefore produce a *continuous spectrum* of X-rays, with a sharp cut-off at the shortest wavelength possible for the voltage

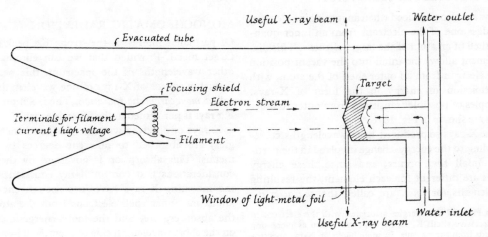

Fig. 135. Diagram of an X-ray tube.

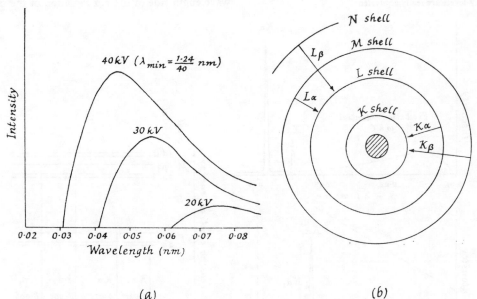

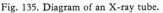

(a)   (b)

Fig. 136. (a) Effect of excitation potential on minimum wavelength. (b) Diagram of electron infall from outer to inner shells.

applied to the tube (Fig. 136). By analogy with light, this is called the *white radiation*.

CHARACTERISTIC RADIATION

X-rays are also produced when an impinging electron dislodges one of the electrons from an inner quantum shell of an atom of the target element. This dislodgement allows the infall into the vacant position of an electron from an outer shell of the atom, with the emission of energy in the form of X-rays. Examples of possible transitions from outer to inner shells are shown in Fig. 137b.*

The X-rays so produced have wavelengths corresponding to the energy-change involved in the particular infall that occurs, and since these energy values are different for each element, the resulting wavelengths, or *lines*, are called the *characteristic*

---

* In Fig. 137 the energies associated with the electrons in the successive shells K, L, M, ..., are shown as increasing outwards from the nucleus. In some books on X-ray spectroscopy the energy-level diagram is turned upside-down with K at the top. In this kind of diagram it is the energy of the *K-state* that is being graphed, i.e. the excited energy of the atom when a K-electron has been knocked out. When the lost electron is replaced by one from an outer shell, the atom falls to a less excited state, having lost some potential energy in the form of an X-ray quantum. The two ways of looking at energy-levels are really equivalent.

---

lines of the element concerned. They stand out as peaks in the intensity of the radiation produced, above the continuous spectrum, as shown in Fig. 137a.

MONOCHROMATIC RADIATION

Of particular interest is the strong $K\alpha$ peak of any target metal. Provided that we can eliminate the other wavelengths of the spectrum, this peak can furnish a source of X-rays of one wavelength only, called *monochromatic radiation*. The isolation of the $K\alpha$ rays is performed in the following way.

X-rays of differing wavelengths (and hence energies) are absorbed to different degrees by other metals. This absorption is governed by the same considerations that control X-ray emission. When the incoming X-rays have just enough energy to knock out inner shell electrons from the atoms of the absorber, they and the more energetic X-rays on the short wavelength side of them, will have their energy mostly absorbed in such interactions. X-rays of longer wavelength than this threshold, having insufficient energy for such interaction, will pass through without being much absorbed. If a thin foil of a metal with its *absorption edge* just on the short wavelength side of the $K\alpha$ radiation of the target is

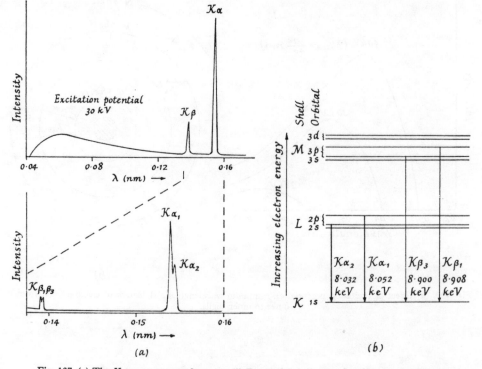

Fig. 137. (*a*) The X-ray spectrum of copper. (*b*) Energy-level diagram for electron transitions.

placed in the X-ray beam, its effect will be to *filter out* the unwanted Kβ line and the short wavelength part of the continuous spectrum. As is shown in Fig. 138 the insertion of a foil of nickel in the beam from a copper target will give an X-ray beam in which the wavelength 0·154 nm is overwhelmingly dominant.

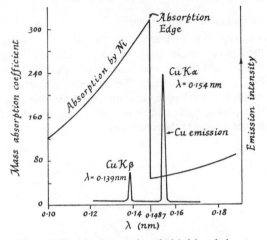

Fig. 138. The absorption edge of nickel in relation to the emission spectrum of copper.

DIFFRACTION OF X-RAYS BY CRYSTALS

The wavelength of X-rays used in crystallography is less than the distance between lattice points, but it is of the same order of magnitude as this distance. Hence, just as light is diffracted from a grating of very closely spaced lines, so X-rays are diffracted from the three-dimensional array of equivalent points of the crystal lattice. In the diffraction of light each aperture in the grating functions as a new source of rays: so in the crystal each atomic electron cloud set into vibration by the X-rays acts as a source of secondary rays of the same wavelength.

W. L. Bragg was able to show, in 1912, that the X-rays behaved, in these circumstances, exactly as though they were *reflected* from the planes of atoms making up the crystal structure. The whole topic can therefore be discussed in the more familiar terms of reflections from planes, and this is a masterly simplification of the situation.

Imagine a set of atomic planes (Fig. 139) and a train of X-rays striking them at an angle θ. The X-rays penetrate the layers, but are also reflected by them. Ray *a* is reflected at the first plane, ray *b* at the second. The reflected rays from all planes of the set take the same direction, and if they are out of phase they will interfere with one another and be destroyed. Only if the difference in path between rays reflected from successive planes is an even number of wavelengths, will they be able to reinforce one another and form a viable reflected train. The path difference between rays reflected at successive planes is seen to be *ef+fg* in Fig. 139. Further, *ef = fg = d* . sin θ where *d* is the perpendicular distance between the planes. Hence the condition for successful reflection is given by the *Bragg Law*

$$n\lambda = 2d . \sin \theta$$

where *n* is any whole number, *d* is the interplanar spacing and θ is the glancing angle of incidence. (Note that θ is the complement of the angle of incidence used in optics.)

From the Bragg equation we see that a family of planes of a particular spacing can reflect X-rays of a given wavelength *at one angle of incidence only*.

We wish to use Bragg's Law to find out the interplanar spacings of a crystal. As we have seen, X-rays of a known wavelength can be produced by using a suitable target and the proper filter. The angular positions of the reflected X-rays are recorded, in

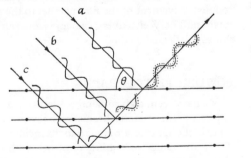

(a) *Condition for reflection*

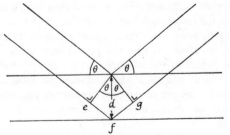

(b) *Path difference = 2d sin θ*

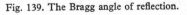

Fig. 139. The Bragg angle of reflection.

relation to the direction of the undeviated beam, most usually on photographic film placed around, or in front of, the crystal. The film is darkened where the reflected beams strike it. As is seen from Fig. 142b the angle between the undeviated and reflected beams is 2θ. Knowing λ and θ the interplanar spacing, d, may be found.

TYPES OF DIFFRACTION PHOTOGRAPHS

The photographs used in X-ray crystallography fall into two groups, single crystal photographs and powder photographs. Single crystal photographs are the basis of complete determinations of crystal structure. Their interpretation is a special study falling outside the scope of this book, and they therefore receive only brief mention below. The powder photograph, on the other hand, though limited as a means of working out a structure, is ideally suited and extensively used for identifying crystalline substances. It is therefore of great value in mineralogy, and will be described more fully.

SINGLE CRYSTAL PHOTOGRAPHS

In these methods a single crystal or crystal fragment, usually a small one a millimetre or two across, is mounted in a known orientation (determined optically or by trial and error) and irradiated by X-rays, which are reflected by appropriately angled planes to yield spots on a photographic film.

LAUE PHOTOGRAPHS

These were the earliest type. A stationary crystal is irradiated with *white radiation* and the reflections recorded on a flat plate placed beyond the crystal (Fig. 140), or between the crystal and the X-ray source, the beam in the latter case passing through a hole in the plate.

Because the white radiation contains a wide spectrum of wavelengths, each of the various sets of planes of different d-spacing in the crystal is able to pick out the wavelength suited to the angle at which it lies to the beam, and reflect that wavelength. Thus many planes reflect, and the photograph gives a general picture of the symmetry of the crystal; but because we do not know which wavelength is being reflected by any set of planes we cannot deduce the d-spacing, and cannot find out the dimensions of the unit cell.

ROTATION PHOTOGRAPHS

In this type the single crystal is rotated about a zone-axis in a beam of monochromatic X-rays. In this way successive crystal planes are brought into a position to reflect the wavelength being used, and reflections are recorded on a cylindrical film placed around the specimen. By taking three photographs with rotation about three suitably chosen axes at right angles to one another a complete picture of the interplanar spacings and lattice dimensions is obtained.

OSCILLATION PHOTOGRAPHS

The crystal is oscillated about an axis of rotation, through an angle of about 15°, instead of being rotated, and the reflections recorded on a cylindrical film around the crystal. This type of photograph avoids the overlapping of spots encountered in rotation photographs and facilitates the assignment of Miller indices to the reflections.

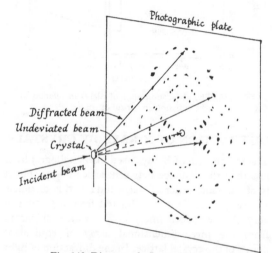

Fig. 140. Diagram of a Laue photograph.

MOVING FILM METHODS

Here, in addition to rotation of the crystal, there is a coupled movement of the film parallel to the axis of rotation. The Weissenberg camera is an example of this sophisticated method.

POWDER PHOTOGRAPHS

As we have seen, a given family of lattice planes can reflect a beam of monochromatic X-rays only when it lies at the appropriate angle θ to the incident beam. There are numerous families of planes, with a sufficiently dense population of atoms to reflect X-rays, transecting a crystal lattice. In the powder method we seek to record reflections from all these planes.

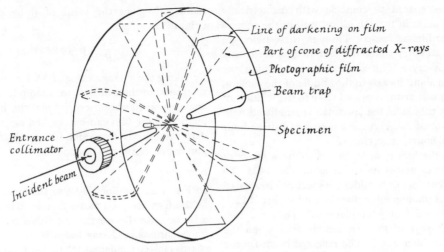

Fig. 141. The arrangement for a powder photograph.

This is done by first grinding the mineral to an impalpable powder (less than 240 mesh). The powder may then be treated in one of the following ways:

(*i*) placed in a fine capillary glass tube of 0·2 mm bore,

(*ii*) coated on a fine glass fibre by dipping the fibre in alcohol and rolling it in the powder,

(*iii*) mixed with a little gum arabic and rolled between slips of glass into a fine spindle or a tiny ball.

The objective is a specimen not more than 0·3 mm diameter.

In such a specimen, if the powder is fine, there will be some grains in every conceivable orientation. Furthermore, to ensure that grains in all orientations are exposed to the X-ray beam it is usual to rotate the specimen while it is being irradiated.

The rod- or ball-shaped mass of powder is now placed at the centre of a shallow cylindrical camera (Fig. 141) the wall of which is lined with 35 mm wide film. The film has previously been punched with

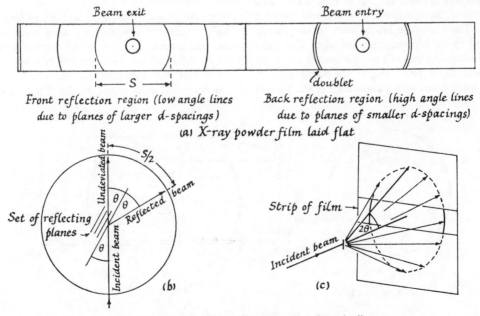

Fig. 142. Diagram to illustrate the formation of powder lines.

two holes spaced to coincide with diametrically opposed holes in the cylindrical wall of the camera when the film is loaded. Through these holes are inserted an entry collimator and an exit beam trap for the X-rays. With a small motor turning the specimen about its axis in the middle of the camera, the entry collimator is placed close to the window of an X-ray tube, and the specimen is irradiated with monochromatic X-rays for a period of half an hour to several hours, as necessary.

When the film is developed it shows a series of arcuate lines across its width as in Fig. 142a. To explain these, let us consider the specimen irradiated with a flat photographic plate behind it (Fig. 142c). Consider one family of planes of spacing d that reflects X-rays of the wavelength used when the angle of incidence is θ. The reflected beam from a crystal fragment at this angle to the beam will reflect a ray at an angle 2θ to the undeviated beam (Fig. 142b). There are crystal fragments in the specimen in all conceivable orientations. Therefore, X-rays will be reflected from this family of planes in all possible directions at an angle 2θ to the undeviated beam—that is, along a conical fan of directions, coalescing into a conical surface of semi-angle 2θ with its apex at the specimen (Fig. 142c). To record the presence of a family of planes of this d-spacing we need only photograph a portion of the trace of this cone of reflected rays. On the other hand, we wish to record all the cones that arise from the different d-spacings present. A strip of film is therefore placed round the specimen, at a uniform distance from it. The curvature of the lines on the film is explained by the fact that they are parts of circular sections of these cones of rays. Measurement of the separation, S, of two parts of the same cone of reflections, coupled with a knowledge of the camera

radius, R, gives the value of θ by the relation (Fig. 143a)

$$\frac{S}{4R} = \theta \text{ (radians)}.$$

By taking the average value of $\frac{1}{2}S$ for several reflections around the exit hole for the undeviated beam, and around the beam-entry hole, the half-circumference of the film can be checked against its standard value, and a correction for any shrinkage of the film in processing can be applied.

DOUBLETS

In Fig. 142a the lines close to the beam-entry hole in the film are seen to be double. These doublets arise because the Kα radiation that constitutes the 'monochromatic' X-ray beam is, in fact, made up of two closely similar wavelengths, produced by the Kα_1 and Kα_2 transitions shown in Fig. 137. In the front reflection region these two wavelengths are not separated in reflection, and produce only a single line. In the back reflection region, however, θ is large, approaching 90°. From the curve of $\sin \theta$ against θ we see that in this region a small change in $\sin \theta$ makes a relatively large change in θ and hence in 2θ, the angle of reflection. A small change in λ in the Bragg equation, leading to a small change in $\sin \theta$, thus produces a large change in 2θ. As there are two values of λ in the incident beam, two powder lines appear for one value of d.

THE MEASUREMENT OF θ

The measurement of distances on the powder film is usually carried out by laying it over a millimetre scale, and moving a cursor with a vernier attached to it from line to line with the aid of a low-power (4× or less) travelling microscope or a lens. The measurement is made to the middle of the arc form-

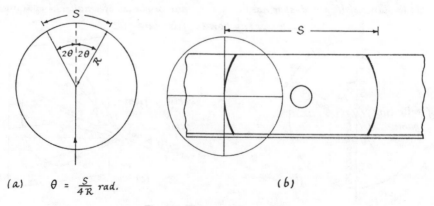

(a) $\theta = \frac{S}{4R} \text{ rad.}$

(b)

Fig. 143. The measurement of θ.

ing the powder line (Fig. 143b) and may be done, on good lines, with an error not exceeding 0·05 mm. The angle θ is derived from the separation of corresponding arcs and the camera radius (see above), and $\sin \theta$ and λ are substituted in the Bragg equation to give d. The appropriate values of λ for $K\alpha_1$ and $K\alpha_2$ radiation are used for each line of a doublet. For single lines $\lambda K\alpha_1$ may be used, or a weighted mean of the two wavelengths, $(2\lambda K\alpha_1 + \lambda K\alpha_2)/3$, may be taken. d may be obtained from θ without calculation by the use of tables published for the wavelengths commonly employed.

Direct reading scales of d may also be used for rough work, each being constructed for a specific camera diameter and radiation wavelength. Film shrinkage and the rapidly changing magnitude of the scale divisions precludes great accuracy in their use.

The relative intensities of the lines may be estimated visually on a scale where the strongest line is taken as 10.

THE POWDER DIFFRACTION FILE (formerly the ASTM index)

The immediate use of the powder photograph is to act as a finger-print in mineral identification. Because the variation in possible cell dimensions is infinite, and also because any change in the kinds of atoms alters the reflecting powers of the atomic planes, it is virtually impossible for two different substances to have the same X-ray powder pattern. A comprehensive and ever-expanding index of powder patterns of crystalline substances, including not only minerals but artificial inorganic and organic compounds, is maintained by the Joint Committee on Powder Diffraction Standards (formerly the American Society for Testing Materials), for the identification of crystals.

Each card gives primarily the d-spacings of the three strongest lines, which forms the key to the system, followed by the largest d-spacing recorded. Then follows a complete list of the d-spacings of lines recorded, with their intensities, and their indices (where known), together with experimental details, optical properties, the density and a reference. Fig. 144 is an example of a card from the File. The cards are classified in order of spacing of the three strongest lines of the pattern (Hanawalt Index) and on a system employing the eight strongest lines (Fink Index). It is not necessary to detail the procedure for using these indexes, as the volumes themselves provide this. It is, of course, useful to establish a collection of powder photographs of minerals for direct comparison, as well, in a laboratory where the method is in continual use.

9-427 MAJOR CORRECTION

d	2.57	1.54	2.87	4.04	$FE_3 AL_2 (SIO_4)_3$		$3FEO.AL_2O_3.3SIO_2$	
I/I_1	100	50	40	30	IRON ALUMINUM ORTHOSILICATE		ALMANDITE	

Rad. CuKα λ 1.5405 Filter NI Dia. 114.6MM	d Å	I/I_1	hkl	d Å	I/I_1	hkl
Cut off I/I_1 VISUAL	4.04	30	220	0.947	5	12.2.0
Ref. L.G. BERRY, QUEEN'S UNIVERSITY, KINGSTON, CANADA	2.873	40	400	.935	10	12.2.2
	2.569	100	420	.869	5	12.4.4
	2.447	5	332	.860	10	12.6.0
Sys. CUBIC S.G. O_H^{10} - IA3D (230)	2.348	20	422	.850	5	12.6.2
a_0 11.53 b_0 c_0 A C	2.257	20	510	.835	5	888
α β γ Z 8 Dx 4.313	2.102	20	521	.792	20	14.4.0
Ref. IBID.	2.043	10	440	.785	30	14.4.2
	1.866	30	611			
ε α n $\omega\beta$ 1.815 ε γ Sign	1.660	30	444			
2V D 4.29 mp Color RED	1.599	40	640			
Ref. IBID.	1.540	50	642			
	1.441	20	800			
SAMPLE FROM CAVENDISH TP., PETERBOROUGH CO., ONTARIO.	1.287	20	840			
	1.257	30	842			
	1.228	10	664			
	1.167	5	941			
	1.070	20	10.4.0			
	1.051	10	10.4.2			
	1.019	10	880			

Fig. 144. A card from the JCPDS Powder Diffraction File (reproduced by permission).

LINE SHIFTS AND INTERNAL STANDARDS

Atomic substitution, an almost universal phenomenon in minerals, leads to small changes in unit cell dimensions, and hence in the positions of powder lines. Such shifts may be accurately measured between different specimens of the same mineral group by mixing a known substance with the mineral powder before photographing it, and measuring the position of a particular line (or lines) in the test sample with respect to one of the lines of the internal standard. Charts of line shift against composition, based on chemically analysed material, provide a valuable means of rapid composition determination.

INDEXING POWDER LINES

To extend the use of powder photographs to the investigation of changes in unit cell size, and similar matters, we must determine the Miller indices of the planes represented by the powder lines. This can be done straightforwardly for orthogonal systems (cubic, tetragonal, hexagonal, trigonal and orthorhombic) *when the unit cell is known*, and, with more difficulty, for the monoclinic and triclinic systems also. The difficulties of indexing increase as the unit cell dimensions become larger. When the unit cell is not known, the difficulties of indexing from a powder photograph alone are greatly increased, and it may be impossible.

These preliminary requirements before indexing might be thought to be severely limiting, but the situation is not as bad as it appears to be. Some knowledge of the symmetry is often obtainable from optical study under the microscope. Moreover, unless one is dealing with an entirely new mineral, it ought to be possible to identify the substance, or some analogue of it, from the ASTM index, and this will give some approximate cell dimensions to start with. We shall first consider the procedure when this sort of preliminary information is available, and shall limit ourselves to the orthogonal systems. The nonorthogonal systems require the use of a new concept—that of the reciprocal lattice—and so will not be considered here.

INDEXING ORTHOGONAL SYSTEMS WHEN THE UNIT CELL IS KNOWN

This problem involves calculating the Bragg angles, θ, for all possible values hkl of Miller indices for the cell, and comparing these with the θ values read from the photograph. Having assigned probable indices to reflections from some key planes, using assumed cell dimensions which are likely to be only

approximately correct, it is then possible to revise the dimensions and repeat the procedure to obtain a more refined result.

The first step is to find an expression relating d, the interplanar spacing, to the indices hkl. Fig. 145 shows the geometry involved. A plane with Miller indices hkl has intercepts on the crystallographic axes of $1/h$, $1/k$, $1/l$. In an orthogonal unit cell with sides a, b and c, the family of such planes divides the sides into parts of length a/h, b/k and c/l respectively; d is the perpendicular distance between the planes.

In the triangles dissected out in Fig. 145 the following relations hold:

$$\frac{OA}{OS} = \frac{AB}{OB} \text{ (similar triangles)}$$

$$\frac{a/h}{s} = \frac{\sqrt{(b^2/k^2 + a^2/h^2)}}{b/k}$$

$$\frac{1}{s} = \frac{hk}{ab} \sqrt{\left(\frac{b^2}{k^2} + \frac{a^2}{h^2}\right)}$$

$$= \sqrt{(h^2/a^2 + k^2/b^2)}$$

$$\frac{OS}{OD} = \frac{SC}{OC}$$

$$\frac{s}{d} = \frac{\sqrt{(c^2/l^2 + s^2)}}{c/l}$$

$$\frac{1}{d} = \frac{l}{cs} \sqrt{\left(\frac{c^2}{l^2} + s^2\right)}$$

$$= \sqrt{(1/s^2 + l^2/c^2)}$$

Substituting for $1/s^2$

$$\frac{1}{d} = \sqrt{(h^2/a^2 + k^2/b^2 + l^2/c^2)}.$$

Hence

$$d = 1/\sqrt{(h^2/a^2 + k^2/b^2 + l^2/c^2)}.$$

Formulae for the orthogonal systems

(*i*) *Cubic system*

For the cubic system, where all cell edges are the same, the formula above becomes

$$d = a/\sqrt{(h^2 + k^2 + l^2)}. \tag{1}$$

Combining this with the Bragg equation

$$\lambda = 2d \cdot \sin \theta \tag{2}$$

we have

$$\sin \theta^2{}_{hkl} = \frac{\lambda^2}{4a^2} (h^2 + k^2 + l^2). \tag{3}$$

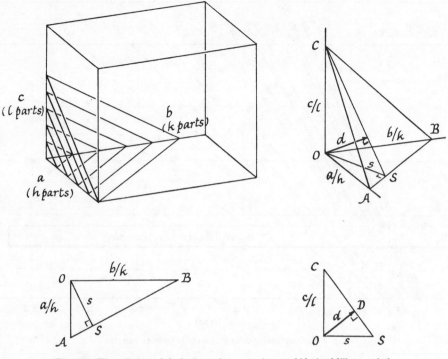

Fig. 145. The relation of d, the interplanar spacing, to hkl, the Miller symbol.

The term $(h^2 + k^2 + l^2)$ is a whole number, designated N. The term $\lambda^2/4a^2$ is evaluated from the known wavelength and the assumed value for a, and is multiplied by the possible values of N, the results then being compared with the values of $\sin^2 \theta$ obtained from the powder photograph. Exact agreement is not expected at the first trial; but probable values of hkl will be suggested for a number of reflections. These can be fed back into equation (1) (above) to give a new value of a that can be used in a fresh trial, and so on until all the lines can be indexed.

(ii) Tetragonal system

The formula here becomes

$$\sin^2 \theta_{hkl} = \frac{\lambda^2}{4a^2} (h^2 + k^2) + \frac{\lambda^2}{4c^2} l^2.$$

(iii) Hexagonal and Trigonal systems

For hexagonal crystals, and trigonal crystals indexed on hexagonal axes, the formula is

$$\sin^2 \theta_{hkl} = \frac{\lambda^2}{3a^2} (h^2 + hk + k^2) + \frac{\lambda^2}{4c^2} l^2.$$

Note the change to $\dfrac{\lambda^2}{3a^2}$.

(iii) Orthorhombic system

The corresponding formula now is

$$\sin^2 \theta_{hkl} = \frac{\lambda^2}{4a^2} h^2 + \frac{\lambda^2}{4b^2} k^2 + \frac{\lambda^2}{4c^2} l^2.$$

The amount of calculation involved in applying these formulae is reduced by making use of published tables that give (i) d for changing values of θ for commonly-used wavelengths; (ii) $\sin^2 \theta$; (iii) values of $(h^2 + k^2 + l^2)$, $(h^2 + k^2)$ and $(h^2 + hk + k^2)$. These may be found in books listed at the end of this chapter.

Indexing when the unit cell is unknown

In the first place, of course, every effort should be made to acquire information about the symmetry from optical examination and to assign the substance to one of the groups isotropic, uniaxial, biaxial.

Cubic patterns may be recognized by the fact that the $\sin^2 \theta$ values have a common factor, namely $\lambda^2/4a^2$, which we may call x. If this is found, then $a = \lambda/2\sqrt{x}$, and a start may be made on this basis.

Graphical methods are also available; the simplest again is for cubic crystals, based upon $d = a/ \sqrt{(h^2 + k^2 + l^2)}$. Lines giving d for various values

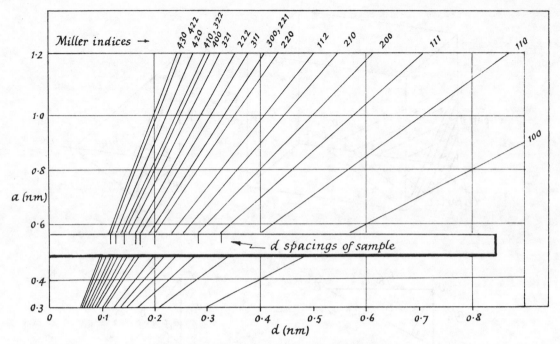

Fig. 146. Graphical indexing of a cubic powder pattern.

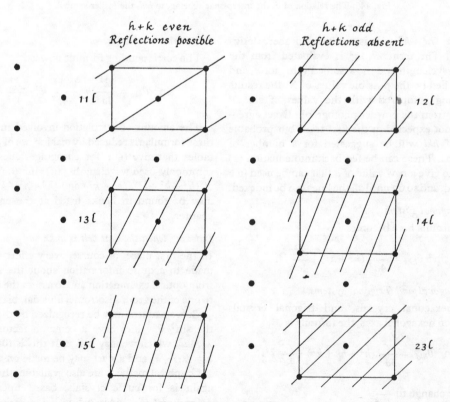

Fig. 147. Absent reflections.

of a, for some simple Miller indices, are drawn on a graph of a in nm, against d in nm (Fig. 146). The d-values for the lines of the powder photograph are plotted on a strip of paper on the same scale as d of the diagram, and the strip is moved up and down parallel to the d axis until the d values on the strip coincide with lines on the graph. The level at which this occurs gives a for the substance.

Analogous but more complicated methods for systems of lower symmetry will not be dealt with here.

ABSENT REFLECTIONS

Only when the lattice is primitive (Fig. 26, p. 32) will reflections be able to occur for all possible values of hkl. When the lattice is face-centred or body-centred, families of planes with certain indices will not contain all the lattice points. For example, Fig. 147 shows a plan view of a C face-centred ortho-rhombic lattice, with the traces of some representative families of planes upon it. It will be seen that those on the left-hand side of the diagram contain all the points of the lattice, while those on the right-hand side omit the centre point in the C face. Viable reflections will not arise from a family of planes that omits centring points in this way, because the omitted and included points together make up a parallel family of planes with half the d-spacing shown on the diagram. This smaller interplanar spacing is the one that will be effective in satisfying the Bragg equation. The wider-spaced set of planes cannot reflect independently, because the reflection from them would be extinguished by another $\frac{1}{2}\lambda$ out of phase, from the planes of lattice points half way between them.

In the case illustrated, the requirement that a family of planes includes all lattice points is met when $(h+k)$ is an *even number*. When $(h+k)$ is odd, centre points are omitted and no reflection will be produced.

This principle can be generalized into a set of rules that allows the lattice type to be identified from the combinations of index numbers that are found on a photograph. Those not conforming to the restrictions will be systematically absent.

Lattice type	Condition for reflection
P (primitive)	no restriction
F (all-face-centred)	h, k, and l all even or all odd
I (body-centred)	$(h+k+l)$ even
A face-centred	$(k+l)$ even
B face-centred	$(h+l)$ even
C face-centred	$(h+k)$ even
R rhombohedral indexed on hexagonal axes	$(-h+k+l)$ a multiple of 3 or $(h-k+l)$ a multiple of 3.

FURTHER READING

The account given above of the interpretation of X-ray photographs is, of course, limited to the simpler procedures. Nevertheless, it should enable the student to make good use of powder photography, which is probably the most widely-used technique. The general account of the interaction between X-rays and crystals should serve to show how our knowledge of mineral structures has been derived, and it is hoped that it will be helpful to those who wish to study the subject in greater depth in one of the works listed below.

AZAROFF, L. V., and BUERGER, M. J. 1958. *The powder method in X-ray crystallography*. New York, McGraw-Hill Co.

BUERGER, M. J. 1942. *X-ray crystallography*. New York, Wiley & Sons.

HENRY, N. F. M., LIPSON, H., and WOOSTER, W. A. 1951. *The interpretation of X-ray diffraction photographs*. London, Macmillan & Co.

For reference:

Tables for conversion of X-ray diffraction angles to interplanar spacing. 1950. U.S. Dept. of Commerce, National Bureau of Standards, Applied Mathematics Series 10. Washington, U.S. Govt. Printing Office.

6

MINERAL ASSOCIATIONS

CHEMICAL CONSTITUTION OF THE EARTH

From the study of meteorites and from geophysical data on the mass of the earth, its moment of inertia, geomagnetism and seismology it is believed that the earth has a *core* (the *siderosphere*) of nickel-bearing iron extending from the centre (6371 km from the surface) out to the Gutenberg-Weichert Discontinuity at 2900 km from the surface.

Outside this lies the *mantle* extending to a distance of only 40 km from the surface under the continents and to less than 10 km from the surface under the deep oceans. Its upper boundary is the Mohorovicic Discontinuity. Again from meteorite study and geophysics, supplemented by the petrography of denser surface rocks and ejecta from volcanoes, the mantle is thought to be made of the high pressure equivalents of the magnesium silicates olivine and pyroxene that make up stony meteorites and ultra-mafic rocks. With these there may well be some nickel-iron and sulphides.

In the *crust* are concentrated the silicates of Al, Fe, Ca, Na, K and Mg, forming rocks of broadly granitic composition in the continents and basaltic composition in the oceanic regions.

The *lithosphere*, in the chemical sense, includes those parts of the earth made principally of silicate minerals. So defined, it embraces the crust and the mantle.*

If we suppose that the earth began as an accumulation of more or less uniform interstellar dust, there has been, during its evolution, a sorting out of the chemical elements into these three zones or geo-

* In geophysics, the term lithosphere means the rigid outer layer of the earth down to perhaps 100 km, which lies above a weaker asthenosphere, in which creep in the solid state allows the possibility of slow circulation of material.

spheres, with the hydrosphere, atmosphere and biosphere (the ever-changing living material of the earth's surface) in addition (Fig. 148). This chemical sorting appears to be irreversible.

Measurements of the relative amounts of the less abundant elements in the metal, sulphide and silicate minerals of meteorites led V. M. Goldschmidt to classify the elements into the groups shown in Table 15 according to their affinity for the metal phase, the sulphide phase, or the silicate phase. He introduced the terms siderophile ('iron-loving'), chalcophile (for the sulphide-formers), lithophile (elements concentrating in silicates or other oxygen compounds), and corresponding terms for the elements concentrated in the other geospheres.

ELEMENT SEPARATION IN THE CRUST

Considering now the crust rocks, we find that the chemical sorting that gave them their present overall composition is carried further by processes within the crust and upon it. These processes are (*i*) melting and subsequent crystallization of the mineral aggregate to form *magmatic* rocks, (*ii*) weathering, partial solution and mechanical breakdown, coupled with washing and grading by surface waters to form *sediment*, and (*iii*) redistribution of materials within the new deposits and all types of rocks by subsurface circulating waters heated by the general rise of temperature with depth. From being a low temperature process with relatively abundant weak solutions near the surface, this leads on to (*iv*) a high temperature recrystallization at depth, less dependent upon water circulation, called *metamorphism*. Metamorphism may lead on to partial melting to complete the cycle.

Throughout this cycle the sorting of the chemical

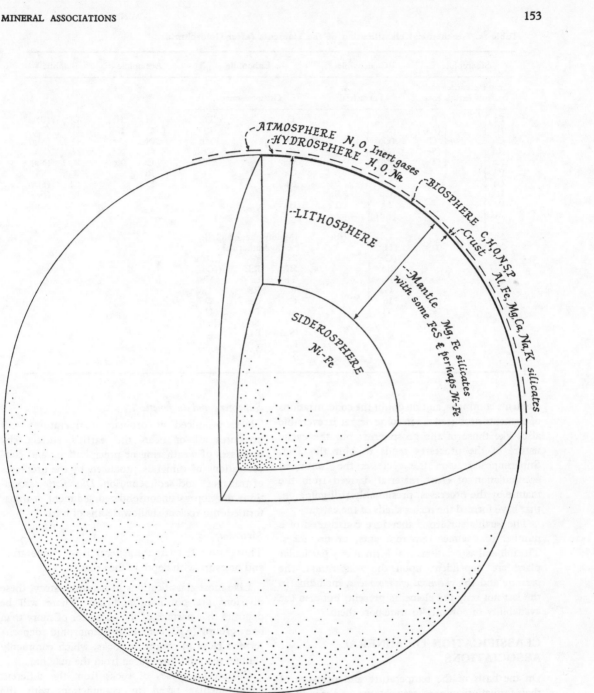

Fig. 148. The geospheres.

Table 15. Geochemical classification of the elements (after Goldschmidt).

Siderophile	Chalcophile	Lithophile	Atmophile	Biophile
Iron meteorites and possibly earth's core	Terrestrial	Orthomagmatic		

Siderophile		Chalcophile		Lithophile		Atmophile		Biophile	
Fe	P	(Fe)		O	Ti	H	He	C	(B)
Ni	(As)	(Ni)		Si	(P)	C	Ne	H	(Ca)
Co	C	(Co)		Al	Mn	N	Ar	O	(Mg)
Ru	(Se)	(Mo)		(Fe)	Cr	O	Kr	N	(K)
Rh	(Te)	(Pd)		Mg	(S)	Cl	Xe	P	(Na)
Pd		(Pt)		Ca		Br		S	(V)
Os			S	Na		I		Cl	(Mn)
Ir		Cu	Se	K				I	(Fe)
Pt		Ag	Te	(H)					(Cu)
Au		(Au)	As						
Ge		Zn	Sb						
Sn		Cd	Bi						
Mo		Hg		Mainly pneumatolytic-hydrothermal					
		(Ge)		Zr	Y	((Ni))			
		(Sn)		Hf	R.E.	((Co))			
		Pb		Th	Li	Nb			
		Ga		F	Rb	Ta			
		In		Cl	Cs	W			
		Tl		Br	Be	U			
				I	Sr	((C))			
				(Sn)	Ba				
				B	V				
				Sc					

elements continues, and no doubt the concentrations of elements produced are in some degree irreversible also, like those of the geospheres; but the cyclic nature of the processes tends to blur this fact. Superimposed upon the cycle is the continued accumulation of crust material derived from the mantle by the processes, presumably still going on, that have formed the major shells of the earth.

The earth's surface is therefore constituted of a number of chemical environments, or provinces. The mineral associations that form at any particular place are dependent upon the *temperature*, the *pressure* and the *chemical environment*, including in the last not only the elements present, but also the availability of solvents to transport them.

CLASSIFICATION OF MINERAL ASSOCIATIONS

On the basis of the temperature and pressure of their formation, mineral associations, or *parageneses*, can be divided into three groups.

Magmatic

Those associated with local centres of heat sufficient to melt part of the rocks. Included here are the products of crystallization of melts, and of gases and solutions derived from them.

Supergene and diagenetic

Those produced at ordinary temperatures and pressures at, or near, the earth's surface, by processes of weathering or precipitation; and concentrations of minerals produced by the processes of transport and sedimentation. *Diagenetic* associations develop as unconsolidated sediment is transformed into rock at shallow depths of burial.

Metamorphic

Those associated with the general rise of temperature and pressure with depth in the crust.

Like most classifications in natural history, these divisions are not hard and fast. There will be deposits that partake of the character of more than one group (e.g. contact metamorphic deposits formed close to magmatic rocks, which commonly contain material introduced from the magma).

Chemical analyses of rocks from the different environments, taken in conjunction with the amounts of these rocks, furnish estimates of the abundance of the elements in the crust. These are summed up in Fig. 149. The most obvious point of the figure (bearing in mind that it is on a logarithmic scale) is the great relative abundance of the eight elements O, Na, Mg, Al, Si, K, Ca, and Fe. These are the *major elements* of the crust. The picture that

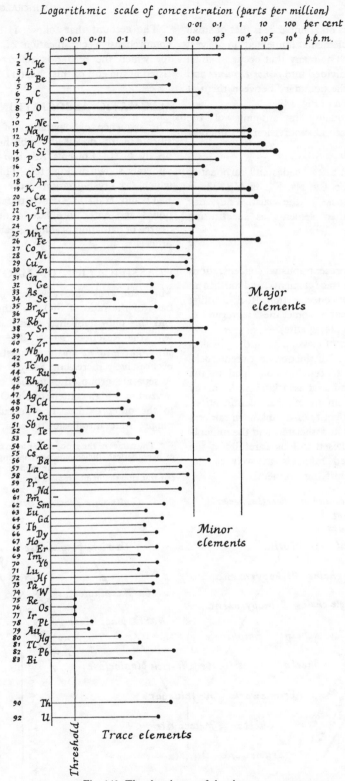

Fig. 149. The abundances of the elements.

emerges from the data on abundance is that of small amounts of many elements immersed in a vast abundance of a few. The many that occur in small amounts may be subdivided into *minor elements* and *trace elements*, with the boundary between them at 0·01 per cent, as shown in Fig. 149.

The marked inequality in amounts of the elements leads to a second subdivision of minerals.

Rock-forming minerals

Chiefly silicates, with a few oxides and carbonates, of the elements Al, Fe, Ca, Na, K and Mg. These build the common rocks of the crust. They may contain traces of other elements as substituting atoms in the lattice.

Ore minerals

All other minerals represent unusual concentrations of one of the minor or trace elements, and must form by special processes of concentration. Rock-forming minerals may themselves sometimes segregate to form ores (e.g. magnetite, apatite).

These two elements of classification, the physical environment and the distinction between rock-forming minerals and ores, are combined in the following survey of mineral associations. A survey of this kind is of help in mineral identification, because clues to the identity of an unknown mineral are often provided by its associates, and the mineralogist will be well advised to look carefully at the matrix and associated minerals as well as the substance itself in identifying a mineral.

The account that follows is not intended as a comprehensive classification of mineral deposits, for which the reader is referred to other works listed at the end of this chapter.

MAGMATIC ASSOCIATIONS

The sequence of separation of minerals from a silicate melt is now well-established, and is summed up by N. L. Bowen's *Reaction Series* of igneous minerals shown in Fig. 150. This series is one in which, as crystallization proceeds, SiO_2, Na_2O, K_2O and H_2O are concentrated in the residual liquid and enter into late-formed minerals. The degree of oxidation becomes greater as the sequence proceeds.

CLASSIFICATION OF MAGMATIC ROCKS

Magmatic rocks are classified chemically on the basis of the ratio of SiO_2 to other oxides present, and this old-established classification links closely with the Reaction Series. The groups *ultrabasic*, *basic*, *intermediate* and *acid* constitute a series with progressively increasing SiO_2 content* (Table 16). A separate group, the *alkaline* rocks, are characterized by a high content of Na_2O and K_2O relative to the other oxides. They occur throughout the range from ultrabasic to acid, but have their

* The names of the groups refer to an old-fashioned concept of hypothetical silicic acids participating in the formation of silicate minerals. Though the acids have disappeared from our thinking the names remain in use.

Fig. 150. Bowen's reaction series of the igneous rock-forming minerals.

Table 16. Classification of non-feldspathoidal magmatic rocks.

(Members of the main calc-alkali series heavily underlined
PLUTONIC rocks in capitals, Volcanic rocks in lower case)

Ratio of light to dark minerals	Group chemistry	Feldspar mainly alkali (K, Na)	Alkali feldspar $\simeq$ plagioclase	Feldspar mainly plagioclase
Felsic—light minerals (quartz and feldspar) > 60% by volume	Acid	GRANITE Rhyolite	ADAMELLITE Rhyo-dacite	GRANODIORITE Dacite
	—66% SiO$_2$—			
Mesotype—light minerals < 60% > 30% by volume quartz < 10%, plagioclase has Ab/An > 50/50	Intermediate	SYENITE Trachyte	MONZONITE Trachyandesite	DIORITE Andesite
	—52% SiO$_2$—			
Mafic—rich in dark (ferromagnesian) minerals (> 60% by volume) plagioclase has Ab/An < 50/50	Basic	ALKALI GABBRO Alkali basalt	SYENO-GABBRO Trachybasalt	GABBRO Basalt (and dolerite)
	—45% SiO$_2$—			
Ultramafic—almost feldspar-free	Ultrabasic			DUNITE PERIDOTITE PYROXENITE

Table 17. Leading types of alkaline rocks.

(PLUTONIC rocks in capitals, Volcanic rocks in lower case)

	With feldspathoid but without feldspar	With feldspar and feldspathoid	With feldspar but without feldspathoid
Felsic—rich in light minerals, but without quartz	URTITE		
Mesotype	MELTEIGITE	NEPHELINE-SYENITE Phonolite	ALKALI-SYENITE Alkali-trachyte
Mafic—rich in dark (ferromagnesian) minerals	IJOLITE Nephelinite	THERALITE Nepheline tephrite (no olivine) Nepheline basanite (with olivine)	ALKALI-GABBRO Alkali-basalt
Ultramafic	JACUPIRANGITE KIMBERLITE		

strongest expression in the acid-intermediate part of this range (Table 17).

Magmatic rocks are also classified into two groups related to the depth of burial during crystallization.

Plutonic rocks are those crystallized at depth within the crust. Deep burial retards heat loss and loss of volatile constituents, and leads to slow cooling, coarse grain-size (>1 mm) of the minerals, and a relatively low temperature of final consolidation. Some later-formed minerals retain H_2O in their lattices and opportunity is given for post-consolidation changes, such as exsolution of crystals that were formed as homogeneous solid solutions and alteration to lower temperature assemblages of early-formed high-temperature phases.

Volcanic rocks have crystallized from magma poured out at the surface of the earth or introduced at shallow depth. They have cooled relatively rapidly, the grain-size of their crystals is small (<1 mm), some part of the melt may be chilled to a glass, volatiles are lost, and anhydrous minerals of high temperature of crystallization are prevalent.

TYPES OF MAGMATIC ORE-DEPOSITS

The ore-deposits associated with magmatic rocks may occur either as segregations of the minerals separating from the silicate melt itself, or as minerals crystallized from the residual fluids after the main rock-forming minerals have precipitated. The following succession of stages can be recognized.

(*i*) magmatic—crystallizing earlier than, or at the same time as the principal rock-forming minerals;

(*ii*) pegmatitic—crystallizing from vapours or supercritical fluids during the later stages of crystallization of the rock-forming minerals;

(*iii*) hydrothermal—crystallizing from watery solutions remaining after the rock-forming minerals have crystallized. The last group may be divided into hypothermal deposits (close to the magmatic source), mesothermal deposits, and epithermal deposits (distant from the magmatic source). These classes contain minerals of progressively lower temperature of crystallization.

The mineralogy of the principal magmatic rock groups and their associated ores is given in the following sections.

ULTRABASIC ROCKS (Table 18)

These are the rocks called dunite (essentially olivine only), peridotite (olivine+pyroxene) and pyroxenite. They are believed to be formed by accumulation of early-formed crystals from magma of basic composition. Their bulk composition does not represent that of any liquid melt and they are plutonic rocks only, without volcanic representatives.

Table 18. The Ultrabasic group of rocks.

Rock types: dunite, peridotite, pyroxenite, kimberlite

Rock-forming minerals: olivine, hypersthene, augite, hornblende, phlogopite, subordinate bytownite or labradorite, chromite, magnetite (garnet, ilmenite in kimberlite)

Ore minerals:

Stage	Magmatic	Hydrothermal
Element		
Cr	chromite	
Pt, Pd, Ir, Os	native Pt iridosmine sperrylite	
Ni	pyrrhotite, pentlandite (also in veins)	
C	diamond (in kimberlite)	
Mg		chrysotile-asbestos talc magnesite

ROCK-FORMING MINERALS

Olivine, pyroxene and minor Ca-rich plagioclase feldspar are the essential minerals of ultrabasic rocks. Amphibole and biotite occur in some. Chromite and other spinel minerals are accessories. The olivine and pyroxene are commonly altered to serpentine, talc and carbonates.

ORE MINERALS

A number of ore minerals are entirely confined to ultrabasic rocks. Segregations of chromite, often as layers, are found only in these rocks. Chrysotile asbestos is restricted to serpentinized ultrabasics in which it forms veins. Native platinum, cooperite (PtS) and iridosmine are only found in basic and ultrabasic rocks, and the rarities awaruite (Ni_3Fe) and wairauite (CoFe) belong to the ultrabasic association only. Nickel-bearing pyrrhotite and pentlandite occur in sulphide segregations from ultrabasic and basic plutonic rocks. Talc, talc-magnesite and quartz-magnesite veins are found associated with some serpentinized peridotites.

Diamond, except when obtained from gravels, is exclusively associated with the special ultrabasic rock kimberlite. The rock-forming minerals here are olivine (and serpentine), the mica phlogopite, calcite, ilmenite, magnetite, perovskite and pyrope garnet. Ilmenite, perovskite and pyrope in soils are a guide to kimberlite below.

Corundum is found associated with some peridotites. The associated rock-forming minerals are notable for being Al-poor, and some introduction of aluminium must be involved in these occurrences. Magnetite segregations are found in this association also, but are not confined to it.

In general, the elements concentrated in the mineral association of ultrabasic rocks are those that find ready acceptance into the lattices of high-temperature precipitates from silicate melts. They include also elements, like Ni and Pt, known to be relatively concentrated in the metal rather than the silicate phase of meteorites—that is, elements with a low affinity for oxygen.

BASIC ROCKS (Table 19)

The plutonic representatives here are the gabbros; the volcanic rocks are the basalts and dolerites.

Table 19. The Basic group of rocks.

Rock types: gabbro, basalt
Rock-forming minerals: olivine, hypersthene, augite, labradorite, magnetite, ilmenite, apatite
Ore minerals:

Stage	Magmatic	Hydrothermal
Element		
Ti	ilmenite	
Ti, Fe	titanomagnetite (concentrated as placers)	
Cu		native Cu (in amygdales)
Na, K, Ca		pumpellyite epidote calcite zeolites
Cl		scapolite

ROCK-FORMING MINERALS

Pyroxene and Ca-rich plagioclase feldspar are the essential minerals. Olivine is often present. The accessory minerals are ilmenite, magnetite and apatite.

ORE MINERALS

Ilmenite occurs as segregations from gabbroic magmas, with some associated magnetite. Although magnetite and titanomagnetite occur in gabbros and basalts as widespread and conspicuous accessory minerals, these do not generally form concentrated deposits in the magmatic rock. Weathering of the rocks and sorting by wave or river action may, however, concentrate them in *placer deposits* as in the black sand beaches of the world.

Native copper is found in amygdales (gas-bubble holes) in a few basalts and dolerites, along with pumpellyite and epidote. Zeolites and calcite are also minerals characteristic of amygdales in basalts. Though quartz is not a rock-forming mineral of this association, secondary silica in the form of agate occurs in the amygdales of rocks of basaltic to andesitic type.

CALC-ALKALINE INTERMEDIATE AND ACID ROCKS

The petrographic classification of intermediate and acid rocks is shown in Table 16.

As far as the broader mineral associations are concerned, it is more appropriate to divide them into a series from diorite to granite (and andesite to rhyolite), heavily underlined in the Table; and a syenite-trachyte group which is associated with alkali syenites.

The diorite-granite series, called the calc-alkaline series, is characteristic of the belts of mountain folding around the earth. Both the rocks and the ores associated with them are linked by continuous transitions and are dealt with together below. The syenites and alkali-syenites form, on the whole, restricted centres in regions subject to block-faulting and rifting of the crust: they are brought together in a separate section.

ROCK-FORMING MINERALS

Amongst the iron- and magnesium-bearing (*ferromagnesian*) minerals, pyroxene, amphibole and biotite characterize this group. Pyroxene-granites are few, this mineral being mainly confined to the intermediate plutonic types (diorites). It occurs more commonly in acid volcanics, however, because their rapid cooling freezes in high temperature minerals. An iron-rich olivine occurs in some granites, but generally olivine is absent from the group. Speaking broadly, pyroxene and hornblende are the dark minerals of diorites and andesites, while hornblende and biotite occur in granodiorites and dacites, granites and rhyolites.

The feldspar includes plagioclase ranging from andesine (with about 40 mol. per cent $CaAl_2Si_2O_8$) in the diorites and andesites, to oligoclase and albite in the granites. With increase in SiO_2 content, potash feldspar (orthoclase, microcline or sanidine) enters: it is found particularly in the granites and rhyolites.

The entry of quartz in amount greater than 10 per cent is taken as the dividing line between the intermediate and acid types.

Accessory minerals are apatite and magnetite, with zircon and sphene entering in granodiorites and granites.

ORE MINERALS (Table 20)

A wide range of ores is found in this association. With the higher content of volatiles in intermediate and acid magmas, there is a change in the character of the associated mineral deposits. They no longer consist of elements entering the lattices of high-temperature minerals, but include many that do not find a place in early precipitates, and hence concentrate in the residual fluids (gases, or liquids) of crystallization. The chief residual fluid is H_2O, but others including F, Cl and CO_2 are important, and they are collectively known as *mineralizers*. Borne by the mobile fluids, the elements that have not found a place in the high-temperature minerals precipitate as *vein minerals* in fissures traversing the magmatic rocks and radiating outwards from the magmatic centres into the surrounding *country-rocks*, which may be magmatic, metamorphic or sedimentary in origin.

With the useful ore-minerals of a vein are associated others of no commercial value. The latter are grouped by miners as the *gangue*. With changes in demand, the gangue of one generation may become the ore of another, as has been the case with fluorite in the Northern Pennine Orefield of England.

The deposition of the transported elements as mineral compounds is greatly influenced by temperature, which falls as the solutions move upwards and outwards from the magmatic centre. The minerals with the lowest temperatures of crystallization (e.g. stibnite, cinnabar) may migrate so far that it is difficult, or impossible, to point with confidence to a particular parent magmatic rock-body. Deposition is also influenced by the nature of the country-rocks. The fluids react with these in varying degree, to produce *replacement deposits* in the neighbourhood of the vein fissures.

In Table 20 the minerals listed as belonging to the pegmatitic stage, together with cassiterite, molybdenite, wolframite and scheelite, are particularly associated with granites. The other minerals are found in association with both intermediate and acid rocks.

SECONDARY ENRICHMENT OF VEINS

When erosion exposes mineral veins at the ground surface, they are often drastically modified by weathering. This extensive alteration is due in part to their open texture but largely to the presence of pyrite, which oxidizes and hydrates to yield acid solvents. The metals move downwards and are deposited at lower levels often as carbonates, or as native metal. At ground water level oxidation is checked and secondary sulphides are precipitated. These processes result in the formation of a leached, porous mass of iron oxide at the surface, called *gossan*, or the *iron hat*, succeeded downwards by a zone of secondary enrichment, consisting of carbonate, sulphate and oxide minerals in its upper part, and of secondary sulphides below the water table (Fig. 151). These minerals are shown in the right-hand column of Table 20.

SYENITES, NEPHELINE-SYENITES, TRACHYTES AND PHONOLITES

Although small amounts of trachyte and phonolite occur in many of the volcanic islands of the ocean basins, the main plutonic masses of syenite, alkali-syenite (syenite rich in soda, but without nepheline) and nepheline-syenite occur in the stable continental shield regions. They are often associated in time with episodes of block-faulting, and are accompanied by trachytes and phonolites. The total bulk of syenites and nepheline-syenites is small compared with that of the calc-alkaline series.

ROCK-FORMING MINERALS

Augite, hornblende and biotite are found in syenites, the sodic pyroxene aegirine and various sodic amphiboles being characteristic of the alkali-syenites. Iron-rich olivine is found in some trachytes.

Potash feldspar as orthoclase or microcline is dominant amongst the feldspar of syenites, while anorthoclase or sanidine occurs in trachytes. Albite may also be important. The nepheline-syenites and phonolites may carry nepheline, sodalite, analcime, cancrinite or natrolite. Leucite occurs only in volcanic rocks rich in potash, being unstable on slow cooling.

Quartz does not exceed 10 per cent in syenites, and is absent from the feldspathoidal (nepheline-, leucite- or sodalite-bearing) types.

ORE MINERALS

The notable concentration of magnetite and apatite at Kiruna in Sweden is associated with rocks

Table 20. Classification of minerals according to temperature at time of deposition.

Falling temperature: increasing dilution →

Element	Pegmatites	Hypothermal	Mesothermal	Epithermal	Secondary
Li	lepidolite amblygonite spodumene petalite				
Be	beryl				
B	tourmaline axinite				
F	topaz apatite / fluorite				
R.E., Y, Th	monazite				
Nb, Ta	samarskite pyrochlore				
(P)	(monazite apatite amblygonite)				
Fe	magnetite				jarosite copiapite limonite
Sn		cassiterite			
Mo		molybdenite			
As		arsenopyrite		realgar orpiment	realgar orpiment
W		wolframite scheelite			tungstite
Bi		native bismuth bismuthinite		tetradymite	
Au		native gold (with glassy quartz)	native gold (with milky quartz)	Au selenides and tellurides native gold electrum (with chalcedony)	
Cu		chalcopyrite tetrahedrite bornite			native Cu cuprite chalcocite covellite bornite malachite azurite chrysocolla
Co, Ni		niccolite smaltite-chloanthite linnaeite			nickel and cobalt blooms smithsonite hemimorphite willemite
Zn			sphalerite		
Pb			galena (argentiferous)		cerussite anglesite linarite pyromorphite mimetite wulfenite crocoite
Ag			argentite / ruby silver	hessite	native Ag cerargyrite stephanite
Sb				stibnite	cervantite senarmontite valentinite
Hg				cinnabar native Hg	
Gangue minerals	microcline albite / biotite muscovite	albite garnet pyroxene / biotite muscovite magnetite / ilmenite calcite / pyrite	calcite dolomite / ankerite barytes	calcite dolomite barytes / adularia alunite sericite / kaolin chlorite zeolites	

quartz changing from glassy through milky to chalcedonic and opaline

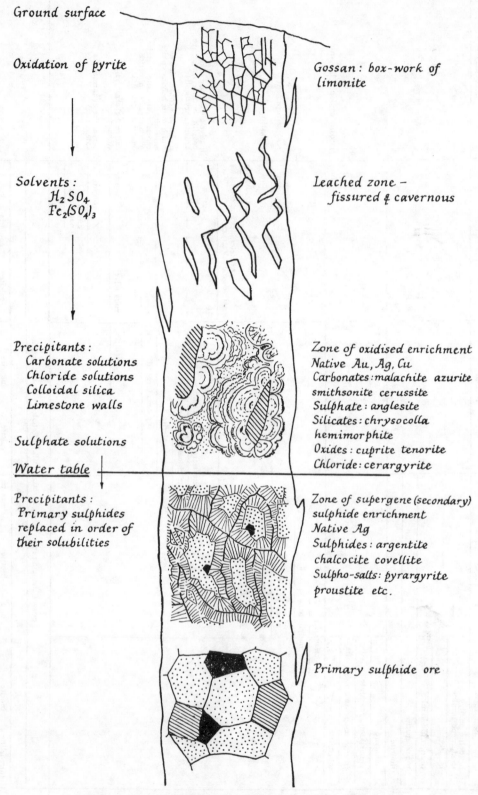

Ground surface

Oxidation of pyrite — Gossan: box-work of limonite

Solvents:
H_2SO_4
$Fe_2(SO_4)_3$

Leached zone – fissured & cavernous

Precipitants:
  Carbonate solutions
  Chloride solutions
  Colloidal silica
  Limestone walls

Sulphate solutions

Zone of oxidised enrichment
Native Au, Ag, Cu
Carbonates: malachite azurite
  smithsonite cerussite
Sulphate: anglesite
Silicates: chrysocolla
  hemimorphite
Oxides: cuprite tenorite
Chloride: cerargyrite

<u>Water table</u>

Precipitants:
Primary sulphides
replaced in order of
their solubilities

Zone of supergene (secondary)
sulphide enrichment
Native Ag
Sulphides: argentite
  chalcocite covellite
Sulpho-salts: pyrargyrite
proustite etc.

Primary sulphide ore

Fig. 151. Zones of weathering and enrichment of a sulphide mineral vein.

described as syenites; but other ores related to simple syenites are few.

Alkali-syenites and nepheline-syenites, on the other hand, possess a characteristic and important suite of ore minerals, in which many of the rarer elements are concentrated. These concentrations clearly represent the product of long and thorough segregation of the elements concerned from large volumes of the more common materials of the earth. Some of the minerals associated with these rocks are shown below (Table 21).

Table 21. Minerals found in association with alkaline rocks.

Stage	Magmatic	Pneumatolytic	Hydrothermal
Element			
Al	nepheline	nepheline corundum	
Nb Ta Y Zr	pyrochlore	pyrochlore	
P Cl F	apatite	fluorite	villiaumite
	pyrochlore	pyrochlore	(NaF)
Ti Nb R.E. Y	sphene	sphene	anatase
	loparite	various Na Ti Nb silicates	rutile
Zr Y	eudialyte	eudialyte Na Zr silicates	
Th R.E.		thorite	
R.E.			R.E. carbonates

CARBONATITES

These bodies of eruptive carbonate rock are restricted to small ring-complexes in the stable, unfolded regions of the continents. The rock types are søvite and alvikite (chiefly calcite), rauhaugite and beforsite (chiefly dolomite).

ROCK-FORMING MINERALS

Calcite, dolomite, ankerite, siderite, together with the silicates orthoclase, albite, forsterite, pyroxene, amphibole, biotite and melanite. Magnetite and apatite are important associates.

ORE MINERALS

Fe	magnetite
P, F	fluor-apatite, monazite, fluorite
Nb, Ta, Ti	pyrochlore, koppite, microlite, sphene, rutile, anatase, brookite, perovskite, knopite
Ce, R.E.	monazite, R.E. fluocarbonates
Sr	strontianite
Zr	zircon,
Th, U	thorianite

SUPERGENE AND DIAGENETIC ASSOCIATIONS

SEDIMENTARY ROCK-FORMING MINERALS

The common sedimentary rocks, except conglomerates, can be represented as mixtures of three fractions, quartz sand, clay minerals and calcite (with dolomite).

(*i*) Quartz sand is a resistant product of breakdown by weathering of earlier rocks. Ultimately the quartz has its origin in igneous or metamorphic rocks or veins.

(*ii*) Clay minerals are formed by the breakdown of high-temperature silicates under weathering, and recombination of their constituents.

(*iii*) Calcite is precipitated from solution either as the hard parts of organisms or by rise of temperature and loss of CO_2 in shallow tropical waters.

The weathering and sedimentation processes lead to segregation of the elements making up the weathered rock, so that some sedimentary rocks are composed very largely of a single mineral e.g. quartz in sandstone, or calcite in limestone. These types naturally have a much more extreme chemical composition than common igneous rocks.

In the sedimentary rocks, movement of circulating waters may concentrate elements dissolved from the rocks themselves. The commonest result of this is the production of *concretions*, some examples of which are given below. The process by which they form is called *lateral secretion*. In some cases workable deposits of valuable metals have been formed in this way. The extent to which lateral secretion, without supply from a magmatic source, may be responsible for major ore-bodies in sedimentary rocks is, however, a matter of debate.

SANDSTONES

These are composed of quartz, with variable amounts of clay and sometimes with a cement of calcite, or of silica as quartz, chalcedony or opal. Other resistant minerals of the rocks weathered to yield the sediment constitute the *heavy mineral suite* of a sandstone and may include such minerals as zircon, rutile, tourmaline, garnet, ilmenite, staurolite and kyanite.

Pure windblown sandstone, or strongly leached quartz sandstone (ganister) below coal seams may provide a source of very pure silica for glassmaking. In sandstones accumulated on land, or in shallow waters, far from any igneous rocks, there is sometimes found the association galena, pyromorphite, chalcocite, malachite and azurite. The chalcocite may replace fossil wood fragments. The uranium minerals carnotite and autunite, associated with some rare vanadium minerals are found in a similar situation in Jurassic sandstones of Colorado and Utah. These minerals have presumably been introduced from without, but their source is unknown.

SHALES

Shales consist of clay minerals, with varying amounts of quartz, carbonates and iron oxides. The crystallization of the clay minerals and iron oxides is often poor, there may be much colloidal matter, and it is necessary to use X-ray methods to obtain information about the minerals.

Concretionary minerals include calcite, siderite and barytes, which at times form ovoid nodules and discontinuous layers in the rock.

Unique amongst shales is the organic mud deposit forming the Kupferschiefer of Mansfield in Germany. The mineral association here includes bornite, chalcocite and rarely galena, sphalerite and tetrahedrite. The presence of these minerals is believed to be connected with the richly organic nature of the shale and its special environment of deposition but the source of the metals is not known.

The association of clay, or shale, with gypsum, sulphur, celestine, barytes, opal or quartz, and hydrocarbons is one that occurs in many parts of the world in sedimentary deposits of shallow waters that repeatedly dried up.

LIMESTONES

Dominantly of calcite, these may contain dolomite formed by replacement of Ca by Mg from sea-water or circulating waters. Aragonite, present in some sea-shells, tends to invert to calcite during consolidation of the rock. Flint and chert (impure forms of crypto-crystalline silica), pyrite and marcasite are concretionary minerals in limestones. Occasionally concentrations of celestine and strontianite occur. Magnesite may replace dolomitic limestones near igneous intrusions.

Because of their reactive nature, limestones assist the precipitation of minerals from vein-forming solutions, and veins often expand sideways and contain rich ore-deposits where they cut limestones.

IRONSTONES

Iron deposits of sedimentary origin are widespread in the geological succession. The characteristic parageneses are given below.

Bog iron ore

Limonite, siderite and minor vivianite of swamps and lakes in high latitudes, in Asia, Europe and North America.

Black band ironstone

Concretionary layers of siderite with small amounts of pyrite, arsenopyrite, galena, sphalerite and chalcopyrite in shales of deltaic coal-bearing successions.

Marine oolitic iron ores

Oolites (rounded, concentrically-layered granules a millimetre or so across) of chamosite, hematite and siderite in a matrix of chamosite, siderite or clay. This material forms beds interstratified with other types of lagoonal sedimentary rocks. Pyrite, sphalerite, galena and barytes are minor associates.

Banded siliceous hematite-magnetite ores

These consist of interbanded fine-grained quartz or jasper and layers rich in hematite or magnetite. Siderite and greenalite are also important minerals. The occurrences are found amongst the ancient rocks of the continental shields, and local concentrations of the original sedimentary iron ore have been produced by circulating waters and metamorphic processes.

MARINE EVAPORITE DEPOSITS

The evaporation of sea-water in basins partly cut off from the open sea has led in the past to the formation of salt deposits. The dissolved salts of the sea-water are deposited in a regular order as the brine is progressively concentrated: but subsequent burial leads to recrystallization and replacement amongst the minerals, which may modify the original succession and introduce minerals whose temperatures of formation are higher than would be expected to prevail at the surface. This constitutes a kind of low-temperature metamorphism.

The common mineral associations in a typical evaporite succession are given below, with the lowest stratum at the bottom of the list and the latest at the top. Not all the minerals listed are primary—sylvine, for example, may be secondary, resulting from the metamorphism just mentioned. There are many salts not listed below that may be present also. Details of complete successions are referred to in F. H. Stewart's survey of the subject.*

The short list is:

> carnallite–kieserite–halite
> sylvine–kieserite–halite
> halite–kieserite–anhydrite
> halite–polyhalite–kieserite
> halite–polyhalite
> halite–anhydrite–polyhalite
> halite–anhydrite
> anhydrite
> dolomite

Dolomite is consistently the basal member, halite occurs in most assemblages after its initial entry, and the potassium-bearing salts, sylvine and carnallite, occur as the latest precipitates from the most highly-concentrated brines.

Comparable successions are formed in the evaporite deposits of salt lakes and inland seas.

TERRESTRIAL EVAPORITES

In desert areas, where temporary lakes of enclosed basins dry out, or dissolved salts are brought to the surface by capillarity, deposits of halite, trona, thenardite, borax and gypsum may be formed. They are deposited on the floors of playa lakes and as incrustations and cements in the soil or sand and gravel.

The soda nitre deposits of the Atacama Desert in Chile are unique. Here the nitrate forms a cement in superficial beds of gravel, associated with halite, sulphates and borates of sodium and calcium and a small amount of sodium iodate.

SEDIMENTARY PHOSPHATE DEPOSITS

The minerals of these deposits include dahllite and francolite (carbonate apatites), and amorphous hydrous phosphates called collophane. Acid hydrous phosphates and various complex phosphates of magnesium, sodium and ammonium occur in guano deposits formed from the excrement of sea birds.

RESIDUAL DEPOSITS

The regolith, or crust of weathering, has its greatest importance in providing the mineral matter of soils. The principal soil minerals are:

(*i*) The *resistates*, minerals resistant to chemical breakdown by weathering, of which quartz is by far the most important.

(*ii*) The layer silicates kaolinite, halloysite, illite, montmorillonite, chlorite and their relatives, which are grouped together as *clay minerals*. These do not all occur together: different environments of parent rock, drainage, aeration and acidity favour different members or assemblages.

(*iii*) The oxides of iron and aluminium, goethite, hematite, gibbsite and diaspore.

Where chemical weathering and leaching under tropical or warm-temperate conditions has proceeded very thoroughly, to great depth, deposits of bauxite or laterite may form. These are respectively mixtures of aluminium and iron oxides and hydroxides.

Bauxite is the ore of aluminium and is composed of gibbsite, boehmite* or diaspore, or of mixtures of gibbsite and boehmite.

Laterite contains hematite and sometimes maghemite. It is widespread in tropical countries: it may sometimes furnish low-grade iron ore and is used for building purposes.

Transport of weathered material leads to the concentration of heavy resistate minerals in the gravels and sands of rivers and beaches, by the winnowing action of water-currents. The concentrations so formed are called *placer deposits*. Minerals concentrated as placers include gold, platinum, diamond, cassiterite, rutile, monazite, zircon, ruby, sapphire, columbite-tantalite, ilmenite and magnetite. Ancient gold-bearing conglomerates, including the famous Witwatersrand banket, are believed by some to be fossil placers.

METAMORPHIC ASSOCIATIONS

Sedimentary assemblages of minerals undergo progressive reorganization and recrystallization from the time of their deposition onwards. The earlier stages of this process are grouped together as *diagenesis* and *lithification*, terms embracing the changes that transform unconsolidated sediment into sedimentary rock.

After lithification, the rocks may undergo further recrystallization with the production of new minerals. When the outlines of original sedimentary grains are obliterated, and the texture or *fabric* of the rock is decisively altered, the changes may be said to have

---

* 1963. Marine evaporites. *Data of geochemistry*, 6th edn *U.S. Geological Survey Prof. Paper* **440**, Chapter Y.

* Boehmite is possibly formed by later baking of bauxite by lava flows or dykes.

passed beyond the stage of diagenesis and to have entered that of *metamorphism*. Though at times the boundary between these two processes may be vague, in practice sedimentary and metamorphic rocks are usually quite distinct.

Magmatic rocks may also undergo metamorphism, recognized by a change in fabric. Since they are originally formed at high temperatures, and often at high pressures, their minerals are likely to be changed in the first instance to an assemblage stable at a lower temperature than that at which the original association crystallized.

Metamorphism may be of two kinds. *Contact metamorphism* (or *thermal metamorphism*) is due to local heating of the rock by the intrusion of magma nearby. *Regional metamorphism* extends over wider areas and is the result of deep burial with consequent rise of temperature and static pressure, usually with the help of the folding movements that accompany the formation of mountain ranges.

In both types of metamorphism the mineral association that forms will depend upon the original composition of the affected rock. In contact metamorphism however, especially that by granitic magma rich in water and other volatile components, there may be much material added to the metamorphosed rock, carried in solution from the magma,

and original constituents of the metamorphosed rock may be removed. This process is called *metasomatism* and produces rocks called *skarns* (from the Swedish for wreckage).

Leaving aside skarns for the moment, the mineral associations developed depend upon original rock composition, temperature and pressure, and they can be predicted fairly accurately. A great deal of detailed information has been amassed about these associations and their temperatures of formation. In what follows, only brief lists of some of the commoner assemblages are given. For further details the student is referred to textbooks of metamorphic petrology, some of which are listed at the end of this chapter.

CONTACT METAMORPHISM

In contact metamorphism the purer sediments produce little variety of minerals. Pure sandstones recrystallize to quartzites composed of quartz with perhaps a little biotite derived from any impurities. Pure limestones yield calcite marbles. Highly aluminous material, such as laterite, will yield the association hercynite plus corundum.

When a greater variety of elements is present in the rock a larger number of minerals can form. The resulting metamorphic rock is given the general

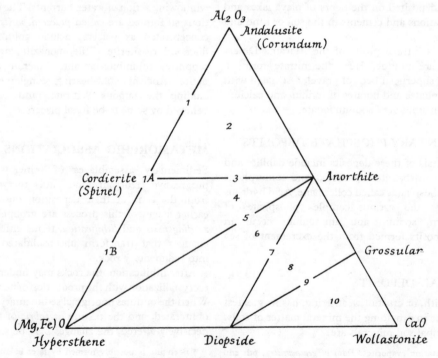

Fig. 152. Goldschmidt's classification of hornfelses.

name of hornfels. In considering the mineral assemblages of hornfelses, we may use the diagram (Fig. 152) due to V. M. Goldschmidt, on which the compositions of the common types of sediment can be represented, provided it is assumed that there is more than enough SiO_2 present to combine with the oxides shown to form silicates. The excess of SiO_2 will be present as quartz. FeO is taken as substituting without restriction for MgO in the minerals containing the latter. The diagram does not, however, show K_2O which will be an important constituent of the muddy part of the sediment, and this omission must be allowed for by adding biotite as a possible member of all the associations shown below, except numbers 8, 9 and 10, while orthoclase may occur in all of them. A further proviso is that vesuvianite may replace grossular in lime-rich rocks.

The list of assemblages is as follows:

1   andalusite-cordierite-(quartz-biotite-orthoclase)

1A   cordierite-(quartz-biotite-orthoclase)

1B   cordierite-enstatite-(quartz-biotite-orthoclase)

2   andalusite-cordierite-anorthite-(quartz-biotite-orthoclase)

3   cordierite-anorthite(quartz-biotite-orthoclase)

4   cordierite-anorthite-enstatite-(quartz-biotite-orthoclase)

5   anorthite-enstatite-(quartz-biotite-orthoclase)

6   anorthite-enstatite-diopside-(quartz-biotite-orthoclase)

7   anorthite-diopside-(quartz-biotite-orthoclase)

8   anorthite-diopside-grossular-(quartz-orthoclase)

9   diopside-grossular-(quartz-orthoclase)

10   diopside-grossular-wollastonite-(quartz-orthoclase)

The bracketed members of these assemblages may be reduced to nothing in amount; but if silica is reduced beyond the point where quartz fails to appear, its deficiency will result in the appearance of the oxides spinel and corundum, in place of silicates. The following possibilities then arise as substitutes for the numbered assemblages above:

1   andalusite-cordierite-corundum
cordierite-corundum
cordierite-corundum-spinel
corundum-spinel

1A   cordierite-spinel

1B   cordierite-enstatite-spinel

2   andalusite-cordierite-corundum-anorthite
andalusite-corundum-anorthite

3   cordierite-spinel-anorthite

REGIONAL METAMORPHISM

Typical rock-forming minerals of regionally metamorphosed shales are chlorite, epidote, the micas biotite and muscovite, chloritoid, garnet, staurolite, kyanite and sillimanite. These minerals enter successively into the constitution of the rocks as the temperature of metamorphism rises in roughly the order listed, being accompanied by quartz and albite in the early stages, quartz and oligoclase in the middle stage, and quartz, plagioclase and orthoclase at the highest grades. This sequence is found in areas where folding movements and consequent high shearing stresses have accompanied the metamorphic recrystallization. In regions where static load has been high, but less shearing stress has been applied, assemblages more akin to those of contact metamorphism may develop.

In regionally metamorphosed limestones zoisite, grossular, vesuvianite and diopside are common. Tremolite and forsterite appear in dolomitic limestones. In calcareous shales epidote and green hornblende develop.

Amongst regionally metamorphosed basic igneous rocks, hornblende with plagioclase is a common association, often accompanied by garnet. High temperatures and pressures of metamorphism of rocks of this composition produce hypersthene-augite-plagioclase-(orthoclase) granulites, in which quartz may appear if the original composition is more acid.

Assemblages containing lawsonite, pumpellyite and glaucophane, along with albite, epidote, micas and garnet, are being discovered in increasing numbers and represent a distinctive type of metamorphism of shales, thought to be associated with high pressures but relatively low temperatures.

Another series of assemblages found in the earliest stages of metamorphism of suitable parent rocks, particularly volcanic ashes, comprises analcime, the zeolites heulandite and laumontite, prehnite and pumpellyite. Analcime and heulandite appear at an early stage, accompanied by original lime-bearing plagioclase, while laumontite, pumpellyite and prehnite develop with greater depth of burial, and the plagioclase becomes albitized.

SKARNS

These deposits characteristically occur at the contact of granites with limestones. They contain a wide variety of minerals including often magnetite, garnet, diopside, enstatite, forsterite, minerals of the humite group, small amounts of sulphides and various rarer minerals containing elements such as fluorine and boron derived from the granite. Cassiterite and tourmaline are sometimes found in this association.

Further Reading

GOLDSCHMIDT, V. M. 1954. *Goechemistry*, London, Oxford Univ. Press.
HARKER, A. 1950. *Metamorphism*, 3rd edn, London, Methuen & Co.
HATCH, F. H., RASTALL, R. H., and GREENSMITH, J. T. 1965. *Petrology of the sedimentary rocks*, 4th edn, London, Allen & Unwin Ltd.
HATCH, F. H., WELLS, A. K., and WELLS, M. K. 1961. *Petrology of the igneous rocks*, 12th edn, London, Allen & Unwin Ltd.
LINDGREN, W. 1933. *Mineral deposits*, 4th edn, New York, McGraw-Hill Co.
NIGGLI, P. (trans. H. C. Boydell). 1929. *Ore deposits of magmatic origin*. London, Murby & Co.
PARK, C. F., and MacDIARMID, R. A. 1964. *Ore deposits*. San Francisco, Freeman & Co.
WINKLER, H. G. F. 1967. *Petrogenesis of metamorphic rocks*, 2nd edn, Berlin, Springer-Verlag.

PART II DESCRIPTIONS
OF MINERALS

FORM, SYMBOLS AND ABBREVIATIONS USED FOR MINERAL DESCRIPTIONS

The mineral descriptions take the following form:

Name x [Formula] expressed in accordance with the structure. The number (x) outside the square brackets gives the number of formula units in the unit cell.

Crystal System   Lattice-type symbol (see p. 31)
Symmetry class symbol (see Fig. 68)
a, b, c the cell edge-lengths in nanometres
α, β, γ the interaxial angles (see Table 12)

Structure

Habit, *twinning* and *cleavage* using:

x, y, z for the crystallographic reference axes and Miller symbols (see p. 53) for the crystal faces.

$\wedge$ angle between

H the hardness on Mohs's scale

D the density in g/cm^3

Optics. Under this heading are given optical data using:

n refractive index (occasionally *R.I.*)

$R\%$ percentage reflectance

ω, ε refractive indices of uniaxial substances

α, β, γ refractive indices of biaxial substances

$\alpha = x$, or $\gamma \wedge z$ etc. expresses the relation of vibration directions to crystallographic axes. Dichroic or pleochroic colours are also shown with the symbol α, β or γ and the numerical value of the refractive index.

$\omega - \varepsilon$, $\varepsilon - \omega$, or $\gamma - \alpha$ as appropriate is the birefringence

$2V_\alpha$, $2V_\gamma$ optic axial angle, with bisectrix

Bx_a, Bx_o acute and obtuse bisectrix, respectively

O.A.P. optic axial plane

$+ve$, $-ve$ positive or negative optic sign

Other symbols and abbreviations used are:

$\|$ parallel to

$\perp$ perpendicular to

$\simeq$ approximately equal to

$>$ greater than

$<$ less than

d interplanar spacing

D.T.A. differential thermal analysis

M.P. melting point

R.E. Rare Earth elements

U.V. ultra-violet

nm nanometre (10^{-9} metre)

The *axial ratio* and the *angles between principal faces*, information often quoted in mineralogy books, are seldom given in this book. The axial ratio is effectively given by quoting the cell dimensions. Where the axial ratio originally chosen is some multiple of that given by the cell edges, there is a tendency nowadays to change it and bring it into conformity with the cell edge-lengths, assigning new indices to the faces. This seems to be a sensible move.

Given the cell dimensions, it is a simple matter to derive the angles between important faces, and indeed these have been quoted in the past principally in support of the axial ratios derived from them.

CLASSIFICATION

The minerals are grouped for description into the following chemical classes, on the basis of the anionic part of the formula:

Native Elements

Sulphides and Sulpho-salts

Halides

Oxides and Hydroxides

Carbonates

Nitrates

Tungstates and Molybdates

Phosphates, Arsenates and Vanadates

Sulphates

Borates

Silicates

NATIVE ELEMENTS

The salient fact about the native elements is that there is only one kind of atom in the structure. In metals the atoms are arranged in one of the forms of close-packing of equal spheres. In some non-metals they are so arranged that the environment of every atom in the structure is the same, when the whole crystal can be considered to be a single molecule. In other non-metals the atoms are linked into separate molecules of fixed shape, and these molecular groups are stacked in a regular way.

The following native elements are considered here:

Metals

Copper	Cu
Silver	Ag
Gold	Au
Platinum	Pt

Semi-metals

Arsenic	As
Antimony	Sb
Bismuth	Bi

Non-metals

Diamond	C
Graphite	C
Sulphur	S

GOLD GROUP **Copper, Silver, Gold**

                                    4[Cu] 4[Ag] 4[Au]

Cubic *F m3m*   *a* Cu 0·3608; Ag 0·4077;
                                                   Au 0·4070

Structure: These native metals all have cubic close-packing of atoms (Fig. 153). They all have the characteristic metal properties of malleability, softness, high density, metallic lustre and high electrical and thermal conductivity.

Habit: Cu is usually dendritic or wiry; sometimes as cubes or dodecahedra. Ag: octahedra, dodecahedra; wiry, dendritic, or as coatings. Au: octahedra, dodecahedra, cubes; wiry, dendritic, sometimes as scales and rounded masses (nuggets). The largest lump recorded is one of 472 lb (214 kg) from Holterman's Reef, Hill End, New South Wales, Australia (1872). The 'Welcome Stranger' nugget, Tarnagulla, Victoria, Australia, weighed 142 lb (64 kg) (1869).

Twins: All three metals twin on {111}, the twin being repeated, especially in Ag, to give branching aggregates.

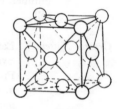

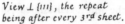

View ⊥ {111}, the repeat being after every 3rd sheet.

Fig. 153. Face-centred cubic closest-packing of equal atoms.

Cleavage absent. All are malleable. $H\ 2\frac{1}{2}$–3·ₐ *D* Cu 8·95; Ag 10·5; Au 19·3. M.P.: Cu 1083°; Ag 960°; Au 1092°C. Lustre metallic.

Colour and streak: Cu: light rose on fresh surface rapidly changing to copper red. Ag: white. Au: yellow.

Optics: Opaque (though the thinnest films of gold transmit green light). Isotropic. *R*% for orange light: Cu 81; Ag 95 (the best known reflector); Au 74.

Tests: Cu fuses readily and becomes coated with black oxide on cooling. Soluble in HNO_3. Ag fuses readily to a white globule. Soluble in HNO_3. Au fuses readily to a yellow malleable globule. Insoluble in acid except aqua regia ($HCl + HNO_3$). Colour, malleability and melting behaviour characterize all three. Pyrite and chalcopyrite, which may be mistaken for gold, are brittle and have brownish- or greenish-black streak. Weathered biotite is readily cleavable.

Alteration: Copper and silver tarnish, the first to black oxide or green carbonate, the second to black sulphide. Gold is immutable.

Occurrence: Cu: in the zone of oxidation of sulphide veins. In amygdales in ancient basalt lavas, with prehnite, pumpellyite, epidote, chlorite, calcite and zeolites, a low temperature assemblage. Ag: in the oxidized zone of hydrothermal sulphide veins, and as primary ore in such veins. Au: in hydrothermal veins in hypothermal, mesothermal and epithermal environments (Table 20, p. 161). As placer deposits in both ancient gravels and recent alluvium.

Varieties: *Electrum* is alloyed gold and silver of a pale yellow colour. At 20% Ag the cell edge of the intermetallic compound is 0·4067 nm, the only known case in which a solid solution has smaller cell dimensions than either of its components.

Use: The usefulness of these metals needs no elaboration here.

Platinum 4[Pt]

Cubic *F m3m* a 0·3916

Structure: Copper type (see under Gold group).

Habit: Rarely as cubes; usually in grains, scales or nuggets.

Cleavage absent. Highly malleable. H 4–4½, increasing with iron content. D 21·46 when pure; natural material 14–19. M.P. 1774°C. Lustre metallic. Colour grey-white to light grey.

Optics: Opaque; in reflected light, white, isotropic. R% 70.

Tests: Infusible; soluble only in hot aqua regia; sometimes magnetic (with polarity) because of dissolved iron.

Occurrence: As segregations in ultrabasic magmatic rocks, especially dunites, along with chromite. With the immiscible sulphide fraction in basic magmatic rocks (norites). As placer deposits derived from erosion of rocks of the foregoing types.

Important production comes from the nickel- and copper-bearing sulphide segregations of Sudbury, Ontario, from segregations in the Bushveld igneous complex, Transvaal, from placer deposits in the Nizhne Tagilsk region of the Ural Mountains, U.S.S.R., and from Colombia.

Varieties: Platinum is host to a variety of other metals, osmium, iridium, rhodium, palladium, iron, copper, gold and nickel, and richness in various of these gives materials of differing hardness. The refining of platinum is the principle source of Os, Ir, Rh and Pd. *Iridosmine*, a hexagonal mineral and an alloy of Ir and Os occurs in placers with platinum and gold. It is notable for its comparative hardness, H 6–7, unusual amongst metals.

Use: In jewellery, for laboratory crucibles, tongs and supports, for non-corroding electrical contacts, in thermocouples and in dentistry. The hardness, thermal stability and freedom from corrosion are its valuable properties. It is also used in finely divided form as a catalyst in industrial chemistry. Palladium, because of its relative lightness (D 11·9), is preferred in some of these applications.

Iron (α form) 2[Fe]

Cubic *I m3m* a 0·286

Structure: A body-centred cubic packing of equal atoms, giving an arrangement a little less dense than the copper type.

Iron is rare as a terrestrial mineral. A considerable mass occurs where basalt cuts coal in Disco Island, Greenland, and small amounts occur in basalts elsewhere, under reducing conditions, as when lava-flows have engulfed and carbonized tree-trunks.

Two rare alloys, awaruite (Ni_3Fe) and wairauite (CoFe), are reported from peridotites, or sands derived from them, in New Zealand and elsewhere.

The special interest of iron is its occurrence, alloyed with nickel, in meteorites. Iron meteorites are made principally of this alloy, and it is an important constituent of the stony irons and the chondritic stony meteorites. In the iron meteorites the nickel-iron is often exsolved into lamellae of Ni-poor iron called kamacite, which contains 6 to 9% Ni, and of Ni-rich iron called taenite with up to 48% Ni. These two kinds of lamellae are set in a ground-mass called plessite, which is an undifferentiated mixture of the other two. The lamellae are parallel to {111} of the iron crystals and, on etching of the cut surface of the meteorite with weak acid, they show up as a characteristic pattern of intersecting strips of differing lustre called Widmanstätten figures.

ARSENIC GROUP Arsenic, Antimony, Bismuth

	2[As] 2[Sb] 2[Bi]	
	a	c
Trigonal *R 3m* As	0·376	1·055
Sb	0·431	1·127
Bi	0·455	1·186

Structure: Based on cubic close-packing (c.f. copper type), but the sheets of atoms ∥{111} separate into pairs (Fig. 154) members of each pair being drawn together by shorter stronger bonds of covalent type (interatomic distance As 0·251, Sb 0·287, Bi 0·310), compared with bonds between the pairs (interatomic distances As 0·315, Sb 0·337, Bi 0·347). Each pair of sheets an indefinitely extended molecule. Symmetry lowered to trigonal.

Habit: Natural crystals of all the species are rare. Artificial ones are pseudocubic. Arsenic is generally granular massive, or in concentrically-layered reniform masses. Antimony is massive or lamellar with a cleavage through the mass. Bismuth is in net-like or branching forms or in leaves. Repeated twinning on {10$\bar{1}$4} is common in antimony and bismuth, rare in arsenic.

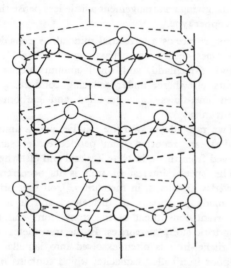

Fig. 154. The structure of arsenic.

Cleavage perfect {0001}, in conformity with the sheet structure. Bismuth also cleaves well on {10$\bar{1}$1}. Fracture uneven.

As is brittle, Sb very brittle, Bi sectile, a little malleable on heating.

Hardness: As 3½, Sb 3–3½, Bi 2–2½.
Density: As 5·7, Sb 6·68, Bi 9·75.
Lustre: Metallic.

Colour and streak: As tin-white tarnishing to dark grey;
Sb tin-white, streak grey;
Bi silvery with reddish tinge, sometimes with iridescent tarnish.

Optics:	As	Sb	Bi
Transmission	opaque	opaque	opaque
Colour	bright white	bright white	creamy-white to yellowish
Pleochroism	weak	very weak	very weak
Anisotropy	distinct (3%)	distinct (5%)	distinct
R%	49½	74½	68

Tests: As: volatilizes without melting in the blow-pipe flame giving dense white fumes with the odour of garlic. Heated in a closed tube gives an As mirror on the upper side of the tube.

Sb: M.P. 670°C. Heated on charcoal fuses very readily; the globule glows after blowing stops until the surface is crusted over with Sb_2O_3 crystals. The cool globule is visibly crystalline. Does not volatilize in the closed tube.

Bi: M.P. 271°C. Heated on charcoal fuses and volatilizes giving an orange-yellow oxide coating, lemon yellow on cooling.

Because the distorted close packing in the crystal has greater volume than the statistical close packing of atoms in the liquid these metals are amongst the unusual substances that expand on crystallizing.

Occurrence: In mesothermal hydrothermal veins with the ores of silver cobalt and nickel, and in mixed sulphide ores. Bismuth occurs also in a higher temperature environment in veins and pegmatites with tin. The native metals are not commercially important.

As and Bi are produced as a by-product of smelting ores containing their sulphides. Sb comes from stibnite deposits.

As is used as a pesticide and preservative.

Sb is used as an alloy in lead for type-metal and bullets. It lowers the melting point and hardens the lead, while the alloy expands slightly when it solidifies. Alloyed with tin and copper, it gives a low-friction metal (Babbitt's metal) for bearings. Bi forms low-melting point alloys with tin, lead or cadmium (e.g. Woods metal) useful in electrical fuses and fire protection devices. Like Sb it is used in metals for fine castings.

Diamond   8[C]

Cubic *F* *m3m*   *a* 0·356

Structure: Carbon atoms linked in tetrahedral groups with a C—C distance of 0·154 nm (cf. resonance bond length of aliphatic organic compounds). There is a C atom at the centre of each tetrahedral box in Fig. 155 as well as at its corners (see also Fig. 6, p. 10).

Habit: Octahedral crystals often with curved faces, sometimes dodecahedral, rarely but perhaps significantly tetrahedral.

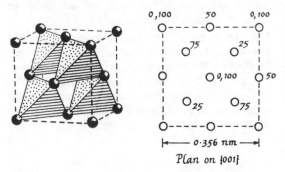

Fig. 155. The structure of diamond.

Twinning on {111} common, simple or multiple. Cleavage perfect {111}.
Fracture conchoidal. Brittle. *H* 10. *D* 3·511 (range 3·50–3·53).
Lustre adamantine to greasy.
Colour white or blue-white through pale yellow to darker shades of brown. Impure varieties grey to black. Sometimes strongly fluorescent in ultraviolet light (λ250 nm) in blue and green, and phosphorescent. Triboelectric.
Optics: Translucent, isotropic except near inclusions and cavities when it may be birefringent. High dispersion giving the 'fire' of the gemstone.

λ_{nm}	687·6	589·3	486·1
n	2·4076	2·4175	2·4354

Tests: The hardness and refractive index are distinctive. Clear quartz grains have often been mistaken for alluvial diamonds. The quickest test is to drop the grains into a tube of bromoform (*D* 2·9) when diamond sinks and quartz floats.
Occurrence: Formed only at high pressures (see Fig. 84, p. 82).

The primary source is an alkaline mica-peridotite called kimberlite, brecciated, often serpentinized and rich in inclusions of other deep-seated rocks, that forms roughly cylindrical 'pipes' and also dykes, located in the stable continental shields. The near surface kimberlite is weathered to friable 'yellow ground'. Deeper in the pipes is the harder less altered 'blue ground'.

The richest pipes occur near Kimberley in the Transvaal, but others occur in Tanzania, Yakutia U.S.S.R., India, and Brazil. From these come the best quality diamonds.

Most of the diamond production is from placer deposits, particularly in southern-central and west Africa and Brazil. The sea-bed off South West Africa is also worked for diamond.

Varieties: Granular cryptocrystalline and rounded aggregates are called *bort*. *Carbonado* is a tough, impure, black diamond with inclusions.

Use: When pure and clear, or pale coloured, the most favoured of gems. The largest diamond recorded is the Cullinan from the Premier Mine, Transvaal, weighing 3106 carats (a carat = 0·2 g) when found.

The most important use of diamond is as an abrasive.

Graphite   4[C]

Hexagonal   *C*   6/*mmm*   *a* 0·246   *c* 0·680

Structure: Hexagonal arrays of C atoms in sheets normal to *z*-axis. Each sheet superposes exactly on the sheet two below it. The intervening sheet is displaced by half a hexagonal ring width. This gives rows of atoms (*A*–*A'* of Fig. 156) that have

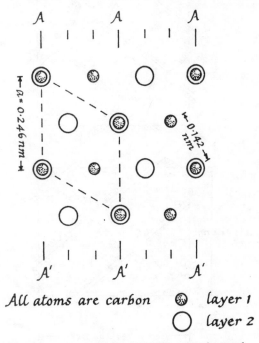

All atoms are carbon   layer 1   layer 2

Fig. 156. The structure of graphite, viewed normal to 0001.

atoms above and below them in the next sheets; between these rows come two rows with atoms above and below in the second sheet away. The distance between the sheets is 0·34 nm.

Habit: Six-sided crystals tabular parallel to {0001}, with prisms {10$\bar{1}$0} and pyramids {10$\bar{1}$1}, {10$\bar{1}$2}. Usually as scales and foliated masses or compact lumps.

Cleavage eminent {0001} parallel to the atomic sheets, conferring on it its softness and lubricant properties. Scales flexible, not elastic, sectile. *H* 1–2, with greasy feel. Marks paper. *D* 2·09–2·23, conspicuously light. Lustre metallic, shining, to dull and earthy. Colour black. Streak black and shining. An electrical conductor.

Optics: Opaque, except in the thinnest flakes, when it is deep blue. In reflected light strongly pleochroic and bireflecting; *R*% ω 17, ε 6.

Tests: Infusible, unattacked by acids. Burns slowly at 900–1000°C. Softness, low *D*, black colour and streak are distinctive.

Occurrence: Chiefly in metamorphosed sedimentary rocks, marbles, schists and gneisses, formed from original organic matter; also in metamorphosed coal. Sometimes associated with basic igneous rocks and with pegmatite dykes and quartz veins, the origin then being obscure; but perhaps it is from volcanic gases. It also occurs in meteorites.

Production comes chiefly from U.S.S.R., Mexico, Korea, Austria, Ceylon and Madagascar, the two last countries producing most of the 'flake' or coarsely crystalline graphite.

Use: In making crucibles, electrodes, generator brushes and lubricants, as well as the familiar lead-pencil. Very pure graphite is used as a moderator to slow down neutrons in atomic reactors.

The polymorphism between graphite and diamond is discussed on p. 82 where Fig. 84 gives the stability relations between the two minerals.

Sulphur (α form)   16[S$_8$]

Orthorhombic *F mmm*   *a* 1·044   *b* 1·284   *c* 2·437

Structure: Eight *S* atoms form a ring-shaped molecule. Sixteen of these are stacked face to face in staggered rows that run parallel to the diagonals

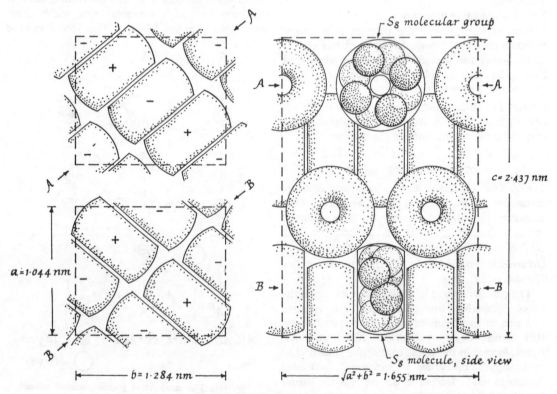

Fig. 157. The structure of α-sulphur.

of the base of the cell. The plane of every ring-molecule is parallel to c of the unit cell (Fig. 157).

Angles: 001 $\wedge$ 101 66° 52′ Common forms {001},
        001 $\wedge$ 011 62° 17′   {111}, {110}, {011},
        100 $\wedge$ 110 39° 7′   {013}

Habit: In bipyramids {111} or thick tablets ||{001}. Also massive, incrusting, stalactitic or as powder. Cleavage poor. Fracture conchoidal to uneven. Brittle to somewhat sectile. H $1\frac{1}{2}$–$2\frac{1}{2}$. D 2·07. Lustre resinous.

Colour: Yellow to yellow-brown. Streak white. Triboelectric, negative.

Optics: Transparent $\alpha(= x)$ 1·958 $(\beta = y)$ 2·038 $\gamma (= z)$ 2·245 $\gamma - \alpha$ 0·287 $2V_\gamma$69°. α and β are much the same, γ considerably greater, reflecting the fact that the first two vibrations are in a plane normal to that of the S_8 molecules while γ is parallel to their plane. The polarizability of the molecules is greater parallel to the plane of the rings, hence light vibrating ||z interacts more strongly with the molecules and is more retarded in its passage giving strong +ve birefringence.

Pleochroic in shades of pale yellow to yellow-brown.

Birefringence extreme, interference colours pale.

Tests: Melts easily (113°C) and burns with a bluish flame. Insol. in water and acids, sol. in CS_2. (N.B. Dissolves in methylene iodide raising its refractive index to 1·8). The colour, softness and combustibility are distinctive.

Varieties: Of the several polymorphs α-sulphur, described here, is the important natural form. β- and γ-sulphur have been recorded from fumaroles.

Occurrence: As a sublimate around volcanic vents, or as a precipitate of hot springs, the immediate source being gaseous H_2S and SO_2. In association with gypsum and bituminous material in shales, sandstones and limestone of Tertiary age, and in the cap-rocks of salt domes. Sulphate-reducing bacteria may play a part in its formation here. Also from oxidation of pyrite with restricted access of air.

The main production is from U.S.A. (salt domes in Louisiana and Texas) and Sicily (Tertiary sediments); sulphur of volcanic origin is mined in Japan, Indonesia and the Andes.

SULPHIDES AND SULPHO-SALTS

This group comprises compounds in which the large atoms S, Se, Te, As, Sb and Bi are combined with Fe, Co, Ni, Cu, Zn, Mo, Ag, Cd, Sn, Pt, Au, Hg, Tl, Pb, and Bi.

The bonding in these minerals is only in part ionic. There is a marked degree of covalency in many of them and a few are in the nature of inter-metallic compounds. They have a special chemistry, different from that of the oxygen compounds that constitute most other minerals. Amongst the large atoms listed above S, Se and Te consistently play an 'anionic' or non-metallic role; but As and Sb may either adopt the same role as S, or act as a metal combining with S. This makes the group difficult to classify.

In this book they are arranged broadly in order of decreasing ratio, metal atoms (A): non-metal atoms (X), but also in such a way as to bring structurally similar species together. The structures of the sulpho-salts are not completely understood, but it is probable that a structural classification will, in the end, be found to be the most satisfactory.

Members of the group form the bulk of the *ore-minerals* as this term is usually understood. They are opaque, dark, heavy, relatively soft, with metallic or submetallic lustre and relatively high thermal and electrical conductivity.

A_2X

Argentite	Ag_2S
Chalcocite	Cu_2S

A_3X_2

Bornite	Cu_5FeS_4

AX

Galena	PbS
Sphalerite	ZnS
Chalcopyrite	$CuFeS_2$
Wurtzite	ZnS
Greenockite	CdS
Enargite	$(Cu_3As)S_4$
Niccolite	$NiAs$
Breithauptite	$NiSb$
Pyrrhotite	$Fe_{1-x}S$
Pentlandite	$(Fe, Ni)_9 S_8$
Millerite	NiS
Covellite	CuS
Cinnabar	HgS

AX_2

Pyrite	FeS_2
Sperrylite	$PtAs_2$
Ullmannite	$NiSbS$
Cobaltite	$CoAsS$
Gersdorffite	$NiAsS$
Marcasite	FeS_2
Loellingite	$FeAs_2$
Gudmundite	$FeSbS$
Molybdenite	MoS_2
Krennerite	$(Au, Ag)Te_2$
Calaverite	$AuTe_2$
Sylvanite	$(Au, Ag)Te_2$

AX_3

Skutterudite	$(Co, Ni)As_3$
Smaltite	
Chloanthite	$(Co, Ni)As_{3-x}$

A_3BX_3

Pyrargyrite	Ag_3SbS_3
Proustite	Ag_3AsS_3
Tetrahedrite	$(CuFe)_{12}Sb_4S_{13}$
Tennantite	$(CuFe)_{12}As_4S_{13}$

Other ratios

| Bournonite | $PbCuSbS_3$ |
| Boulangerite | $Pb_5Sb_4S_{11}$ |

Semi-metal sulphides

Realgar	AsS
Orpiment	As_2S_3
Stibnite	Sb_2S_3
Bismuthinite	Bi_2S_3

Argentite $2[Ag_2S]$

Cubic I $m3m$ above 179°C $a\,0.488$
Monoclinic $P\,2/m$ below 179°C (acanthite $8\,[Ag_2S]$)
but pseudocubic. $a\,0.423\ b\,0.691\ c\,0.787\ \beta\,99°\,35'$

Structure: At ordinary temperatures monoclinic
with S atoms approximately on a cubic body-centred
lattice. Ag atoms in two positions, Ag_I midway
between two S atoms, Ag_{II} equidistant from three
S (Fig. 158). Above 179°C structure uncertain, but
may be analogous to cuprite.

Habit: Cubes, octahedra, dodecahedra; in parallel
groups, arborescent, as a coating.

Twins on {111} interpenetrant. A mass of sub-
microscopic twins at ordinary temperatures due to
inversion.

Cleavage poor. Fracture subconchoidal. Sectile.
$H\,2–2\frac{1}{2}$. $D\,7.04$. Lustre metallic. Colour black to
dark grey. Streak black, shining.

Optics: Opaque, grey-white, faintly anisotropic.
$R\%\,29$.

Tests: On charcoal, before the blowpipe, swells up
and fuses giving off sulphurous fumes and yielding
a globule of silver.

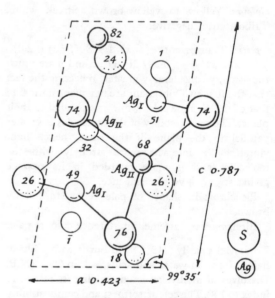

Fig. 158. The structure of argentite.

Occurrence: The most common primary ore of silver.
It occurs in epithermal sulphide vein deposits along
with the ruby silvers (pyrargyrite and polybasite)
and native silver. It occurs in most silver-producing
regions particularly in Peru, Mexico, Colorado (San
Juan district), Nevada (Tonopah, Comstock Lode),
in Saxony (Freiberg), Harz Mountains (Andreasberg), Czechoslovakia, Norway (Kongsberg).

Chalcocite Cu_2S

Orthorhombic mmm $a:b:c = 0.582:1:0.970$

Structure: At room temperature orthorhombic
perhaps with $a\,1.190\ b\,0.675\ c\,1.341$ for $96\,[Cu_2S]$
or $a\,0.394\ b\,0.675\ c\,0.670$ for $4\,[Cu_2S]$. Above
105°C hexagonal (see Fig. 159) with $a\,0.396\ c\,0.672$
for $2\,[Cu_2S]$. Perhaps cubic above 460°C.

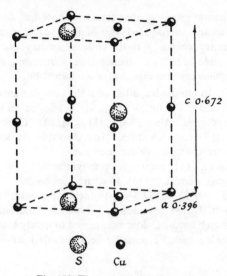

Fig. 159. The structure of chalcocite.

Habit: In crystals tabular ||{001} showing {001}, {110}, {010}, {012}, {021}, {111}, {113}. Also in fine-grained sooty masses. Angles between common faces are

110 ∧ 1$\bar{1}$0 60° 25′; 001 ∧ 111 62° 35′; 001 ∧ 011 44° 8′.

Twins common on {110} giving pseudohexagonal forms. Also cross forms from twins on {032} and {112}. Microscopic lamellar twinning.

Cleavage poor. Fracture conchoidal. Brittle to slightly sectile. *H* 2½–3. *D* 5·77. Lustre metallic. Colour and streak black to dark grey.

Optics: Opaque. Bluish-white, slightly anisotropic. *R*% 32. In etched polished sections may show a triangular pattern of lamellae.

Tests: Fuses easily. After heating in the oxidising flame gives a globule of copper on charcoal in reducing flame.

Occurrence: One of the common secondary copper minerals of the zone of supergene enrichment of sulphide veins, or of disseminated chalcopyrite and pyrite extending over wide areas, then giving chalcocite 'blankets'. Because of its high ratio of metal to sulphur it is a very valuable ore. The associated primary sulphides are chalcopyrite, bornite, covellite, tetrahedrite and enargite, pyrite, sphalerite and galena. It may also be associated with the oxidized copper ores: malachite, azurite, cuprite and native copper. It sometimes contains valuable amounts of silver.

Chalcocite may also be a primary mineral of sulphide veins, replacing bornite.

Important deposits occur at Rio Tinto, Spain, at various places in Nevada, Arizona and New Mexico and at Tsumeb, S.W. Africa. Good crystals have come from Redruth, St. Just, St. Ives and Camborne in Cornwall and from Bristol, Conn., U.S.A.

Bornite (Peacock Copper Ore)   8[Cu_5FeS_4]

Cubic *F* *m*3*m*   *a* 1·093

Structure: S atoms on a face-centred cubic lattice. This has 8 tetrahedral and 4 octahedral spaces in the cell. The metal atoms lie in the tetrahedral spaces, so distributed as to form a supercell. Belov suggests Fe is regularly distributed in tetrahedra as in Fig. 160, Cu being arranged in 2-fold co-ordination along the edges of unoccupied tetrahedra (an example in broken line in Fig. 160). This gives 48 spaces for Cu of which only 40 are (randomly) occupied.

Habit: Rarely in cubes or dodecahedra. Usually massive.

Twins on {111}. Cleavage unimportant. Fracture subconchoidal to uneven. Brittle. *H* 3. *D* 5·07. Lustre metallic. Colour purplish and iridescent, but coppery red or bronze on fresh fracture. Streak grey black.

Optics: Opaque, isotropic. Pinkish-brown when freshly polished. *R*% 22. Often has exsolved lamellae of chalcopyrite (*R* 44%) ||{111}.

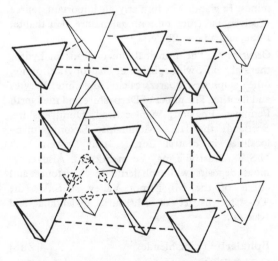

Fig. 160. Belov's suggested arrangement for bornite. The tetrahedra are FeS_4, and the distribution of Cu along edges of a vacant tetradedral space is shown in broken line.

Tests: On charcoal in reducing flame fuses easily to a brittle magnetic globule. Dissolves in HNO_3 with separation of sulphur. The colour and iridescence coupled with softness are characteristic.

Occurrence: A common primary ore mineral of mixed sulphide veins. It is occasionally a product of supergene alteration. It may sometimes occur in high temperature environment in pegmatites. Well-known localities are Redruth, Cornwall (crystals); Butte, Montana; Mt Lyell, Tasmania; and various mines in the Andean belt.

Galena   4[PbS]

Cubic   *F   m3m*   *a* 0·5963

Structure: NaCl type.

Habit: Cubes, cubo-octahedra, more rarely octa-hedra, often massive. Twins on {111}. Cleavage perfect ||{100} giving characteristic stepped or terraced surfaces of fracture. Brittle. *H* 2½. *D* 7·58. Lustre metallic, splendent. Colour and streak lead grey. M.P. 1114°C.

Optics: Opaque, pure white in reflected light, isotropic. *R*% 43.

Tests: Fuses easily on charcoal giving a coating of yellow (PbO) near the assay and white with a bluish border farther away, and yields a globule of lead. Stibnite melts much more easily (550°C) is less dense (4·63) and lacks cubic cleavage.

Varieties: Argentiferous galena may contain up to more than 1% Ag. Galena from the N. Pennine mines, England, was formerly an important source of silver. Isostructural with galena are clausthalite, PbSe, and altaite, PbTe.

Occurrence: The principal ore of lead: in hydro-thermal veins, with sphalerite, chalcopyrite, pyrite, other sulphides, quartz, calcite, dolomite, barytes and fluorite. The mines of Derbyshire and the North Pennines, England; Wanlockhead, Dumfriesshire, Scotland; the Harz Mountains, Saxony; many localities in central Europe; and Broken Hill, N.S.W., Australia may be mentioned. Also enor-mous deposits with sphalerite in limestones and cherts of the Mississippi Valley, U.S.A. (the Tri-state district of Missouri, Oklahoma and Kansas).

Sphalerite (Zinc blende)   4[ZnS]

Cubic   *F   4̄3m*   *a* 0·5409

Structure: The atoms are arranged on a cubic face-centred lattice (Fig. 161) as in diamond. Each

tetrahedral group of that structure now has Zn at the centre surrounded by four S, so that the symmetry planes ||{100} (glide symmetry planes with glide $(a+b)/4$ in diamond) are eliminated, and the symmetry becomes that of a tetrahedron.

Habit: In tetrahedra, often with the complementary {111} and {11̄1} forms unequally developed on the same crystal. Also {100}, {110}, {113}. Crystals distorted and complex. In cleavable masses, granular compact, or concretionary.

Twins on {111} sometimes polysynthetic.

Cleavage perfect ||{110}. Fracture conchoidal.

Brittle. *H* 3½–4. *D* 3·9–4·1 increasing with Fe content. Lustre resinous to submetallic, the latter in Fe-rich kinds. Colour orange-red to nearly black, darkening with Fe. Usually yellow to dark brown.

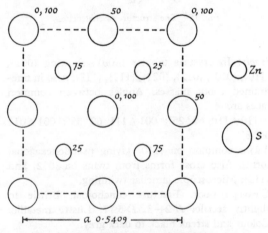

Fig. 161. The structure of sphalerite.

Streak brown to yellow, sometimes white. Fluores-cent in orange under UV light (λ250 and 360 nm) and X-rays. Triboluminescent. Pyroelectric.

Optics: Translucent, isotropic, $n = 2·36–2·47$ in-creasing with Fe content. In reflected light isotropic, with white, yellow and brown internal reflections. *R*% 17. Often intergrown with chalcopyrite.

Tests: Dissolves in HCl giving off H_2S. Fuses with difficulty. On charcoal in oxidizing flame gives a ZnO coating which is yellow while hot and white when cold. Cleavage and lustre are characteristic; the variable colour is a difficulty which must be allowed for.

Varieties: Usually contains some iron which produces variation in properties. Cadmium also substitutes for Zn in amounts up to 0·3–0·5 wt. % and spha-lerite is the commercial source of Cd.

Occurrence: The chief ore of zinc. Found in mesothermal veins along with galena and other sulphides. It also accompanies galena in the replacement deposits found in limestones, as in the Mississippi Valley. Sometimes found with chalcopyrite and other sulphides in hypothermal vein deposits in a higher temperature environment than galena.

Wurtzite. Sphalerite inverts at 1020°C to wurtzite 2[ZnS]. Wurtzite is hexagonal with S atoms in hexagonal close-packing and Zn atoms again in tetrahedral spaces between them. The six-fold symmetry axis is polar and the crystals hemimorphic (i.e. differently terminated at either end). Wurtzite is uncommon.

Greenockite. CdS, also rare, has a similar structure.

Chalcopyrite   4[CuFeS$_2$]

Tetragonal   *I* $\bar{4}2m$ *a* 0·524 *c* 1·030

Structure: Essentially two unit cells of sphalerite one on top of the other, the doubled *c* dimension being necessitated by the alternation of Cu and Fe atoms in sites essentially of one kind (which in sphalerite are all occupied by Zn). It is, hence, an example of a *superlattice* (Fig. 162).

Habit: In sphenoids {112} so near to tetrahedra in angle as to be indistinguishable from them by eye. Also massive and compact.
Twins on {112}. Cleavage on {011} poor. Fracture uneven. Brittle. *H* 3½–4. *D* 4·28. Lustre metallic. Colour brass-yellow often with brown or iridescent tarnish. Streak greenish black.
Optics: Opaque. Weakly anisotropic, pale yellow, *R*% 44. Often shows lamellar twins. May have parallel intergrowths of sphalerite (*R*% 17) or tetrahedrite.
Tests: Heated in a closed tube it decrepitates and gives a sublimate of sulphur. Heated on charcoal gives a magnetic globule. Soluble in HNO$_3$. Distinguished from pyrite by being softer with usually a deeper yellow colour. It differs from pyrrhotite in being non-magnetic. The brittleness, streak and solubility in HNO$_3$ distinguish it from gold.
Occurrence: The commonest copper mineral, and one of the commoner sulphides. As a primary ore mineral it is characteristic of hypothermal and the higher temperature mesothermal veins; but it also forms under epithermal conditions both in veins and as disseminated crystals. Pyrite is a common

associate in the hypothermal and mesothermal veins. At the high temperature end of its range it may occur with cassiterite (Cornwall) and at the low temperature end with galena and sphalerite. It is associated with pyrrhotite and pentlandite in nickeliferous sulphide deposits (Sudbury, Ontario). Rio Tinto, Spain, Mt. Lyell, Tasmania, and Ducktown, Tennessee are a few of the many important localities. It occurs in the richly organic shale deposit of the Kupferschiefer at Mansfield, Germany.

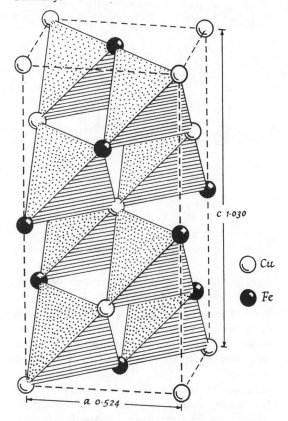

Fig. 162. The structure of chalcopyrite.

Enargite   2[Cu$_3$AsS$_4$]

Orthorhombic   *P* *mm*   *a* 0·646 *b* 0·743 *c* 0·618

Structure: A hexagonal close-packing of S atoms with Cu and As atoms at the centres of tetrahedral groups of 4 S atoms, distributed so as to produce orthorhombic symmetry. In Fig. 163 line-shaded tetrahedra contain As. All tetrahedral S groups have the tetrahedral apex pointing the same way. The structure is hence hemimorphic (terminated differently at opposite ends of *z*-axis).

G

Habit: Tabular ||{001} or prisms elongated ||*z*, showing {001}, {100}, {010}, {110}, {120}; but usually massive and granular.

Twins on {320} sometimes repeated to give pseudohexagonal trillings.

Cleavage {110} perfect. Fracture uneven. Brittle. *H* 3. *D* 4·45.

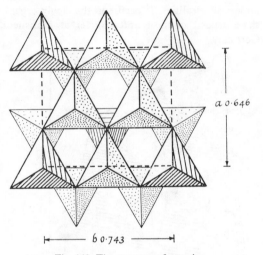

Fig. 163. The structure of enargite.

Lustre metallic. Colour and streak grey-black to black.

Optics: Opaque, anistropic, weakly pleochroic, grey to pinkish-brown. *R*% 25–28.

Tests: The powdered mineral heated gently in an open tube gives sulphurous and arsenical fumes. Roasted and heated with a flux (sodium carbonate) on charcoal yields a globule of copper. Dissolves in aqua regia.

Occurrence: Not common, but an abundant ore in a few places, along with other copper sulphides, sphalerite, galena, and pyrite. Sierra de Famatina, Argentina, localities in Chile and Peru, Butte, Montana and Tintic, Utah are important occurrences.

Famatinite is its antimony analogue and occurs with it.

Niccolite (Kupfernickel)   2[NiAs]

Hexagonal *C* 6/*mmm*   *a* 0·361   *c* 0·503

Structure: Ni atoms are on simple hexagonal lattice (Fig. 164) with As atoms halfway between the sheets of Ni atoms, also in a hexagonal array,

approximately that of hexagonal close packing. Each As lies between 6 Ni atoms, and every Ni between 6 As and 2 near neighbour Ni atoms.

Niccolite-type minerals show departures from their ideal compositions, usually due to a deficiency of metal atoms in the structure.

Habit: Usually massive.

Fracture uneven. Brittle. *H* 5–5½. *D* 7·78. Lustre metallic. Colour pale copper red, sometimes with tarnish. Streak pale brownish black.

Optics: Opaque, pleochroic in pinkish or yellowish colours; anisotropic; highly reflecting, *R*% 52–58. Shows hexagonal zoning. Polishing hardness > pyrrhotite < pyrite.

Tests: Heated before the blow pipe it fuses easily and gives arsenical fumes. Soluble in HNO_3 giving an apple-green solution. Precipitation of Ni-dimethylglyoxime is a highly specific test for Ni. The colour of niccolite is distinctive. Alters to a green coating of 'nickel bloom'.

Occurrence: In hydrothermal veins with other nickel, cobalt, silver and copper ores. Important localities are Cobalt, Ontario, and mines in Cornwall, Saxony (Freiberg, Annaberg), the Harz Mountains (Andreasberg), Bohemia (Joachimstal).

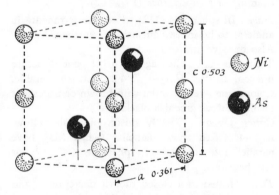

Fig. 164. The structure of niccolite.

Related mineral: *Breithauptite*, NiSb, is the antimony analogue. The two form solid solutions at high temperatures, but at low temperatures they exsolve to form intergrowths.

Pyrrhotite   2[$Fe_{1-x}S$]

Hexagonal *C* 6/*mmm*   *a* 0·343 *c* 0·568

Structure: Niccolite type. Illustrates the metal deficiency mentioned under niccolite.

Habit: Crystals tabular ||{0001} or with steep pyramids. Usually massive granular.

Cleavage absent. Fracture uneven. Brittle. $H\ 3\frac{1}{2}-4\frac{1}{2}$. $D\ 4\cdot58-4\cdot65$ decreasing as Fe becomes increasingly deficient. Lustre metallic. Colour bronze-yellow to brownish yellow tarnishing brown or reddish brown. Streak grey black. Magnetic.

Optics: Opaque, light yellow. Anisotropic with good reflectivity, $R\%$ 38–45. Moderate polishing hardness between that of sphalerite and pyrite. Shows intergrowths with pentlandite.

Tests: The colour and magnetic property are distinctive. It dissolves in HCl giving off H_2S.

Occurrence: The nickeliferous pyrrhotite deposit, a magmatic segregation from basic-ultrabasic rocks at Sudbury, Ontario, is the most important deposit. Also at Bushveld Complex, Transvaal. Pyrrhotite is also found in hypothermal veins, as at Botallack, Cornwall and Andreasberg, Harz Mountains. The nickel of the pyrrhotite body at Sudbury occurs in intergrown pentlandite $(Fe, Ni)_9S_8$, exsolved along {10$\bar{1}$0}. Se substitutes for S in the pyrrhotite and can be recovered.

Troilite, FeS, representing the ideal formula for the pyrrhotite structure, is found in meteorites. It has $D\ 4\cdot79$.

Pentlandite $4[(Fe, Ni)_9S_8]$

Cubic $F\ \ m3m$ $a\ 1\cdot009$ to $0\cdot991$, decreasing as Fe decreases.

Structure: A cubic close-packing of S atoms, with (Fe, Ni) in two types of space, tetrahedral and octahedral. Fig. 165 shows part of the unit cell, the tetrahedra in the figure having (Fe, Ni) at their centres and S at the angles. Dot-dash lines indicate the octahedral arrangement of S around the other type of (Fe, Ni) position.

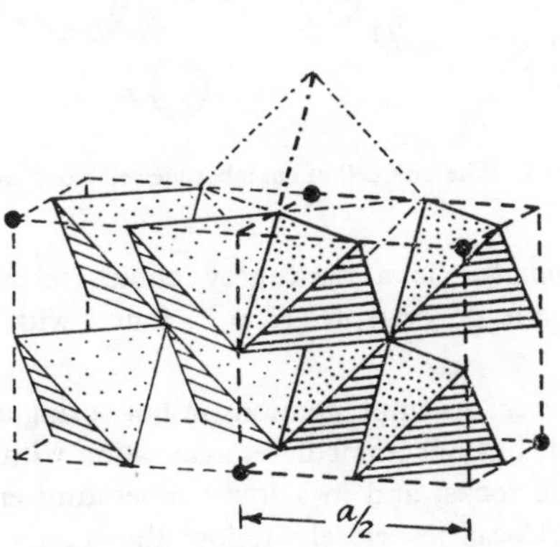

Optics: Opaque, yellowish, isotropic. $R\%$ 52. A little softer in polishing than its host.

Composition: The Fe : Ni ratio is usually near 1 : 1, but may rise to 2 : 1. Co is also present, substituting for Ni.

Occurrence: Usually in intergrowth with pyrrhotite, the triad axes of the cube coinciding with the hexad of the pyrrhotite.

Millerite (Capillary Pyrites) $9[NiS]$

Trigonal $R\ \ 3m$ $a\ 0\cdot961$ $c\ 0\cdot326$
                           (hexagonal axes)

Structure: Not well established. When prepared artificially said to be unstable, inverting to the niccolite structure.

Habit: Slender hair-like twisted crystals and radiating groups. Cleavage perfect on {10$\bar{1}$1} and {01$\bar{1}$2}. Hair-like crystals somewhat elastic. $H\ 3-3\frac{1}{2}$. $D\ 5\cdot5$. Lustre metallic. Colour pale brass-yellow. Streak greenish black.

Optics: Opaque, pale yellow, anisotropic, strongly reflecting, $R\%$ 54–60.

Tests: Easily fusible, M.P. 797°C (a difference from pyrite, 1171°C). Soluble in HNO_3; Ni-dimethylglyoxime can then be precipitated.

Occurrence: A low-temperature mineral forming by alteration of other nickel minerals in nickel-cobalt-silver ores (Cobalt, Ontario; Gap Mine, Pennsylvania). In cavities of carbonate veins. With siderite nodules in clay ironstones of the Coal Measures in South Wales.

Covellite $6[CuS]$

Hexagonal $C\ \ 6/mmm$ $a\ 0\cdot38\ \ c\ 1\cdot64$

Structure: Sandwich-type layers normal to z-axis, the bread of the sandwiches being sheets of tetrahedra with their bases in one plane and their apices pointing inwards, wherein Cu is co-ordinated to 4 S. The sandwich filling is a sheet of Cu atoms at the centres of triangles of S atoms. Two sandwich layers to the unit cell. At half height of the cell are two adjacent sheets of S atoms. The motif in Fig. 166 is drawn on a framework of the upper half of an 'orthohexagonal' cell.

• *Fe, Ni in octahedral positions*

Fig. 165. The structure of pentlandite.

Habit: Crystals thin hexagonal plates. Usually massive. Cleavage perfect ||{0001} in conformity with structure. Cleavage flakes flexible. H 1½–2. D 4·6. Lustre submetallic to resinous. Colour indigo blue, sometimes with purple tarnish. Purple when wet. Streak grey to black.

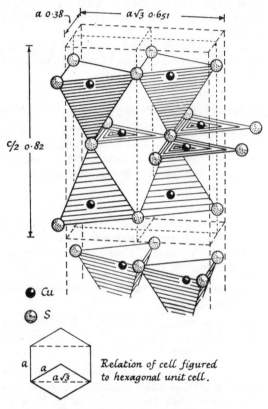

Fig. 166. The structure of covellite.

Optics: Translucent and pleochroic in greens in very thin flakes. Uniaxial positive. ω1·45 in Na light with very high dispersion. In reflected light pleochroic, ω dark blue in air (red-violet in cedarwood oil), ε blue-white. Strongly anisotropic. $R\%$ ω 22, ε 7. Forms intergrowths with chalcocite. Polishing hardness less than that of galena.

Tests: Gives off much S on heating in a closed tube. Roasted and moistened with HCl gives an azure blue flame. The colour of covellite is distinctive.

Occurrence: As a product of secondary enrichment of copper-bearing veins, associated with chalcopyrite, chalcocite, bornite, enargite. Butte (Montana), Bor (Yugoslavia) and Sardinia are well-known localities. Rarely as a volcanic sublimate (Vesuvius).

Cinnabar   [HgS]

Trigonal C 32 a 0·415 c 0·950

Structure: Like that of NaCl slightly extended along one triad axis. Hg and S atoms form two sets, each arranged as in cubic close packing (shown for Hg in Fig. 167): but the S atoms are displaced from above Hg positions by a small rotation of each triangle of S atoms in the projection.

Habit: Rhombohedra or thick tablets with dominant {0001}. Often granular, massive or as an earthy coating.

Twins on {0001}. Cleavage perfect ||{10$\bar{1}$0}. Somewhat sectile. H 2–2½. D 8·09. Lustre of crystals adamantine; but dull in earthy varieties. Colour cochineal red to brownish red, vermilion. Streak scarlet.

Optics: Translucent in thin flakes, red in colour, ω 2·905 ε 3·256 (Na light) varying rapidly with wavelength. Optically positive. Strongly rotates the plane of polarized light.

Tests: Heated in the open tube decomposes giving SO_2 and Hg vapour which condenses on the cooler part of the tube. The colour and streak are distinctive.

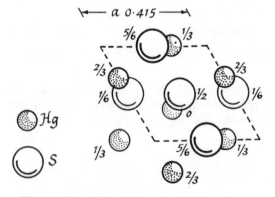

Fig. 167. The unit cell of cinnabar viewed along z-axis.

Metacinnabar is a dark grey cubic (tetrahedral) metastable polymorph (H = 3) found with cinnabar.

Occurrence: As impregnations in hot spring areas in general, but not immediate, association with young volcanic rocks, and in a low-temperature environment. Localities closely follow the Tertiary orogenic and volcanic belts. The most famous locality is Almaden, Spain. Other deposits occur at Monte Amiati, Tuscany; Idria near Trieste; Nikitovka, in the Donetz Basin; Altai Mtns, Outer Mongolia; and in California and Peru. Small amounts occur at Ohaeawai and Ngawha Springs, New Zealand.

Use: The chief ore of mercury. Has been used for 1000 years by the Chinese and others as a red pigment.

Pyrite (Iron pyrites)   4[FeS$_2$]

Cubic   *P   m3*   a 0·5405

Structure: Like NaCl, the Fe atoms alternating with covalently united S$_2$ groups along the edges of the cube. (Fig. 33, p. 38). The S$_2$ pairs form dumb-bell shaped molecules pointing along triad axes, and so arranged that each of the eight sub-cubes of the main unit cell has only one S$_2$ pointing to its centre. In Fig. 168 the arrangement is shown by a sample of the S$_2$ pairs with arrows along their axes, and the body diagonals on which they lie drawn in full line.

Habit: In cubes often striated (Fig. 71, p. 67), pyritohedra {210} and more rarely the diploid {213} (Fig. 65, p. 61). Also massive, or concretionary, sometimes stalactitic.

Twins on {110}, sometimes interpenetrant, giving the 'iron cross' (Fig. 169). Cleavage unimportant.

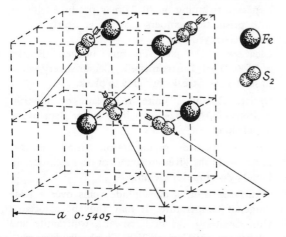

Fig. 168. Part of the structure of pyrite to show the orientation of S$_2$ groups. Cf. Fig. 33.

Fracture uneven. Brittle. *H* 6–6½—high for a mineral of metallic lustre. *D* 5·01. Lustre metallic, splendent. Colour pale brass yellow. Streak greenish black. Pyroelectric. Conducts electricity. Sparks when struck with steel.

Optics: Opaque, pale cream-yellow. Isotropic, *R*% 54. Polishing hardness one of the greatest amongst the ore minerals.

Tests: The colour, streak and hardness are distinctive. Pyrrhotite is softer and magnetic. Marcasite is paler on the fresh surface and orthorhombic, often bladed, in crystallization, but distinction may be difficult. Marcasite alters more readily.

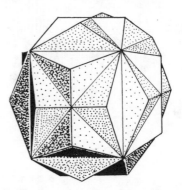

Fig. 169. The iron-cross twin of pyrite. Cf. the pyritohedron, Fig. 65.

Alteration: In a damp atmosphere pyrite may oxidize to iron sulphates and acid liquid. This so-called pyrites disease is a problem for museum curators—susceptibility to it varies in different specimens. In the near-surface parts of veins a similar process is very important in the initiation of supergene enrichment. The heat generated in decomposition of pyrite in coal seams and mine spoil-tips may lead to 'spontaneous' combustion of seams or pit-heaps.

Related minerals: *Bravoite*, (Ni, Fe)S$_2$ with some Co; *sperrylite*, PtAs$_2$; and *hauerite*, MnS$_2$, all rare minerals, have the pyrite structure. *Ullmannite*, NiSbS, is interesting because the SbS replacing S$_2$ groups are not symmetrical end for end, and this lowers the symmetry of the whole crystal to that of point group 23 in which all symmetry planes are lost. *Cobaltite*, CoAsS, and *gersdorffite*, NiAsS, may be similar.

Marcasite   2[FeS$_2$]

Orthorhombic   *P   mmm*   a 0·444   b 0·541   c 0·338

Structure: Fe atoms are octahedrally co-ordinated to six S atoms. The octahedra have two orientations, the one at the centre of the cell differing from those at its corners. Both settings link edge to edge with octahedra above and below to give chains along z-axis. The S atoms fall into pairs whose members are 0·221 nm apart (cf. pyrite 0·214 nm). One such pair is shown by a double-headed arrow on Fig. 170

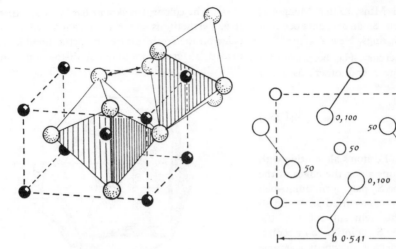

Fig. 170. The structure of marcasite.

and the pairs are joined on the plan. The orientation chosen conforms with that of arsenopyrite.

Habit: Tabular ∥{010}, faces often curved. In cockscomb aggregates with wedge-shaped tips; also concretionary and stalactitic.

Twins on {101}, often repeated, giving the cockscomb form. Cleavage {101} poor. Fracture uneven. H 6–6½. D 4·88. Lustre metallic. Colour pale bronze yellow to silvery on fresh surfaces. Streak grey to very dark brown.

Optics: Opaque, pleochroic in pale cream, yellowish white and pinkish brown. Anisotropic, $R\%$ 49–55. Relatively hard to polish, like pyrite.

Tests: Difficult to distinguish from pyrite, the crystal form and paler colour being the best guides. It inverts to pyrite at 450°C.

Occurrence: Formed at low temperatures and under acid conditions. Pyrite, which is more stable, forms under less acid or alkaline conditions and at higher temperatures. Marcasite is found as nodules in sedimentary rocks, particularly limestones. Also a late mineral in some low temperature veins. *Loellingite*, $FeAs_2$, a rather uncommon silvery to grey mineral of sulphide veins, has the same structure.

Arsenopyrite (Mispickel)   8[FeAsS]

Monoclinic B $2/m$ a 0·951 b 0·565 c 0·642
(pseudo-orthorhombic) β 90°

Structure: The large cell (Z = 8) does not depart measurably from orthorhombic shape. The atomic arrangement is closely similar to that of marcasite,

with As—S pairs replacing S—S pairs. The cell is chosen to conform with that of marcasite.

Habit: Prismatic crystals with striations ∥z on {hk0}; columnar aggregates; compact and granular. Twins on {100} and {001} to give pseudo-orthorhombic crystals; also on {101} sometimes repeated. Repeated twinning on {012} may produce trillings with prismatic members crossing at nearly 60° to give a six-rayed star.

Cleavage {101} fair. Fracture uneven. Brittle. H 5½–6. D 6·1. Lustre metallic. Colour silvery-white to steel-grey. Streak dark grey to black.

Optics: Opaque. White to cream or pinkish. Anisotropic, $R\%$ 52–56. A little softer than pyrite in polishing.

Tests: Fuses easily, giving arsenical fumes. Heated in a closed tube gives an abundant As sublimate. In HNO_3 decomposes yielding S.

Occurrence: Forms in sulphide veins over a wide range of temperature. It may thus be associated with cassiterite (as in Cornwall), wolframite and scheelite (as at Carrock Fell, Cumberland) or with lead-silver veins of mesothermal type. It is the most abundant arsenic mineral.

Glaucodot, (Co, Fe)AsS, and *gudmundite*, FeSbS, are less common analogues.

Molybdenite   2[MoS₂]

Hexagonal C $6/mmm$ a 0·316 c 1·230

Structure: A layer structure, each layer made of two sheets of S atoms separated by a sheet of Mo atoms. Mo in each layer lies at the centre of a triangular prism of 6 S. In the drawing (Fig. 171) the triangular cages with lids contain the Mo. Each layer is staggered $\frac{1}{3}a_1$ and $\frac{1}{3}a_2$ to bring S atoms above a triangle of S atoms in the layer below (as shown by stalks in Fig. 171). This stagger reverses in the next layer, so that the layers treated as units are in hexagonal close-packed arrangement.

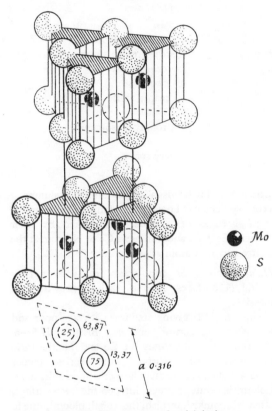

Fig. 171. The structure of molybdenite.

Habit: In shining hexagonal leaves and scales, or massive and foliated. Cleavage {0001} highly perfect in accordance with the layered structure. Scales flexible, not elastic; sectile. H 1–1$\frac{1}{2}$; greasy feel. D 4·7. Lustre metallic, splendent. Colour lead-grey. Streak on porcelain greenish-grey. Makes a bluish-grey mark on paper. Electrical conductivity much greater normal to layers than along them.

Optics: Translucent in extremely thin flakes. Opaque for practical purposes. Strongly pleochroic in reflected light, white, grey-white. Anisotropic,

$R\%$ ω 37 ε 15. Surprisingly, is harder in polishing than chalcopyrite.

Tests: Infusible, M.P. 1185°C. Soluble in aqua regia. Extreme softness, cleavage and bluish-grey mark on paper are distinctive. Distinguished from graphite by its lustre, higher specific gravity and bluish grey, not black, mark on paper.

Occurrence: Characteristic of the pegmatitic phase of igneous action, associated with granite rocks either in the granite or in pegmatites and veins with cassiterite, wolframite, scheelite, topaz, tourmaline, and garnet.

The largest known deposit is at Climax, Colorado, U.S.A., in silicified granite with fluorite and topaz. Chile, Canada, Mexico and Norway are other producers. In small amounts molybdenite is not uncommon.

Use: The chief ore of molybdenum, which is widely used as an alloy in steels, and in non-ferrous stellite alloys. Molybdenite has lubricant properties conferred by its atomic structure (cf. graphite). Tungstenite, WS_2, is a rare analogue.

KRENNERITE GROUP

The structure of the group is characterized by linear $AuTe_2$ molecular groups.

Krennerite occurs in short striated prisms.

Calaverite: bladed lath-like crystals, slender, striated prisms and granular or massive. Twins on {310} give prisms crossing at 123° and on {111} prisms roughly at 93° 40′.

Sylvanite: short prisms, or thick tablets. Often skeletal, roughly columnar and granular. Twins on {100} are common, either simple, lamellar or interpenetrant, giving prisms crossing at 69° 44′, about 55° or nearly 90°. These forms resembling cuneiform writing are characteristic and led to the name *graphic tellurium*.

Calaverite and sylvanite fuse readily and heated on charcoal give a malleable bead of gold (calaverite) or an alloy (sylvanite). All alter to a spongy mass of gold. They are distinguished, one from another, by the absence of cleavage in calaverite and by crystal habit.

Krennerite is less abundant than the other two but all occur in the same environment of low temperature hydrothermal veins along with pyrite, small amounts of tetrahedrite and other sulphides and gold. Famous localities are Transylvania (Nagyag), California (Calaveras County), Western Australia (Kalgoorlie, Coolgardie).

	Krennerite	**Calaverite**	**Sylvanite**
	$8[AuTe_2]$	$2[AuTe_2]$	$4[(Au, Ag)Te_2]$
	Orthorhombic	Monoclinic	Monoclinic
		pseudo-orthorhombic	
	$P\,mm$	$C\,2/m$	$P\,2/m$
	a 1·65	0·72	0·90
	b 0·88	0·44	0·45
	c 0·45	0·51	1·46
		$\beta 90°$	$\beta 145°$
Cleavage	perfect {001}	none	perfect {010}
H	2–3	$2\frac{1}{2}$–3	$1\frac{1}{2}$–2
D	8·62	9·24	8·16
Lustre	metallic	metallic	metallic, splendent.
Colour	silver to yellow	yellow to silver	steel-grey to silver
Streak	—	yellowish grey	steel-grey
Optics	opaque	opaque	opaque
	creamy white	creamy white	creamy white
$R\%$	61	63	48–60
	anisotropic	anisotropic	strongly anisotropic
Polishing			
hardness	soft	soft	very soft

Skutterudite $8[(Co, Ni)As_3]$
Smaltite $8[(Co, Ni)As_{3-x}]$
Chloanthite $8[(Ni, Co)As_{3-x}]$
Cubic $I\ m3\ a$ 0·819 (Skutterudite), 0·824 (Smaltite)

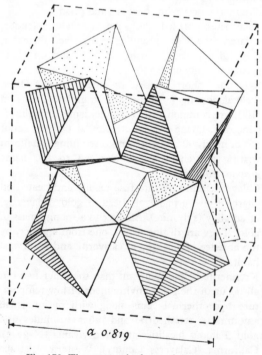

Fig. 172. The structural scheme of skutterudite.

Structure: May be visualized as tottering columns or crumpled chains of octahedra running parallel to the vertical edges of a cube, cross-linked by their corners. Each octahedron has a Co or Ni atom at the centre and an S atom at each corner (Fig. 172).

Habit: Cubes, or cubo-octahedra, with pyritohedral {110} faces. Also as skeletal growths and massive.

Twins on {112} giving sixlings. Cleavage poor and inconstant. Fracture uneven. Brittle. $H\,5\frac{1}{2}$–6. D 6·1–6·9. Fe substitutes for Co and Ni the ratio of which varies through the series giving variation in D. Lustre metallic, sometimes splendent. Colour tin-white to grey-white, on the fresh surface, often with pinkish tarnish, the 'cobalt bloom', due to alteration, or green 'nickel bloom' in chloanthite.

Optics: Opaque, grey-white to yellowish white. Isotropic. $R\%$ 56. Often strikingly zoned, the outer zones richer in As. Softer in polishing than pyrite or arsenopyrite, harder than pyrrhotite.

Tests: Fuses on charcoal giving arsenical fumes and a magnetic globule. Fused in a borax bead the cobalt variety gives a fine blue colour. The 'blooms' mentioned under colour, may be helpful.

Occurrence: In cobalt-silver-nickel veins of mesothermal type as in Saxony and Ontario. At Skutterud (Norway) in veins in gneiss. Arsenopyrite, other cobalt and nickel arsenides, native silver, native bismuth and silver sulpharsenides are the associated minerals.

Use: The complex ore is a source of cobalt, which is used as an alloy in steel and stellite, in alloys for strong permanent magnets, as a binder for abrasives and as a catalyst.

Stibnite $4[Sb_2S_3]$

Orthorhombic *P mmm* *a* 1·130 *b* 1·131 *c* 0·384

Structure: Pyramidal SbS_3 groups are joined in fours by common edges, the foursomes being linked by longer bonds into chains snaking along the *x*-direction (Fig. 173).

Habit: Bundles of slender striated prisms, columnar or somewhat radiating, or granular.

Twinning mainly seen in polished section, {130} and others. Cleavage ||{010} perfect, in conformity with the structure. Flakes flexible, crystals easily bent or twisted about their length, not elastic, slightly sectile. *H* 2. *D* 4·63. Lustre metallic splendent. Colour lead-grey, sometimes tarnishing darker. Streak lead-grey.

Optics: Transparent only in the far red or infra red. In reflected light, white to grey, strongly pleochroic, Anisotropic, *R*% 30–40, the strongest reflection for vibration ||*z*-axis. Rarely with red internal reflections. Zonary growth is common. Shows lamellar twinning. Softer in polishing than galena.

Tests: Easily fusible, M.P. 550°C, colouring the flame greenish blue. With KOH solution a yellow coating at once forms.

Occurrence: In low-temperature veins often far from any igneous body. Large deposits are rare. They occur in the Hunan Province of China; Sabah; Shikoku Island, Japan; Mexico; Bolivia. A small deposit occurs at Glendinning, Dumfriesshire.

Associated minerals are realgar, orpiment, galena, pyrite, cinnabar, carbonates and quartz.

Alteration: Stibnite alters to yellowish or white oxides, cervantite Sb_2O_4, senarmontite Sb_2O_3 and valentinite Sb_2O_3. The Algerian and some other antimony deposits are of these alteration products. Bismuthinite Bi_2S_3 has the same structure as Sb_2S_3. By contrast it is found in high-temperature veins but is rather rare.

Use: Stibnite is the principal ore of antimony, which is used to harden alloys of lead, particularly type-metal. It is alloyed with tin and copper in the manufacture of the low-friction Babbitt metal for bearings.

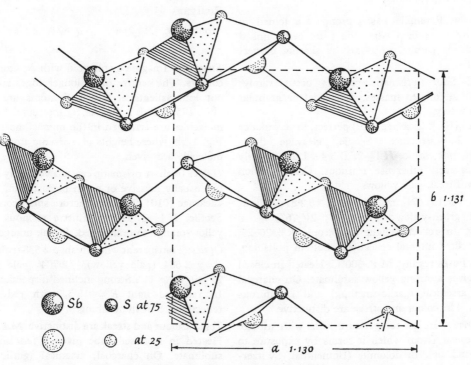

Fig. 173. The structural arrangement in stibnite.

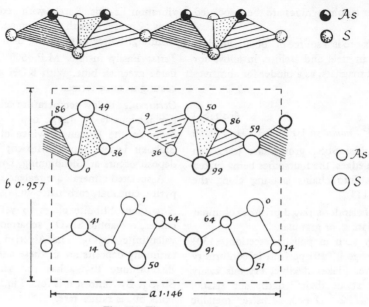

Fig. 174. The structure of orpiment.

Orpiment   4[As₂S₃]

Monoclinic   *P* 2/*m*   *a* 1·146 *b* 0·957 *c* 0·422
                                        β90° 30′

(Pseudo-orthorhombic)

Structure: Pyramidal AsS₃ groups are joined in pairs by a common edge, the pairs being linked into chains parallel to *x*-axis by sharing corners (Fig. 174).

Habit: Short prismatic crystals occur rarely; usually in foliated masses, botryoidal or granular and powdery.

Twins ‖{100}. Cleavage {010} perfect, in accordance with the structure. Sectile, cleavage flakes flexible, not elastic. *H* 1½–2. *D* 3·49. Lustre pearly on cleavages, otherwise resinous. Colour lemon yellow. Streak pale yellow.

Optics: R.I.(λ671 nm) α=y 2·4,   β 2·81 yellow, γ 3·02 green-yellow; γ∧z1½–3°; $2V_\alpha$76°; $r > v$ strong. In reflected light anisotropic *R*% 20–25, with yellow internal reflections. Soft in polishing.

Tests: Fuses readily, M.P. 300°C. Heated in closed tube gives a strong yellow sublimate. On charcoal gives arsenical (garlic-smelling) and sulphurous fumes. The colour and streak are distinctive.

Occurrence: In low temperature veins with stibnite and realgar (from which it forms by exposure to light and air). In dolomite (Binnenthal, Switzerland) and marls with native sulphur, calcite and barytes. As a hot spring deposit (Yellowstone Park, U.S.A.) and at fumaroles (Vesuvius).

Realgar   16[AsS]

Monoclinic   *P* 2/*m*   *a* 0·927 *b* 1·350 *c* 0·656
                                        β106° 37′

Structure: Imagine a sphenoid with As atoms at its corners. The centres of its longer edges are pulled out into salients at which S atoms are located (Fig. 175a). This is the As₄S₄ molecule. 4 of these molecules are arranged in the monoclinic cell as in Fig. 175b, where heights of 3 arsenic atoms in each molecule are given.

Habit: In short prismatic crystals vertically striated, but usually massive or as an incrustation.

Cleavage {010} distinct. Fracture small conchoidal. Sectile. *H* 1½–2. *D* 3·48. Lustre resinous. Colour yellow-orange to orange-red. Streak orange to red.

Optics: Transparent when fresh. α 2·538 (colourless), β=y 2·684 (pale yellow), γ 2·704 (pale yellow), $2V_\alpha$40° α∧z 11°. Strong inclined dispersion $r > v$. In reflected light *R*% 18·5, with red internal reflections. Soft in polishing.

Tests: Colour and streak are distinctive. M.P. 307°C. Heated in a closed tube gives a reddish yellow sublimate. On charcoal, arsenical (garlic odour) and sulphurous fumes.

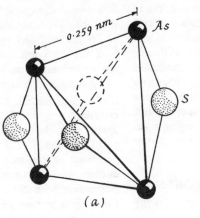

(a)

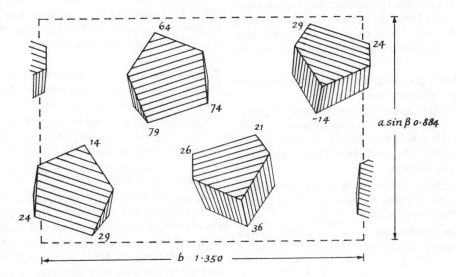

Projection parallel to c : heights given above 001

(b)

Fig. 175. The structural scheme of realgar.

Alteration: On exposure to light and air decomposes to an orange yellow powder (orpiment, As_2S_3 and arsenolite As_2O_3).

Occurrence: In low temperature veins associated with orpiment, stibnite or cinnabar, in some limestones and dolomites (Binnenthal, Switzerland) and clay rocks. Also as a hot spring deposit (Yellowstone Park U.S.A.) and volcanic sublimate (Naples and Vesuvius).

Boulangerite

Monoclinic P 2/m

Structure: no details available.

Habit: Acicular crystals striated $\|z$. Feathery, or in compact fibrous masses.

Cleavage {100} good. Brittle; thin needles, flexible. $H\ 2\frac{1}{2}$–3. $D\ 5.98$–6.23. Lustre metallic. Colour bluish lead-grey, with yellow oxidation spots. Streak brownish-grey to brown.

Optics: Opaque. Grey-white, markedly anisotropic. $R\%\ 37$–44. Softer in polishing than galena or covellite.

Tests: Fuses easily. Soluble in hot HCl giving off H_2S. Often not distinctive, requiring confirmation by X-ray examination.

Boulangerite $8[Pb_5Sb_4S_{11}]$

Monoclinic P 2/m   a 2.152 b 2.346 c 0.807
                              $\beta 100°\ 48'$

Occurrence: In mesothermal or epithermal sulphide veins with galena, arsenopyrite, sphalerite and pyrite. Found at Molières, Gard, France; in the Harz and Silesia; at Sala, Sweden; Přibram, Bohemia; Coeur d'Alene, Idaho; Iron Mountain Mine, Montana. *Zinkenite* $PbSb_2S_4$, *plumosite* $Pb_2Sb_2S_5$ and *jamesonite* $Pb_4FeSb_6S_{14}$, are related minerals of similar general properties and fibrous or acicular crystallization, belonging with boulangerite to the group of *feather-ores*. X-ray study is required to distinguish them.

Bournonite (Wheel Ore) $4[CuPbSbS_3]$

Orthorhombic   *P mmm*   *a* 0·816 *b* 0·871 *c* 0·781

Structure: Cu atoms lie in tetrahedral S groups and each As atom forms the summit of a flat pyramid with 3 S atoms. These elements are linked, with Pb, in a structure with resemblances to that of stibnite.

Habit: Short prisms striated ||*z*-axis or *y*-axis, in parallel growth, or tablets ||{001}.

Twins on {110} commonly repeated, giving cross-like or cog-wheel-like forms. Cleavage poor. Fracture sub-conchoidal. Brittle. $H\ 2\frac{1}{2}$–3. D 5·83. Lustre metallic often splendent. Colour and streak steel grey to blackish.

Optics: Opaque. White, weakly anisotropic. $R\%$ 36–38. Shows lamellar twinning on {110}. Harder in polishing than galena.

Tests: Fuses easily. Decomposed by HNO_3 to give a blue solution and a residue of S and oxides of Sb and Pb.

Occurrence: A common mineral of mesothermal veins with galena, sphalerite, chalcopyrite, tetrahedrite, pyrite and quartz. Occurs as microscopic inclusions in galena, and forms oriented intergrowths with it. Its temperature range is rather wide, however, and it is reported from Cornwall (it was first found at St. Endellion) and Broken Hill, N.S.W., in high-temperature environment as well as, in other places, with low-temperature minerals like stibnite and cinnabar. Found in the mining districts of the Harz, Saxony, Silesia, Romania, in tin-silver veins of Bolivia, and in the western mining areas of the U.S.A.

TETRAHEDRITE SERIES

Tetrahedrite $2[(Cu,Ag)_{12}Sb_4S_{13}]$
Tennantite $2[(Cu, Ag)_{12}As_4S_{13}]$

Cubic   *I* $\bar{4}3m$ *a* 1·033 (Sb) to 1·019 (As)

Structure: Comparable with sphalerite. In a sphalerite cell doubled along the edge, 8 of the 32 Zn are replaced by (Cu, Ag) 24 by (As, Sb). 8 of the 32 S are removed so that the (As, Sb) are co-ordinated to 3 S instead of 4. Chemical analysis suggests 2 extra S (location uncertain).

Habit: Tetrahedral, also massive, coarse or fine-granular.

Twins with the triad axis as twin axis and composition plane || or ⊥{111}. Twins may be interpenetrant or repeated. Cleavage absent. Fracture uneven. $H\ 3$–$4\frac{1}{2}$ (increasing with As). D 4·6–5·1 (increasing with Sb). Lustre metallic, splendent. Colour grey to nearly black. Streak black to brown.

Optics: In very thin splinters translucent, red *n* 2·72 (λ671 nm). In reflected light isotropic, grey to olive brown, rarely with red internal reflections. $R\%$ 30. Moderate polishing hardness > chalcopyrite < sphalerite.

Tests: Fuses easily. Is decomposed by HNO_3 leaving S and Sb_2O_3. The variety of elements present makes bead tests unreliable.

Composition: The simpler formula $8[(Cu,Ag)_3 (Sb,As)\ S_3]$ seems structurally more probable. Besides Cu and Ag there may be notable amounts of Fe, Au, Hg, Pb, Ni, Co, while Bi may enter in place of (As, Sb).

Occurrence: Tetrahedrite is an important copper ore of mesothermal to epithermal veins, associated with other primary copper sulphides and sulphides of lead and silver. Its crystallographic relation to chalcopyrite and sphalerite is expressed by orientated overgrowths on these minerals. It is found in the mines of Saxony, the Harz, Potosi (Bolivia), and many of the mining districts of western U.S.A. Unusual are the occurrences in high-temperature environment in Cornwall and Broken Hill, N.S.W.

RUBY SILVER GROUP

Proustite $6[AgAsS_3]$
Pyrargyrite $6[AgSbS_3]$

Trigonal   *R* 3 *m*   *a* 1·080 *c* 0·869 (proustite)
                          a 1·102 *c* 0·873 (pyrargyrite)
                                      (for hexagonal cell)

Structure: AsS_3 (or SbS_3) groups form triangular pyramids with As(Sb) at the apex. The S atoms at the base are linked to triangular sheets of Ag atoms above and below the pyramid. Towers of these Ag sandwiches with $AsS_3(SbS_3)$ inside them run parallel to *z*-axis (Fig. 176) and are cross-linked

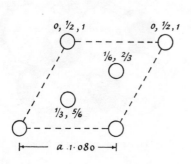

Positions of As atoms in plan

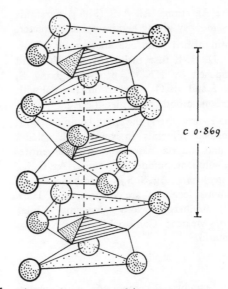

As atoms at summits of low pyramids which represent AsS_3 groups. Circles = Ag.

Fig. 176. The structural scheme of proustite, typifying that of the Ruby Silver group.

	Proustite	Pyrargyrite
Habit	Prismatic ‖z-axis, with rhombohedra and scalenohedra. Massive	Prismatic ‖z-axis, with rhombohedral terminations different at either end. Massive.
Twins	On {10Ī1}, also on {10Ī4} giving trillings.	On {10Ī4} giving lamellae and complex aggregates.
Cleavage	{10Ī1} distinct.	{10Ī1} distinct.
Fracture	Conchoidal to uneven.	Conchoidal to uneven.
Tenacity	Brittle	Brittle
H	2–$2\frac{1}{2}$	$2\frac{1}{2}$
D	5·57	5·85
Lustre	Adamantine	Adamantine
Colour	Scarlet	Deep red
Streak	Vermilion	Purple-red
Optics	Translucent, red, pleochroic in reds Uniaxial negative ω 2·98 ε 2·71 (λ671 nm)	Translucent, deep red Uniaxial negative ω 3·08 ε 2·88 (λ671 nm)
In reflected light	Bluish grey Pleochroic Anisotropic $R\%$ 25–28 Red internal reflections Low polishing hardness ≫ argentite ≪ silver	Bluish grey Pleochroic Anisotropic $R\%$ 28–31 Red internal reflections Low polishing hardness ≫ argentite ≪ silver

through the Ag atoms. All $AsS_3(SbS_3)$ pyramids point the same way giving hemimorphic character.

Tests: Distinguished from one another by colour, streak and *D*. Both fuse easily (M.P. 490°C) and are decomposed by HNO_3, proustite giving S, pyrargyrite S and Sb_2O_3. Heated on charcoal with sodium carbonate both yield metallic Ag.

Occurrence: Pyrargyrite is the more common. Both are late-formed minerals in veins with calcite, dolomite and quartz associated with argentite, other silver sulphantimonides (*polybasite*, $[Ag, Cu]_{16}$ Sb_2S_{11}; *stephanite* Ag_5SbS_4), tetrahedrite and silver. Famous localities are Andreasberg, Harz Mountains; Freiberg, Saxony; Atacama, Chile; Guanajuato, Mexico; Comstock Lode, Nevada; Cobalt, Ontario.

HALIDES

This group comprises compounds with wholly ionic bonding. Though there are some 85 species, some complex, the commoner members are simple in composition and fall into four groups.

Anhydrous halides AX	Halite	NaCl
	Sylvine	KCl
	Cerargyrite	AgCl
Anhydrous halides AX_2	Fluorite	CaF_2
Anhydrous halides of two metals	Cryolite	Na_3AlF_6
Hydrous halides	Carnallite	$KMgCl_3 . 6H_2O$
	Atacamite	$Cu_2Cl(OH)_3$

Halite (Rock Salt) $4[NaCl]$
Cubic *F   m3m*   *a* 0·564

Structure: One type of atom lies at the corners and face centres of the unit cube, the other at the midpoints of the edges and at the body-centre (Fig. 177). Each atom is surrounded octahedrally by 6 of the other kind. (Fig. 10, p. 12). The first structure to be analysed by X-rays (W. L. Bragg, 1914. *Proc. Roy. Soc. A*, 89, 468).

Habit: In cubes often with faces hollowed stepwise (hopper crystals, Fig. 178). Massive.
Cleavage {100} perfect. Fracture conchoidal. *H* $2\frac{1}{2}$.
D 2·16. Lustre vitreous. Colourless, yellowish, reddish, blue or purplish. Colour in part due to lattice imperfections (see p. 87). Also often

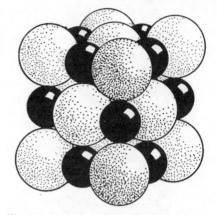

Fig. 177. Atomic packing in halite. Cf. Fig. 10.

coloured by discrete impurities. Streak colourless to white. Taste salty. Sometimes fluorescent in UV light. Has low absorption of heat (diathermanous). Paramagnetic.

Optics: Transparent, isotropic, $n(\lambda589$ nm$)$ 1·544. Microscopic, brine-filled cavities common.

Tests: Taste and ready solubility in water are characteristic. When it is dissolved in water and silver nitrate solution is added a dense precipitate of AgCl forms.

Occurrence: In thick sedimentary successions formed by the evaporation of sea water in former enclosed lagoons. Characteristic associated minerals are basal dolomite, anhydrite and gypsum, polyhalite ($K_2SO_4 . MgSO_4 . 2CaSO_4 . 2H_2O$), kieserite ($MgSO_4 . H_2O$), sylvine and carnallite (see p. 164). Salt deposits occur at many horizons in the geological succession, those of the Permian being especially notable, and of these the Stassfurt

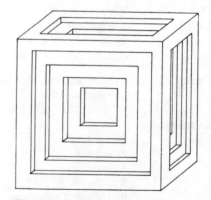

Fig. 178. A crystal of halite with hopper faces, due to more rapid deposition of material at the edges than at the centres of the faces.

deposit of Germany is the most celebrated. In England in Cheshire, S.W. Durham and N.E. Yorkshire, and in Cumberland.

Use: The source of common salt.

Sylvine   4[KCl]

Cubic *F m3m*   *a* 0·6293

Structure: As for halite. Though the two are isostructural there is no solid solution between them.

Habit: In cubes modified by octahedral faces, granular or compact. Cleavage {100} perfect. Fracture uneven. Less brittle than halite. *H* 2. *D* 1·99. Lustre vitreous. Colour colourless, white, greyish, bluish or red. Regarding colour-centres see p. 87. Taste salty but more bitter than halite.

Optics: Transparent, isotropic, $n(\lambda 589$ nm$)$ 1·490.

Tests: Like halite. Distinguished by bitter taste. Colours the flame violet (a blue filter to eliminate the yellow of Na helps this observation). Pressed between glass slips it flattens while halite powders.

Occurrence: As for halite but much less common being found only in the uppermost horizons of the salt sequence where the brine was concentrated to < 1·57% its original volume.

Use: The ore of potassium for fertilizers and other purposes.

Cerargyrite (Horn Silver)   4[AgCl]

Cubic *F m3m*   *a* 0·555

Structure: As for halite. Forms a complete series of solid solutions with bromyrite AgBr. ($a = 0.577$ nm.)

Habit: Usually in crusts and waxy coatings or horn-like masses. Cleavage absent. Tough, sectile and plastic. *H* 2½. *D* 5·55 (bromyrite 6·50). Lustre resinous or horn-like. Colourless when pure. Usually grey or yellowish, becoming purple-brown on exposure to light.

Optics: Translucent, isotropic, n 2·071 (bromyrite 2·253).

Tests: Habit, colour and sectility are characteristic. Placed on zinc and moistened with water, swells, blackens and is reduced to silver.

Occurrence: In the oxidized zone of silver veins especially in arid climates. Associated with native silver, limonite, manganese oxides and also malachite, cerussite and similar secondary oxidation products. Found in many mines of the western

U.S.A. in Colorado, Nevada, New Mexico, and California. Also in Chile, Bolivia and Peru.

Use: An ore of silver.

Fluorite (Fluorspar)   4[CaF₂]

Cubic *F   m3m*   *a* 0·5463

Structure: Ca^{2+} ions lie at the corners and face-centres of a cube. F^- ions are tetrahedrally co-ordinated to $4Ca^{2+}$ (Fig. 179). This means Ca^{2+} is co-ordinated to $8F^-$ which surround it at the corners of a cube. The radius ratio $R_{Ca^{2+}}/R_{F^-}$ of $2·06/1·33 = 0·797$ leads us to expect 8-fold co-ordination for Ca^{2+}.

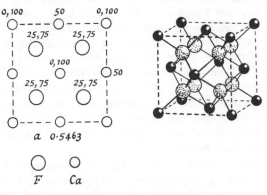

Fig. 179. The structure of fluorite.

Habit: As cubes, rarely with edges bevelled by {110} or less often as octahedra. Massive, cleavable. Twinning on {111} often interpenetrant, is very common (Fig. 72, p. 68). Cleavage {111} perfect. Fracture shallow conchoidal to splintery. Brittle. *H* 4. *D* 3·18. Lustre vitreous. Colour: purple or blue (Blue John), blue-green, and green are the common colours of crystals. Often white when massive. Crystals may also be yellow, colourless or rarely red or pink. The colour may change from green to purple on exposure to sunlight. Colour zones are common. Usually strongly fluorescent in ultra-violet light (especially of $\lambda 360$ nm) or cathode rays and sometimes phosphorescent on heating or irradiation. Sometimes triboluminescent when scratched with a knife. Streak white.

Optics: Colourless to pale purple, often in spots or bands. Isotropic though sometimes showing zones of weak anomalous birefringence. $n = 1·434$ ($\lambda 589$ nm). The rather high relief with $n <$ cement and perfect cleavage are distinctive in thin section.

Tests: The crystal form and twinning, octahedral cleavage and often the colours are distinctive. Softer than feldspar, harder than calcite. Does not effervesce with HCl. M.P. 1360°C. A fluorine test as under cryolite may be applied. The refractive index and isotropic character with the other properties is distinctive. Opal, which is harder and less dense, and alums, which are water-soluble, are the main optically similar minerals.

Occurrence: As an accessory mineral of late hydro-thermal formation in granites, thence through the whole range of temperature to deposition as a cement in sandstones. Most commonly as a vein mineral especially in mesothermal lead-silver veins, where it may be the main gangue mineral. In the Northern Pennine ore-field, England, it occurs in the highest temperature zone near the emanative centre. The blue granular banded material from Derbyshire, called Blue John, deserves mention. The Permian and Trias of Elgin, Scotland, provide examples of sandstones partly cemented by fluorite.

Varieties: Yttrium and cerium may substitute for Ca^{2+} giving *yttrofluorite*, $(Y+Ce):Ca$ reaching $1:6$ in natural material, and other Rare Earths occur in trace amounts.

Use: As a flux in steel-making, for enamelling iron, in making special glasses, and in making hydro-fluoric acid. Optically perfect material is used for making special lenses.

Cryolite $2[Na_3AlF_6]$

Monoclinic P $2/m$ $a\ 0.546$ $b\ 0.560$ $c\ 0.780$
                                                         $\beta 90°\ 11'$

Pseudo-cubic $\sqrt{(a^2+b^2)} \simeq c$

Structure: AlF_6 octahedra centred at the corners and body-centre of the cell are linked through inter-mediate Na^+ ions at or near 0,100 and 50 from the 010 datum-plane (Fig. 180).

Habit: Usually massive, granular. Crystals when they occur are equant, equal development of {001} and {110} giving a cubic aspect.

Twinning common and complicated. Often repeated, on several laws at the same time, made possible by the pseudo-cubic structure. Cleavage absent. Fracture uneven. Brittle. $H\ 2\frac{1}{2}$. $D\ 2.97$. Lustre vitreous to greasy. Colourless to white, less often brownish or reddish. Streak white.

Optics: Colourless. $\alpha\ 1.338$, $\beta=\gamma\ 1.338$, $\gamma\ 1.339$, $\gamma-\alpha\ 0.001$, $\gamma \wedge z\ 44°$, $2V_\gamma\ 43°$, horizontal dispersion $r > v$.

Tests: M.P. 1020°C. Refractive index above that of pure water but below that of a 1% cane sugar solution. Heated with potassium bisulphate in a closed tube, the HF given off etches the glass, the silica being redeposited as a white coating higher up the tube.

Occurrence: At Ivigtut, Greenland, as a pegmatitic mass in granite, associated with siderite, galena, chalcopyrite, sphalerite and fluorite. There is also some cassiterite and topaz. Other occurrences are few: Pikes Peak, Colorado; Miask, Ilmen Mts., U.S.S.R.; and in the Spanish Pyrenees. Cryolite alters to a suite of rare aluminium fluorides.

Use: As a flux in the electrolytic extraction of aluminium from bauxite. The natural mineral is being replaced in this rôle by artificial cryolite made from fluorite.

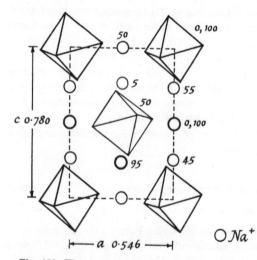

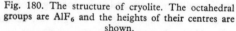

Fig. 180. The structure of cryolite. The octahedral groups are AlF_6 and the heights of their centres are shown.

Carnallite $12[KMgCl_3.6H_2O]$

Orthorhombic P mmm $a\ 0.951$ $b\ 1.602$ $c\ 2.252$

Structure: Not definitely known but its bromine analogue, and $K\ Mg\ F_3$, are like perovskite.

Habit: Pseudohexagonal pyramidal crystals with {111} and {011} equally developed. Sometimes in hexagonal tablets ||{001}.

Cleavage absent. Fracture conchoidal. Brittle $H\ 2\frac{1}{2}$. $D\ 1.602$. Lustre of fractured surface brilliant. Colourless to white or reddish due to hematite inclusions, zonally arranged. Taste very bitter. Deliquescent.

Optics: Transparent, colourless. $\alpha = z$ 1·466, $\beta = y$ 1·475, $\gamma = x$ 1·494, $\gamma - \alpha$ 0·028, $2V_\gamma$ 70°. May show polysynthetic twinning. Strong birefringence, coupled with indices below 1·50 are distinctive.

Tests: Soluble in water. Fuses easily colouring the flame violet (use a blue filter). The refractive indices and birefringence characterize it.

Occurrence: In the upper zone of evaporite successions with halite, sylvine and kieserite, and in overlying marls. The Stassfurt salt deposits of Germany are the most famous locality, but it is found in N.E. England, New Mexico, Texas and other salt successions.

Atacamite $4[Cu_2Cl(OH)_3]$

Orthorhombic   *P   mmm* a 0·601 b 0·913 c 0·684

Structure: Cu, co-ordinated octahedrally to 4OH and 2Cl, is placed at the corners, the top and bottom face-centres and the body-centre of the cell. 4Cu co-ordinated to 4OH, with another OH and a Cl at the apices of the roughly octahedral group, lie in pairs at 25 and 75.

Habit: Slender striated prismatic crystals elongated along *z*-axis. Crystals tabular ||{001}. May be massive, fibrous, or granular.
Twins on {110}, and others. Cleavage {010} perfect, {101} fair. Fracture conchoidal. Brittle. *H* 3–3½. *D* 3·76. Lustre adamantine to vitreous. Colour a handsome bright to dark green. Streak pale green.

Optics: Transparent, pleochroic in greens. $\alpha = y$ 1·831 pale green, $\beta = x$ 1·861 yellow green, $\gamma = z$ 1·880 grass green $\gamma - \alpha$ 0·049, $2V_\alpha$ 75° $r > v$ strong.

Tests: Fuses readily, colouring the flame blue. Heated with sodium carbonate on charcoal yields a globule of copper. Though soluble in acids it does not effervesce like malachite. The green colour is characteristic, and differs from that of malachite.

Occurrence: A secondary oxidation product of copper ores under arid, salty conditions. Mainly found in the Atacama desert in Chile with malachite, cuprite, chrysocolla and gypsum. Sparingly in some mines of arid western U.S.A., and South Australia. Also at St. Just, Cornwall, and as a fumarole deposit at Vesuvius and Etna.

OXIDES AND HYDROXIDES

The oxides fall into two groups. One is a group of hard, dense, refractory minerals on the whole, occurring as accessory minerals of igneous rocks and as resistant detrital grains in sediments. The other is a group of secondary alteration products of sulphide ores. These are softer and may be earthy. The hydroxides by contrast tend to be soft, of moderate or low density, and occur in secondary products of weathering or alteration. They are less well crystallized and harder to characterize than the oxides.

Oxides and hydroxides may be classified as follows. Note that the commonest of all oxides, quartz, SiO_2, and its polymorphs are omitted. These are considered as the ultimate term in the classification of silicates, since the view is here taken that mineral structure is more fundamental than chemical composition.

Type A_2O

| Cuprite | Cu_2O |

Type AO

| Periclase | MgO |
| Tenorite | CuO |

Type A_2O_3

Corundum	Al_2O_3
Hematite	Fe_2O_3
Ilmenite	$FeTiO_3$

Type ABO_3

| Perovskite | $CaTiO_3$ |

Type AB_2O_4

Spinel Group

Spinel	$MgAl_2O_4$
Hercynite	$FeAl_2O_4$
Gahnite	$ZnAl_2O_4$
Galaxite	$MnAl_2O_4$
Magnesioferrite	$MgFe_2O_4$
Magnetite	$FeFe_2O_4$
Maghemite	Fe_2O_3
Ulvospinel	$Ti(Fe, Mg)_2O_4$
Franklinite	$ZnFe_2O_4$
Jacobsite	$MnFe_2O_4$
Trevorite	$NiFe_2O_4$
Magnesiochromite	$MgCr_2O_4$
Chromite	$FeCr_2O_4$
Chrysoberyl	$BeAl_2O_4$

Type AO_2 (excluding SiO_2)

Rutile	TiO_2
Anatase	TiO_2
Brookite	TiO_2
Pyrolusite	MnO_2
Cassiterite	SnO_2
Uraninite	UO_2
Thorianite	ThO_2

Type $A_mB_nO_{2(m+n)}$ of Ni, Ta, Ti and Rare Earths

Pyrochlore–
microlite	$(Na, Ca)_2 (Nb, Ta, Ti)_2$ $O_6(O, OH, F)$
Fergusonite	$(Nb, Ta, Ti)O_4$
Columbite–tantalite	
Euxenite–polycrase	$(Ca, R.E., U, Th, etc.)$
Samarskite	$(Nb, Ta, Ti, etc.)_2$
Tapiolite	$(O, OH)_6$

Hydroxides

Brucite	$Mg(OH)_2$
Gibbsite	$Al(OH)_3$
Diaspore	$AlO(OH)$
Goethite	$FeO(OH)$
Lepidocrocite	$FeO(OH)$
Boehmite	$AlO(OH)$
Manganite	$MnO(OH)$
Psilomelane	$(Ba,H_2O)_2 Mn_5O_{10}$

Cuprite $2[Cu_2O]$

Cubic *P* *m3m* a 0·427

Structure: Refer to the structure of diamond. Let the C atoms become O. Now in each tetrahedral group place a Cu atom halfway between the centre O and each apex of the tetrahedron as in Fig. 181a. This is half the cuprite structure. Into the unoccupied sub-cubes of the diamond cell must now be

introduced a linked set of similar tetrahedra in parallel disposition to the first set. This completes the cuprite structure (Fig. 181b). As Bragg pointed out, the two sets of tetrahedral groups in the structure have no cross-linking bonds between them. One sub-cube constitutes a unit cell (Fig. 181c), the corner oxygens whose heights are underlined belonging to the opposite network from the rest.

Habit: Octahedra, more rarely dodecahedra or cubes. Also massive, or earthy. Sometimes in matted, straight, hair-like crystals (plush ore). Cleavage poor {111}. Fracture conchoidal to uneven. Brittle. H $3\frac{1}{2}$–4. D 6·14. Lustre submetallic or adamantine to earthy. Colour shades of red, especially cochineal-red. Streak brownish red, shining.

Optics: Opaque. Bluish white in reflected light. Shows anomalous pleochroism and anisotropy. $R\%$ 27. Red internal reflections. Harder in polishing than chalcopyrite or copper, softer than sphalerite. It often encloses native copper.

Tests: M.P. 1235°C, colouring the flame green. Heated on charcoal yields a globule of copper. Soluble in HCl: a concentrated solution on cooling and dilution gives a white precipitate of CuCl.

Occurrence: Common in the oxidized zone of copper veins, with native copper, malachite, azurite and relict primary sulphides. Many localities include Redruth, Liskeard and Lizard, in Cornwall; Chessy near Lyons, France; Broken Hill, N.S.W., Australia; Mt. Lyell, Tasmania; Tsumeb, S.W. Africa; and Bisbee, Arizona.

Use: An important ore of copper.

Periclase $4[MgO]$

Cubic *F* *m3m* a 0·421

Structure: NaCl type.

Habit: Octahedra, more rarely in cubes. Often in irregular rounded grains.

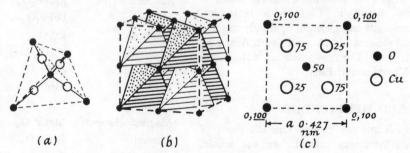

Fig. 181. The structure of cuprite.

Cleavage perfect {100}. Sometimes a {110} parting. *H* 5½. *D* 3·56. Lustre vitreous. Colourless or grey-white, brownish yellow. May be green or black due to inclusions. Streak white.

Optics: Transparent, colourless, $n = 1·735$, increasing with Fe content. High relief and good cleavage {100}. Often anhedral and partly altered to brucite, or hydromagnesite.

Occurrence: In contact-altered magnesian limestone as a result of decarbonation of dolomite.

$$CaMg(CO_3)_2 \rightarrow CaCO_3 + MgO + CO_2$$

If silica is not present in excess, periclase enters after tremolite, forsterite and diopside as metamorphic temperature rises, and thereafter remains stable to the highest temperatures. It is found disseminated and as small clusters of crystals in the marble. Predazzo in the Tyrol is a famous locality, but periclase is widely distributed though rare.

Composition: May contain Fe substituting for Mg up to 8 wt.% FeO, giving a brownish colour and higher refractive index.

Use: Artificial material is used as an electric insulator.

Tenorite (Melaconite)   4[CuO]

Monoclinic *C* 2/*m*   *a* 0·465 *b* 0·341 *c* 0·511
                                                   β99° 29′

Structure: Cu atoms are linked to 4O in an approximately plane square (Fig. 182). The tendency to such co-ordination is a feature of the cupric ion. Each O is surrounded by 4Cu in a distorted tetrahedron.

Habit: As a black powder, also massive or botryoidal, sometimes as thin shining scales.

Twins on {011}. Scales flexible and elastic. *H* 3½. *D* 6.45 as crystals, 4–5 when massive. Lustre metallic. Colour black.

Optics: In reflected light, pleochroic in grey-white and yellowish tints. Anisotropic *R*% 20–27. May show lamellar twinning. Pleochroic in brown in transmitted light.

Occurrence: In massive form in the oxidized zone of copper veins, associated with malachite, chrysocolla, cuprite, limonite and wad. At Bisbee, Arizona; in Cornish mines; at Leadhills, Lanarkshire.

Corundum   6[Al₂O₃]

Trigonal *R* $\bar{3}$ *m*   *a* 0·495 *c* 1·358
                                                    (hexagonal cell)

Structure: An approximately hexagonal close-packing of O^{2-}, with $\frac{2}{3}$ of the octahedral spaces between the oxygens occupied by Al^{3+}. Each octahedron shares a face with one in the layer above

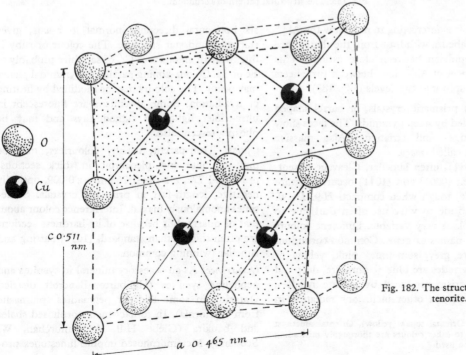

Fig. 182. The structural pattern of tenorite.

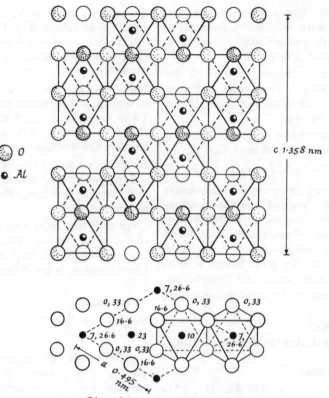

Plan of lowermost ⅓ rd of cell

Fig. 183. The structural pattern of corundum.

or below, and it is interesting to note how the Al^{3+} within the octahedra withdraw from the shared face because of repulsion between close cations. This leads to the sheet of Al^{3+} ions between two sheets of O^{2-} being split into two levels (Fig. 183).

Habit: Rough prismatic crystals, or barrel-shaped crystals bounded by steep pyramids. The faces show horizontal ridges and striations. Also massive granular, then called *emery*.

Twins on $\{10\bar{1}1\}$, often lamellar. Cleavage absent, but partings on $\{0001\}$ and $\{10\bar{1}1\}$ occur. Fracture uneven. Brittle, tough when compact. H 9. D 4·0. Lustre adamantine to vitreous, often dull due to roughness. Colour very variable. Different colours have separate names as gems. Common corundum and emery are grey, sometimes pink, yellow or green. Gem varieties are blue = sapphire, deep red = ruby.* Asterism produced by minute needles, possibly of rutile, or other inclusions radiating in

* Names like Oriental topaz (yellow), Oriental amethyst (purple) applied to other colours are thoroughly misleading and should not be used.

six horizontal directions normal to z-axis, gives star ruby and star sapphire. The colour of ruby is due to small amounts of Cr, of sapphire probably to Fe or Ti. Colour-zoning parallel to pyramidal planes may occur. The colours may be modified by heating or irradiation. Some specimens are fluorescent in ultra-violet light and cathode-rays and may be phosphorescent also.

Optics: Transparent, usually colourless, may be pink or blue and pleochroic in thick sections. ω 1·767–1·772, ε 1·759–1·763, $\omega-\varepsilon$ 0·009, optically negative, length-fast in prismatic crystals. Sometimes anomalously biaxial. Interference colour about that of quartz, but because of its hardness sections are often thicker than standard. $\{10\bar{1}1\}$ parting and lamellar twinning common.

Occurrence: As an accessory mineral in syenites and nepheline syenites (Haliburton-Bancroft district, Ontario), in some feldspar pegmatites (plumasite from California). In contact-metamorphosed shales and bauxite (Glebe Hill, Ardnamurchan, W. Scotland). Metamorphosed impure limestones provide the emery deposits of Naxos Island, Greece, the Aiden region, Turkey and the Central Ural Mountains. Gem corundums come from Mogok, Upper Burma in eluvial deposits from crystalline limestones, and from placer deposits in Thailand, Ceylon, India and elsewhere.

Use: As an abrasive (emery) and a gem-stone. It is now produced artificially for these purposes by fusing bauxite or alumina. Its melting point is about 2015°C.

Hematite $6[Fe_2O_3]$

Trigonal R $\bar{3}m$ a 0·504 c 1·376
                             (hexagonal cell)

Structure: Corundum type.

Habit: Crystals tabular ‖{0001} with {10$\bar{1}$1}, often with triangular striations on {0001}, or rounded into lenticular form; as thin 'micaceous' scales in random orientation; as compact, radially fibrous, reniform masses; oolitic; earthy (ochreous) often mixed with clay. Cleavage absent. Parting on {0001} and {10$\bar{1}$1}. Twinning on {0001} or lamellar on {10$\bar{1}$1} giving striae and a parting. Crystals brittle, pulverulent in earthy varieties, elastic in thin scales. H 5–6. D 5·26–4·9 in fibrous varieties. Lustre metallic splendent in crystals, to dull and earthy. Crystals are steel grey to iron black, sometimes iridescent. Reniform material is red to red-brown. Streak dark red to red-brown.

Optics: Translucent, dark-red in thin scales. Usually dark between crossed polars. $\omega = 3\cdot22$, $\varepsilon = 2\cdot94$ ($\lambda589$ nm), $\omega - \varepsilon$ 0·28. In reflected light white to grey, pleochroic and anisotropic, $R\%$ 28. Red internal reflections. One of the hardest ore minerals to polish, equal in hardness to pyrite and harder than magnetite or ilmenite. It may be intergrown with the two last.

Tests: Infusible, M.P. 1565°C. Becomes magnetic when heated in a reducing flame. Soluble in conc. HCl. The density and streak are helpful.

Occurrence: Widespread in the zone of weathering as an alteration product of iron-bearing minerals. Concentration in the *laterite* of tropical regions may sometimes provide iron ore. Great masses of hematite in Precambrian terrains, worked as 'siliceous hematite ores', are of sedimentary origin later metamorphosed (e.g. Lake Superior, Brazil, Australia). It forms part of Silurian oolitic sedimentary iron ores of the Clinton formation of New York, Pennsylvania, and Alabama. As irregular

bodies in Carboniferous strata, and older Palaeozoic rocks on the borders of the English Lake District, in W. Cumberland and N. Lancashire, whence come fine specimens of specularite and kidney ore.

Varieties: *Specularite* consists of crystalline material of metallic lustre. *Micaceous hematite* is in small shining scales. *Kidney ore* consists of reniform masses. *Red ochre* is the earthy variety.

For solid solution relationships with ilmenite see under that mineral, and under titanomagnetite (p. 204).

Use: An important ore of iron.

Ilmenite $6[FeTiO_3]$

Trigonal R $\bar{3}$ a 0·508 c 1·408
                             (hexagonal cell)

Structure: Corundum-type, with all the metal ions between one pair of oxygen layers Fe, and all those between the next pair Ti. This ordered substitution of Ti for Fe lowers the symmetry from $\bar{3}m$ to $\bar{3}$.

Habit: Thick tablets ‖{0001}; sometimes in acute rhombohedra; massive; as disseminated grains in basic igneous rocks.
Twinning on {0001}, and lamellar on {10$\bar{1}$1}. Cleavage absent. Partings on {0001} and {10$\bar{1}$1}. Fracture conchoidal to sub-conchoidal. H 5–6. D 4·72. Lustre metallic to submetallic. Colour iron black. Streak black.

Optics: Opaque. In reflected light grey-white. Anisotropic $R\%$ 18–21. Shows twin lamellae. Often in regular intergrowth with magnetite or hematite. With magnetite the triad axis is common to both minerals. Rather hard in polishing, harder than magnetite but softer than hematite. In thin section, by the reflected light of a lamp held above the microscope stage, it often shows alteration, along three directions, to yellowish-white *leucoxene* composed of rutile, anatase or sphene.

Tests: Infusible. Distinguished from magnetite by being non-magnetic and by its streak from hematite. Fused in Na_2CO_3 and the pellet dissolved in H_2SO_4, the Ti gives a fine yellow colour on the addition of a drop of H_2O_2.

Occurrence: As disseminated grains in basic igneous rocks or as larger ilmenite-magnetite or ilmenite-hematite segregations in gabbros and anorthosites, Famous deposits occur at Egersund, S. Norway. Taberg in Sweden, Allard Lake, Quebec, and the Adirondacks, New York. Beach placer deposits of ilmenite or titanomagnetite are important in many parts of the world, those of Travancore being well-known. Menaccan Sands, Cornwall, provided one of the early records of the mineral (menaccanite).

Varieties: Ilmenite contains only limited amounts (< 6 wt %) of Fe_2O_3 at ordinary temperatures, but solid solution is complete above 950°C (see further under titanomagnetite, p. 204). Related minerals are *giekielite* $MgTiO_3$ and *pyrophanite* $MnTiO_3$, and both Mg and Mn substitute for Fe in ilmenite lowering its density and causing its colour to vary towards brown and deep red respectively.

Use: An ore of titanium.

Perovskite $4[CaTiO_3]$

Orthorhombic *P   mmm   a* 0·537 *b* 0·544 *c* 0·764

Structure: Pseudocubic. Becomes cubic at 900°C. The idealized cubic structure (which is the type for a large class of compounds) has *a* 0·38 and consists of 8 TiO_6 octahedra centred on the corners of a cube, with a calcium ion at the body centre of the cube (Fig. 184). There is one formula unit $CaTiO_3$

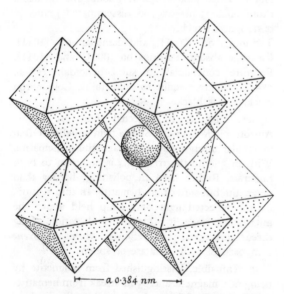

├─── *a* 0·384 *nm* ───┤

Fig. 184. The structure of perovskite, idealized. The octahedra are TiO_6, and the atom at the centre Ca.

in this cell. A small distortion leads to a larger cell and orthorhombic symmetry. The orthorhombic dimensions are roughly $\sqrt{2}a$, $\sqrt{2}a$ and $2a$ of the cubic cell.

Habit: As tiny cubes or octahedra. Rarely as reniform masses. Twinning on {111} interpenetrant; also more complex twins. Cleavage {100} poor.

$H\ 5\frac{1}{2}$. $D\ 4·0$ increasing with Ce and Nb. Lustre adamantine. Colour black, brownish, reddish or yellow. Streak colourless, cinnamon brown.

Optics: Translucent yellow to dark-brown. Colour may be zonal. Nearly isotropic, $n \simeq 2·34$ but with faint interference colours (birefringence 0·002 in larger crystals, with appearance of twinning. In reflected light light-grey or creamy-white (loparite) with brownish-red internal reflections.

Occurrence: A mineral of alkaline basic rocks such as melilite-, nepheline- and leucite-basalts (Eifel, Germany), some syenites (Kola Peninsula, U.S.S.R.) and particularly of carbonatites. Also found in some talc schists.

Varieties: The entry of Ce for Ca gives *knopite*, and of Nb substituting for Ti *dysanalyte*, while $(Ce, Na, Ca)_2(TiNb)_2O_6$ with Y, La, Ta is *loparite*.

SPINEL GROUP $8[XY_2O_4]$

Cubic *F   m3m*

Structure: O^{2-} ions are arranged approximately in cubic close-packing. In a cell of this arrangement about 0.8 nm along the edge there are 64 tetrahedral interstices (A sites) and 32 octahedral interstices (B sites). 8 A sites and 16 B sites are occupied by cations in such a way that ranks of occupied octahedra, joined edge to edge, run along one diagonal of the cube face cross-linked by occupied tetrahedral sites to form a layer (Fig. 185a). The tetrahedra also link them to octahedra of the next layer, whose ranks run along the other diagonal of the cube face. Four such layers make up the unit cell. Fig. 185b is a plan of the lower half of this cell. Each oxygen is common to two octahedra and one tetrahedron. The cations are of two types X^{2+} and Y^{3+} and in *normal spinels* X^{2+} is in A sites Y^{3+} in B sites. There are also, however, *inverse spinels* in which $8Y^{3+}$ are in the A sites and $(8X^{2+} + 8Y^{3+})$ are randomly distributed over the B sites. The choice between the two arrangements is governed by the crystal field stabilization energies of the ions involved, and this results in cases where the larger of the two cations may occupy the tetrahedral sites, in contravention of the usual rule.

Composition and physical properties: Naturally occurring spinels fall into the following series. Solid solution is possible within the series and physical properties vary between the extremes quoted.

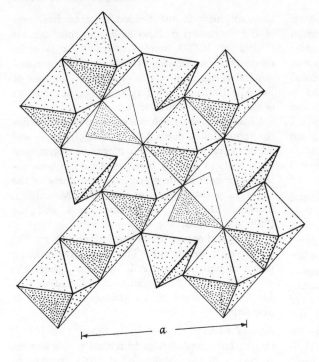

(a)

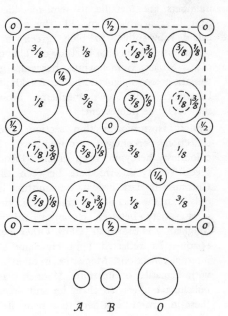

(b)

Fig. 185. The structure of spinel.

Spinel Series (Aluminium spinels). All are normal spinels.

		a (nm)	D	n	H
Spinel	MgAl$_2$O$_4$	0·809	3·55	1·719	7$\frac{1}{2}$–8
Hercynite	FeAl$_2$O$_4$	0·814	4·40	1·83	
Gahnite	ZnAl$_2$O$_4$	0·806	4·62	1·80	
Galaxite	MnAl$_2$O$_4$	0·827	4·03	1·92	

Magnetite Series (Iron spinels). All except franklinite are inverse spinels.

		a (nm)	D	R%	H
Magnesioferrite	Fe^{3+}(Mg, Fe^{3+})O$_4$	0·837	4·51	—	
Magnetite	Fe^{3+}(Fe^{2+}, Fe^{3+})O$_4$	0·8397	5·20	21	5$\frac{1}{2}$–6$\frac{1}{2}$
Franklinite	ZnFe$_2$O$_4$	0·843	5·32	19	
Jacobsite	Fe^{3+}(MnFe^{3+})O$_4$	0·850	5·03	18	
Trevorite	Fe^{3+}(Ni, Fe^{3+})O$_4$	0·841	5·20	—	

Chromite Series (Chromium spinels). All are normal spinels.

		a (nm)	D	n	R%	H
Magnesiochromite	MgCr$_2$O$_4$	0·832	4·43	—	—	
Chromite	Fe^{2+}Cr$_2$O$_4$	0·837	5·09	2·16	14	5$\frac{1}{2}$

Others

		a (nm)	D	R%
Ulvöspinel	Fe^{3+}(TiFe^{2+})O$_4$	0·854	—	< = 21
Maghemite	Fe$_{21\frac{1}{3}}$O$_{32}$ ($\equiv$ Fe$_2$O$_3$)	0·836	4·7	25

Habit: Octahedral. Cleavage absent. Twinning on {111} with this face as composition plane is common (the spinel law Fig. 72, p. 68). Brittle. Lustre: Spinel is vitreous, splendent. The more opaque members are metallic to semi-metallic. Colour: Spinel is variable from red (the gem variety is called ruby-spinel) to blue, green, brown and colourless. Maghemite is brown. The others are black. Streak: Spinel white; hercynite grey-green to dark green; magnetite black; franklinite reddish brown; chromite brown; maghemite brown. Spinel may rarely be fluorescent.

Optics: Isotropic in thin section. Indices and reflection percentages given above. Spinel is colourless, hercynite deep green. (Pleonaste is a green variety intermediate between these two.) Chromite is dark brown, transmitting light often only near the margins of grains. (Picotite, brown, is a variety between chromite and pleonaste.) Maghemite is brown to yellow. Other members are opaque. In reflected light chromite shows red internal reflections. Magnetite is often intergrown with hematite or ilmenite. Magnetite is hard in polishing but less so chromite and ilmenite while these in turn are less hard than hematite.

Tests: Infusible. M.P. spinel 2135°, magnetite 1591°. Members of the magnetite series are magnetic (franklinite and jacobsite only weakly) as is maghemite (see also p. 206). The more nearly opaque members may be distinguished by the colour of thin splinters, magnetism, density and cell size. Fused in borax, chromite gives a green coloured bead.

Occurrence: The spinel series occurs as accessory minerals in basic igneous rocks, in some pegmatites, in metamorphosed aluminous sediments and aluminous xenoliths in igneous rocks, and in contact-metamorphic limestones. Spinel itself is a constituent of gem gravels derived from such rocks in Ceylon and Burma. Hercynite is found in iron-rich, silica-poor aluminous contact hornfelses and xenoliths, in granulites and pyroxenites of metamorphic terrains and in ultrabasic rocks generally. Gahnite is found in granite pegmatites but is uncommon. Galaxite is rare.

Magnetite is widespread as a ubiquitous accessory mineral of igneous rocks and as black sands derived from basic igneous rocks. It also occurs as segregations (with ilmenite) from gabbroic rocks (Egersund, Norway; Bushveld Complex, S. Africa) and with apatite as a segregation from 'syenite' in the famous Kiruna deposit of Sweden. It forms by metamorphism in the bedded siliceous iron ores of the Precambrian. Also as a fumarolic deposit (Valley of 10,000 Smokes, Alaska). It is often titaniferous (see below). Franklinite is the principal mineral of zinc deposits in crystalline limestone at Franklin Furnace, New Jersey. The other members of the magnetite series are rare.

Chromite is invariably associated with peridotites (or serpentinites derived from them), norites, and anorthosites. Its origin as a magmatic segregation often leads to laterally persistent layers of accumulated chromite in the magmatic rock. Some of the larger deposits occur in the Ural Mountains, Turkey, Rhodesia, S. Africa, Cuba and the Phillipine Islands.

Ulvöspinel occurs only as an intergrowth parallel to {100} in magnetite. This intergrowth along planes at right angles may be very fine, giving the so-called box- or cloth-texture in polished section. Maghemite occurs as an oxidation product of magnetite.

Related mineral: *Hausmannite*, $8[Mn^{2+}Mn_2^{3+}O_4]$, though tetragonal, is closely related to the spinels. It occurs in pseudo-octahedral {101} crystals, or massive, has a good {001} cleavage, and is a little softer than magnetite (H 5–5½), with a chestnut streak. Fused in borax, it gives an amethyst-coloured bead, indicating Mn. Found in high-temperature veins and contact-metamorphic deposits, as well as with psilomelane and pyrolusite. It occurs at Ilfeld, Harz Mts, and Långban, Sweden.

Titanomagnetite

Because so much of the magnetite of basic igneous rocks is titanium-bearing, the relations between FeO, Fe_2O_3 and TiO_2 are important. Fig. 186 shows the solid solutions that exist at high temperatures.

(*i*) The trigonal series (α series) hematite-ilmenite shows complete solid solution above 950°C.

(*ii*) The cubic series (β series) magnetite-ulvöspinel shows complete solid solution above 600°C.

(*iii*) The cation-deficient series of spinels on the Fe_2O_3–TiO_2 side of the magnetite-ulvöspinel join is called the γ-series by analogy with maghemite (see below) which is γ-Fe_2O_3.

(*iv*) The pseudobrookite series shows complete solid solution above 1150°C; but $FeTi_2O_5$ is not

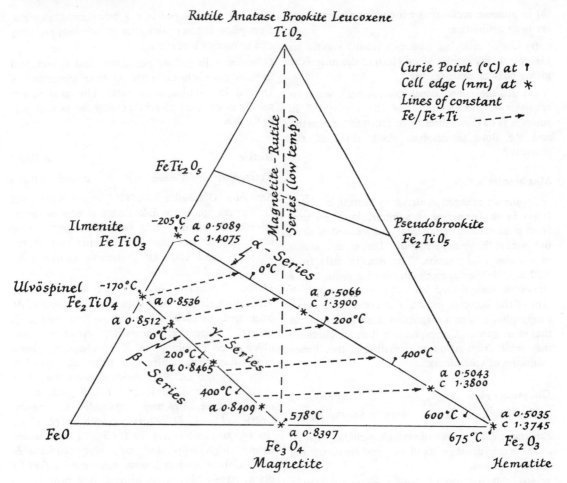

Fig. 186. Solid solution series and some physical properties of iron-titanium oxides.

known to occur naturally, being represented below 1140°C by ilmenite and rutile.

(v) In low temperature metamorphic rocks magnetite and rutile sometimes co-exist showing that the α-series is in part unstable under these conditions.

Most naturally occurring, homogeneous, magnetic spinel minerals fall in group (iii) lying between the magnetite-ulvöspinel and the hematite-ilmenite joins. Experiments show that the solid solution of ilmenite in magnetite is very limited. It is found also that magnetite-ilmenite intergrowths, apparently due to exsolution, cannot be homogenized by heating if the oxygen content of the specimen is kept constant (i.e. if O_2 is not allowed to escape). It is probable, therefore, that the Ti in the magnetite is originally present in solid solution as ulvöspinel. This solid solution may behave in one of two ways.

(a) In volcanic rocks some degree of oxidation takes place changing the composition along lines of constant Fe/Fe + Ti ratio (broken lines with arrowheads in Fig. 186) giving γ-series titanomagnetites which retain the spinel structure. Oxidation takes place by removal of cations from a proportion of the cation sites leading to cation-deficient spinels, a process that has been followed in artificially oxidized magnetite-ulvöspinel solid solutions. The resulting titanomagnetite is probably metastable, since its molecular volume is greater than that of two phases, magnetite-ulvöspinel + hematite-ilmenite.

The unit cell edge-length and Curie Point of these titanomagnetites can be approximately contoured across the γ-series field. Values of these properties may be thus predicted from composition; but the reverse cannot be done.

(b) In plutonic rocks, under higher pressures, there are two possibilities:

(i) Under reducing (water-deficient) conditions ulvöspinel may exsolve along {100} of the magnetite giving box- or cloth-texture.

(ii) In the more usual case, where water is relatively abundant, much of the ulvöspinel in solution is oxidized directly to ilmenite + magnetite and the ilmenite exsolves along {111} of the magnetite.

Maghemite

Magnetite oxidized in air, or by heating in closed tubes in the presence of sulphur, loses Fe ions from octahedral or tetrahedral and octahedral sites, but retains the spinel structure. This is an example of *omission solid solution*. The density falls from 5·20 to < 4·7 and the cell edge-length changes in a somewhat complicated way, depending on which type of site becomes vacant. The resulting mineral, maghemite, retains a magnetism almost as great as that of magnetite. Corresponding titanomaghemites may result from advanced oxidation of the titanomagnetite of volcanic rocks.

Chrysoberyl $4[BeAl_2O_4]$

Orthorhombic P mmm $a\,0{\cdot}547$ $b\,0{\cdot}939$ $c\,0{\cdot}442$

Structure: Like that of olivine (Mg_2SiO_4) with Al atoms occupying the sites of Mg, and Be occupying the Si positions.

Habit: Tablets ||{001}, rarely prismatic. Often striated.

Twins on {130} often repeated to give pseudo-hexagonal crystals. Cleavage {110} distinct. Fracture uneven to conchoidal. Brittle. $H\,8\frac{1}{2}$. $D\,3{\cdot}75$: this is high considering the light elements of which it is composed, and reflects the close atomic packing. Lustre vitreous. Colour shades of green including emerald-green, greenish white, yellow and greenish brown. Sometimes red by transmitted light, or showing opalescence or asterism. Streak colourless.

Optics: Translucent, colourless or pleochroic in pale shades $\alpha = z\,1{\cdot}746$, red; $\beta = y\,1{\cdot}748$ yellow; $\gamma = x\,1{\cdot}756$, green; $\gamma - \alpha\,0{\cdot}010$. $2V_\gamma\,70°$. Fe substituting for Al causes variation in indices.

Tests: Very refractory. Unattacked by acids. The hardness, greater than topaz, is distinctive. Distinguished from beryl by its tabular habit and its biaxial character.

Varieties: *Alexandrite* is red by transmitted light. *Cymophane* shows opalescence or asterism, probably due to minute inclusions.

Occurrence: In granite pegmatites and aplites, and in some mica schists. In the Mourne Mountains of Ireland in cavities in granite. The gem stones mostly come from placers, notably the gem gravels of Ceylon.

Rutile $2[TiO_2]$

Tetragonal P $4/mmm$ $a\,0{\cdot}459$ $c\,0{\cdot}296$

Structure: Octahedra with Ti^{4+} at the centre and O^{2-} at the apices are linked edge to edge in chains ||z-axis, the chains being cross-linked by sharing octahedral corners. Each Ti^{4+} contributes charge to $6O^{2-}$ and each O^{2-} receives charge from $3Ti^{4+}$ (Fig. 187).

Habit: Commonly in long needle-like crystals vertically striated; sometimes granular massive. As orientated needles in corundum (star sapphire), quartz, magnetite, hematite, ilmenite, and phlogopite. Twinning on {011} common, giving knee-twins comparable with those of cassiterite (Fig. 72); also on {031} giving a heart-shaped or spearhead outline. Cleavage {110} distinct, others more obscure. Fracture conchoidal to uneven. Brittle. $H\,6{-}6\frac{1}{2}$. $D\,4{\cdot}23$ rising with entry of Fe (to 4·4) and of Nb and Ta (to 5·6). Lustre adamantine: high dispersion may give considerable 'fire'. Colour reddish brown, sometimes yellow or black. Rarely blue, green. Streak pale brown.

Optics: Transparent, yellow to dark reddish brown depending upon Fe^{3+} content. $\omega\,2{\cdot}612$, $\varepsilon\,2{\cdot}899$, $\varepsilon - \omega\,0{\cdot}287$. Optically $+$ve. Extreme birefringence, but interference colour often masked by absorption colour. High dispersion $n_{red}\text{-}n_{violet} \simeq 0.10$. Straight extinction. Angled twins often characteristic in thin section.

Tests: Infusible. Insoluble in acids. When it is fused in Na_2CO_3 and the pellet is dissolved in H_2SO_4 and diluted, addition of H_2O_2 gives an amber colour (due to Ti).

Composition: Fe and Nb (up to 32 wt %) or Ta (up to 36 wt %) may enter the rutile structure. (Nb, Ta)-bearing members are black and nearly opaque.

Occurrence: The high-temperature form of TiO_2, found as one of the less common accessory minerals in plutonic igneous rocks. It is abundant in eclogites and a well-known rutile-bearing granite comes from

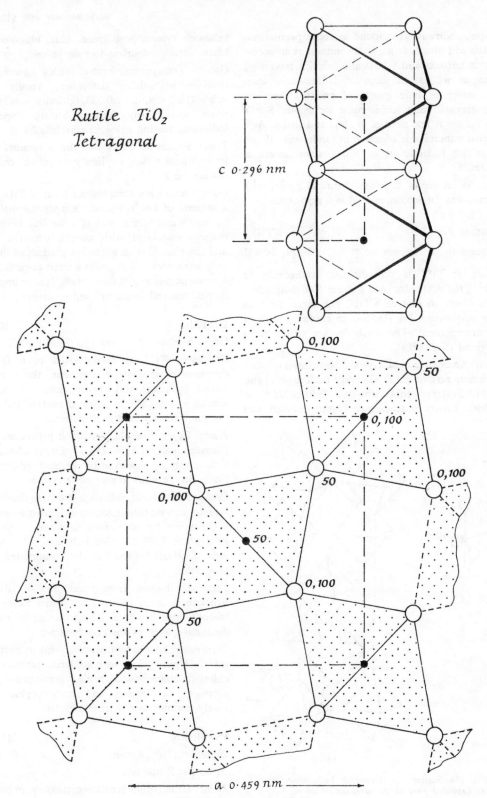

Rutile TiO₂
Tetragonal

c 0·296 nm

0,100
50
0, 100
50
0,100
0,100
50
0,100
50
0,100
50

a 0·459 nm

Fig. 187. The structure of rutile.

Kragero, Norway. It is found in some pegmatites, apatite and quartz veins. It is common as an accessory in gneisses and schists and in veins traversing them, as well as in some crystalline limestones. The milky or blue quartz of gneisses owes its appearance to microscopic rutile inclusions. Rutile also occurs as fine needles in clays and shales. As a detrital mineral it is common in sandstones. Beach sands rich in rutile are worked as titanium ore in Australia.

Use: As an ore of titanium. Artificial rutile, with magnificent dispersion, is used as a gem-stone.

Anatase   4[TiO$_2$]

Tetragonal   *I*   4/*mmm*   *a* 0·379 *c* 0·951

Structure: Ti^{4+} is co-ordinated octahedrally to 6 O^{2-}, the octahedra each sharing 4 of their edges with others so that each O^{2-} has three Ti^{4+} as near neighbours. Fig. 188b is an exploded view of the arrangement. The octahedra are rather badly distorted (Fig. 188a).

Habit: Acute pyramidal crystals. Sometimes tabular. Twinning rare on {112}. Cleavage perfect {001} and {011}. Fracture subconchoidal. Brittle. *H* 5½–6. *D* 3·90. Lustre adamantine. Colour reddish and

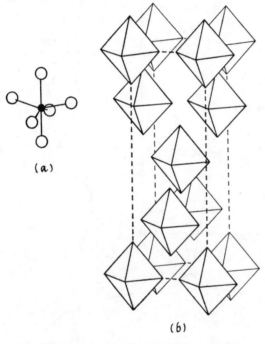

(a)

(b)

Fig. 188. Anatase. (a) Distorted TiO$_6$ octahedron.
(b) Exploded view of the arrangement of octahedra.

yellowish brown. Also green, blue, blue-grey and black. Streak colourless to pale yellow.

Optics: Transparent, brown, yellow, greenish or blue, with variable dichroism, usually weak. ω 2·561, ε 2·488, $\omega - \varepsilon$ 0·073, optically −ve. Sometimes anomalously biaxial. Strong dispersion, indices decreasing at longer wavelengths.

Tests: Similar to rutile but the pyramidal habit, lower density and optically negative character distinguish it.

Occurrence: A low temperature form of TiO$_2$. It is a mineral of low temperature hydrothermal veins in gneiss and schist, and of miarolitic cavities in granite, associated with quartz, adularia, rutile, and chlorite. Also an alteration product of ilmenite in igneous rocks. Sometimes altered to rutile which forms a paramorph of the anatase. It is an important detrital mineral in placers and sediments.

Brookite   8[TiO$_2$]

Orthorhombic   *P*   *mmm*   *a* 0·918 *b* 0·545 *c* 0·515

Structure: Ti^{4+} is co-ordinated to 6 O^{2-} in distorted octahedral arrangement, the octahedra linked by edges and corners into a framework similar in principle to those of anatase and rutile, but more complicated in its pattern.

Habit: Crystals tabular ||{010} or prismatic. Cleavage poor. *H* 5½–6. *D* 4·14. Lustre adamantine. Colour varying shades of brown, and red-brown to black. Streak white to grey or yellowish.

Optics: Translucent, yellowish-brown to dark brown, sometimes with colour zoning in hour-glass pattern. Well-known for its unique dispersion with optic axial plane ||100 for blue light and ||001 for red light. Optically +ve. Fig. 189 summarizes these properties.

Tests: Distinction from rutile may be difficult. Brookite is often tabular, but rutile forms paramorphs after brookite. The twinning of rutile is distinctive, as is its uniaxial character.

Occurrence: In low temperature veins in gneiss and schist (Alpine veins) with adularia, quartz, calcite, chlorite, anatase and rutile. Sometimes as an accessory in igneous and metamorphic rocks. Fairly common as a detrital mineral.

Pyrolusite   2[MnO$_2$]

Tetragonal *P*   4/*mmm*   *a* 0·440 *c* 0·287

Structure: Rutile type.

Habit: Usually reniform, concretionary, in powdery

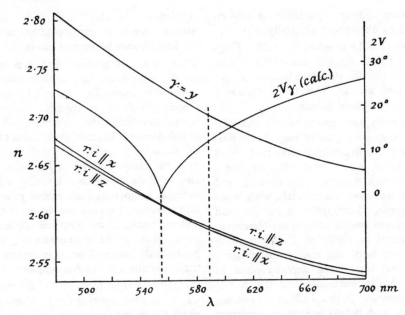

Fig. 189. The optical properties of brookite, showing the effect of extreme dispersion in changing the orientation of the optic axial plane with change in the colour of the light.

form, or as dendritic markings on joints. May be massive, fibrous. Rare, well-formed prismatic crystals are called *polianite*.

Cleavage perfect {110}. Fracture uneven. *H* crystals 6–6½; massive 2–6. Powdery material soils the fingers. *D* crystals 5·06; massive 4·4–5·0. Lustre metallic to dull and earthy. Colour black, or blue-black. Streak black or blue-black.

Optics: Opaque, creamy-white in reflected light. Anisotropic, *R*% 30–42.

Tests: A very small amount fused in a borax bead gives an amethyst colour in the reducing flame. The Na_2CO_3 bead is opaque, blue-green. Distinction from psilomelane may be difficult (see under Varieties, below).

Occurrence: Formed under oxidizing conditions, perhaps partly by the agency of bacteria. It is precipitated as nodules on the sea bed, in shallow waters, and in bogs and lakes. It also occurs as nodules and layers in residual clays from the weathering of manganese-bearing minerals and may be concentrated by circulating near-surface waters along fissures or into concretionary masses. Important deposits occur in the Ukrainian S.S.R. (Nikepol), in the Caucasus, India, Ghana and S. Africa, Cuba and Brazil.

Varieties: Pyrolusite is β-MnO_2. Because of the difficulty of separating the black oxides of manganese

pyrolusite, psilomelane and *manganite*, the general name *wad* is used to cover them all and many deposits are mixtures of them. Much of the material is colloidal and various elements (Co, Cu, Fe, Pb, W, Al and alkali metals) may be adsorbed upon it.

Use: An ore of manganese. Manganese is extensively used in steel-making, chiefly to remove oxygen and sulphur, but also as an alloying metal (in small amount) to improve the properties of the steel.

Cassiterite   2[SnO_2]

Tetragonal *P* 4/*mmm*   *a* 0·474 *c* 0·319

Structure: Rutile type.

Habit: Short prismatic crystals; sometimes long prismatic with steep pyramidal terminations. Also in radially fibrous concretionary masses (wood tin). Twinning very common on {011} producing well-known knee-twins (Fig. 72) and sometimes more complicated repeated twins. Cleavage {100}, imperfect in hand specimen. *H* 6–7. *D* 6·99; may be as low as 6·1 for fibrous masses. Lustre adamantine to submetallic, splendent. Colour reddish brown to brownish black. Sometimes red. Rarely yellow or colourless. Streak grey-white to brownish.

Optics: Translucent, colourless to yellow, reddish or brown, often with colour zoning; rarely dichroic. Cleavage ‖ length. Often twinned. ω 2·006, ε 2·097,

$\varepsilon - \omega$ 0·093, giving high-order interference colours, usually masked by absorption. Optically +ve.

Tests: Infusible, hardly attacked by acids. Fragments placed in a zinc dish with dilute HCl become coated with grey metallic tin which becomes bright when rubbed with a cloth. This test is especially useful on rounded alluvial grains.

Occurrence: In veins with quartz, within and near granites. The associated granite may be hydrothermally altered to a quartz-muscovite rock called *greisen*. Tourmaline, topaz, muscovite, lepidolite, fluorite, wolframite, arsenopyrite, molybdenite, and native bismuth are associated minerals, with lesser amounts of pyrite, chalcopyrite, sphalerite and galena. The deposits associated with the Hercynian granites of Europe especially in Cornwall, were the major source until the nineteenth century. Important alluvial and eluvial deposits occur in Thailand, the Malay Peninsula and the Sunda Islands of Indonesia. Bolivia, China (Yunnan), Nigeria, Congo, and Burma are other important producers. W. Australia, Tasmania and Japan also have deposits. Though small occurrences are known, it is remarkable that no important deposits are known in the N. American continent.

Use: Cassiterite is the only abundant tin mineral, and is virtually the sole source of tin. Minor tin minerals include *stannite* Cu_2FeSnS_4 *teallite* $PbSnS_2$, *frankeite* $Pb_5Sn_3Sb_2S_{14}$ and *cylindrite* $Pb_3Sn_4Sb_2S_{14}$.

Uraninite (Pitchblende) $4[UO_2]$
Cubic F $m3m$ a 0·547

Structure: Fluorite type. Excess oxygen is common, leading to a reduction of cell edge.

Habit: In octahedra, cubo-octahedra or cubes. Pitchblende is massive, or in colloform aggregates (botryoidal, reniform).

Fracture uneven to conchoidal. Brittle. H 5–6. D 8–10 for crystals, 6·5–8·5 for colloform material. Lustre sub-metallic (crystals), pitch-like or greasy and dull (pitchblende). Colour black and brownish black. Streak brownish-black or greenish-black. Radioactive emitting both α and β particles and decaying ultimately to Pb^{206} or Pb^{207} and He.

Optics: Usually opaque. In reflected light, light grey with a brownish tone, isotropic. $R\%$ 17. Almost as hard in polishing as hematite.

Tests: Infusible, M.P. 2176. Slowly soluble in HNO_3 and conc. H_2SO_4. The best test is to scan for radioactivity with a Geiger–Muller counter.

Alteration: Besides radioactive disintegration, natural material is partly oxidized to U_3O_8. Both Th and Pb may be leached out of the mineral.

Occurrence: In granite- and syenite-pegmatites, usually as crystals. Associated minerals are zircon, tourmaline, monazite, mica and feldspar (Bancroft, Ontario; Villeneuve, Quebec; Norway). In high temperature veins, with cassiterite and its associates as colloform pitchblende, as in Cornwall. In the form of pitchblende in mesothermal Co-Ni-Ag veins, as at the classic locality of Joachimsthal, Bohemia and at Great Bear Lake, Canada. The associated minerals are pyrite, chalcopyrite, galena, skutterudite minerals, bismuth and silver, barytes, fluorite and carbonate. In mesothermal sulphide veins with pyrite, chalcopyrite, sphalerite and galena, again as pitchblende. As small grains in ancient placers as in the Witswatersrand banket conglomerates.

Varieties: It often contains Th, and the Rare Earths Ce, Y etc., substituting for U. *Gummite* is a sackname for mixed uranium oxides with Pb, Th, and an indefinite amount of water. It comprehends a number of oxides of uncertain formula in which heavy metal atoms are associated with uranium. *Thorianite*, ThO_2, isostructural with uraninite, is a well-recognised species found mainly in gem gravels and placer deposits.

Columbite—tantalite
                   $4[(Fe, Mn)(Nb, Ta)_2O_6]$
Orthorhombic P mmm a 0·508 b 1·424 c 0·573

Structure: The metal atoms are co-ordinated like Ti in rutile and other TiO_2 polymorphs, the RO_6 octahedra being linked edge to edge in chains that are cross-linked by sharing octahedral corners: only the details of linkage are different. In columbite the chains run along y-axis.

Habit: Varied, prismatic, equant or tabular $\|\{010\}$, in groups. Also massive.

Twins on $\{201\}$ common giving heart-shaped contact twins. Cleavage $\|\{010\}$ distinct. Fracture sub-conchoidal, brittle. H 6, for columbite with Nb to $6\frac{1}{2}$ for tantalite with Ta. D 5·2 (columbite) to 7·95 (tantalite), linearly related to Ta_2O_5 wt %. Lustre sub-metallic to resinous. Colour black to brownish black with iridescent tarnish. Streak dark red to black.

Optics: Red, reddish yellow to brown, sometimes pleochroic. Refractive indices, for example $\alpha = x$ 2·26 $\beta = y$ 2·32 $\gamma = z$ 2·43, $\gamma - \alpha$ 0·17; but variable. $2V_\gamma$ large. In reflected light grey-white with a brown tone. Anisotropic $R\%$ 16–18. Shows red-brown internal reflections. One of the hardest ore minerals in polishing, harder than hematite.

Tests: Fusible only with difficulty. Chemical tests on solutions after fusion may be needed. Crystal form, often with rectangular outlines, distinguishes it from black tourmaline. Cleavage is much poorer than in wolframite.

Occurrence: A mineral of granite pegmatites, and the most widespread Nb–Ta mineral. Associated with albite, lepidolite, spodumene and petalite, and with tourmaline, apatite, cassiterite, wolframite, fluorite and topaz. A particular occurrence is in the cryolite of Ivigtut, Greenland. It also occurs as a detrital mineral.

Related minerals: The *euxinite-polycrase* series (Ca, R.E., U, Th) (Nb, Ta, Ti)$_2$ (O, OH)$_6$ and *samarskite* with even more substituents are structurally and physically similar. These minerals are often metamict and isotropic. They too are found in granite pegmatites with zircon, monazite, xenotime and allanite. As with other radioactive minerals, expansion and cracking of surrounding minerals may accompany metamictization. *Tapiolite* (FeMn) Ta$_2$O$_6$, though tetragonal, is closely comparable also in structure, physical properties and environment of crystallization.

Pyrochlore—microlite

8[(NaCa)$_2$(Nb, Ta)$_2$O$_6$(O, OH, F)]

Cubic F $m3m$ $\qquad a \simeq 1{\cdot}04$

Structure: (Na, Ca) are co-ordinated in cubic arrangement with (O, OH, F) while Nb, Ta are octahedrally co-ordinated.

Habit: Crystals octahedral with subordinate {011}, {100}.

Cleavage {111} seen in thin section. H 5–5½. D 4·2 (pyrochlore with Nb) to 6·4 (microlite with Ta). Lustre vitreous or resinous. Colour brown to black (pyrochlore), pale yellow to brown (microlite). Streak light brown or yellow.

Optics: Colourless to brown, sometimes varying in zones. Isotropic, $n = 1{\cdot}93$–2·18.

Tests: Often radioactive. Because of this it may be metamict and give poor X-ray reflections. Heating may improve the pattern and, by driving off water, raise both D and n. Pyrochlore may turn green on heating.

Occurrence: In the pegmatites of alkaline rocks as in nepheline syenites of Larvik, Norway, in nepheline-rich rocks of Kola Peninsula U.S.S.R. and of Julianshaab, S.W. Greenland, and with carbonatites in various places. Also in alkali granites in Nigeria. It is associated with zircon, apatite, soda pyroxenes and amphiboles and a suite of unusual Zr, Ti, Nb, Ta and Rare Earth minerals.

Varieties: Varieties with U, Th, Ti, Ce, Y, other Rare Earths and La are described under many names. The presence of these substituents is the rule rather than the exception. *Koppite* with Ce and Fe is one of the better known variants. *Fergusonite* in which rare earths wholly replace (Na, Ca) is tetragonal: its formula is Y(Nb, Ta) O$_4$.

Brucite

1[Mg(OH)$_2$]

Trigonal C $\bar{3} m$ $\qquad a$ 0·315 c 0·477

(hexagonal axes)

Structure: Mg^{2+} ions are octahedrally co-ordinated to (OH)$^-$, the octahedra sharing edges to form a layer (Fig. 190b). Successive layers are superposed with (OH)$^-$ in hexagonal close-packing and are held together by weak residual bonding forces between adjacent sheets of (OH)$^-$ ions at the surfaces of the layers.

Habit: Usually massive foliated: crystals when they occur are thin {0001} tablets.

Cleavage perfect ||{0001}, the flakes flexible. Sectile. H 2½. D 2·39. Lustre pearly on cleavage surfaces. Colour white, to pale green or grey blue. Streak white.

Optics: Colourless in transmitted light. Usually appears as a felted mass of curving fibres (actually scales), of low to moderate relief. Often pseudomorphous after equant periclase. ω1·560, ε1·580, $\omega - \varepsilon$ 0·02. Optically +ve. Interference colour orange-yellow, but an anomalous reddish brown due to large variation of birefringence with wavelength is often shown in thicker sections.

Tests: Glows brilliantly before the blowpipe but is infusible. Moistened with cobalt nitrate solution and reheated turns pink.

Occurrence: A low-temperature alteration product. Most common as the alteration product of periclase in metamorphosed dolomitic limestones, where it may give the marble a rasp-like weathered surface. It occurs thus near Ledbeg in Sutherland. Occurs in some serpentinites as at Hoboken, New Jersey, the original locality, and at Swinna Ness, Unst, Shetland.

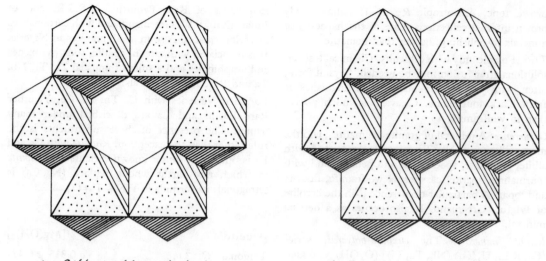

(a) Gibbsite (dioctahedral)   *(b) Brucite (trioctahedral)*

Fig. 190. Contrast between gibbsite and brucite layers.

Gibbsite (Hydrargillite) $8[AlOH_3]$

Monoclinic $P\ \ 2/m$ $a\ 0.862\ \ b\ 0.506\ \ c\ 0.970$
                                              $\beta 85°\ 26'$

Structure: Al^{3+} is octahedrally co-ordinated to $(OH)^-$ and octahedra share edges to form a layer as in brucite, save that only 2/3rds of the possible octahedral spaces contain Al^{3+} (Fig. 190a). This is in accord with its valency compared with Mg^{2+}. The layer is said to have a dioctahedral cation population while brucite is trioctahedral. In gibbsite successive layers are superposed with $(OH)^-$ immediately above $(OH)^-$ in the layer below. This contrasts with brucite, where $(OH)^-$ nests between $3(OH)^-$ in the layer below.

Habit: Usually in foliated masses, earthy, or as pisolitic and concretionary aggregates. Crystals tabular $\|\{001\}$ with pseudohexagonal outlines. Twins on $\{310\}$ with $\{001\}$ as composition plane are common. Cleavage $\{001\}$ perfect. $H\ 2\frac{1}{2}-3\frac{1}{2}$. $D\ 2.40$. Lustre pearly on cleavage surfaces. Colour white, grey or greenish, often stained yellow.

Optics: Transparent, colourless, with moderate relief. $\alpha = y\ 1.568$, $\beta\ 1.568$, $\gamma \wedge z\ 21°\ 1.587$, $2V_y\ 0°$ (effectively uniaxial positive), $\gamma - \alpha\ 0.019$. Interference colours rise to 1st order red, 2nd order blue. Elongation of fibres $+$ve or $-$ve, extinction inclined.

Occurrence: Principally as a constituent of bauxite clays produced by tropical or sub-tropical weather-ing. An alteration product of feldspar and corundum. Also as a low temperature mineral of cavities and veins.

Use: Constitutes part of the ore of aluminium.

Diaspore $4[AlO(OH)]$

Orthorhombic $P\ \ mmm$ $a\ 0.440\ \ b\ 0.943\ \ c\ 0.284$

Structure: O^{2-} and $(OH)^-$ taken together are in hexagonal close-packing with Al^{3+} in octahedral spaces between them. The arrangement is close to that of olivine with the Si^{4+} (tetrahedral) positions vacant and a change in the position of two cations that halves the c-dimension (Fig. 191).

Habit: Usually in disseminated scales, foliated masses, or concretionary forms. Crystals are thin $\{010\}$ plates. Cleavage $\{010\}$ perfect. Fracture conchoidal. Very brittle. $H\ 6\frac{1}{2}-7$. $D\ 3.4$. Lustre pearly, shining on cleavage faces. Colour grey, white or colourless, also pink.

Optics: Transparent, colourless or pale-blue, with high relief and good cleavage. $\alpha = z\ 1.702$, $\beta = y\ 1.722$, $\gamma = x\ 1.750$, $2V_y\ 84°$, $\gamma - \alpha\ 0.048$ giving interference colours of 3rd order (higher than those of olivine). Straight extinction, length-fast.

Occurrence: In metamorphosed aluminous rocks often with emery, chloritoid and spinel. More importantly, it is widespread in bauxites and laterites. It may also form by hydrothermal

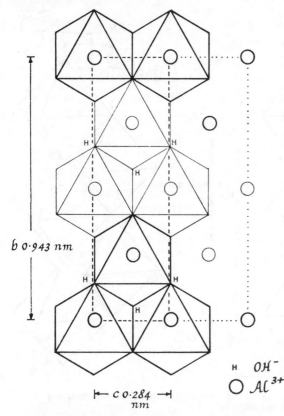

$b\ 0.943\ nm$

$\longmapsto c\ 0.284 \longrightarrow$
nm

H OH^-
O Al^{3+}

Fig. 191. The structure of diaspore. H indicates a hydroxyl group. Other octahedral corners are O^{2-}. A second cell is shown in dotted outline so that the two together may be compared with olivine, Fig. 210.

alteration of corundum and occurs with alunite in altered igneous rocks.

Use: As part of bauxite it is an ore of aluminium.

Goethite $\qquad$ 4[FeO(OH)]

Orthorhombic *P mmm* $a\ 0.464$ $b\ 1.00$ $c\ 0.303$

Structure: Diaspore type.

Habit: Variable. Good crystals rare, prismatic or scaly ||{010}. Usually massive with radially fibrous structure and botryoidal, stalactitic or concretionary form. Also earthy.
Cleavage {010} perfect. Fracture uneven. Brittle. $H\ 5–5\frac{1}{2}$. $D\ 4.28$, but varying to 3.3 in massive material. Lustre submetallic to silky (fibrous sorts) and dull. Colour dark brown to yellow. Streak yellow-brown.

Optics: Transmits red and yellow light; pleochroism is often not well marked. Optical orientation and properties vary rapidly with wavelength.

$\alpha = y\ 2.26$, $\beta = x$ in red, z in yellow light 2.39, $\gamma = z$ in red, x in yellow light 2.4, $\gamma - \alpha\ 0.14$, $2V_\alpha\ 0°$ (at $615\ nm$) to $27°$. Strong dispersion. In reflected light, grey, anisotropic $R\%\ 16–18$. Fairly hard in polishing, between pyrrhotite and magnetite.

Tests: Soluble in HCl. Becomes magnetic on heating in reducing flame.

Occurrence: One of the most widespread of all minerals. A principal constituent of *limonite* (which is a sack-name for all forms of hydrated ferric oxide) formed by the weathering or hydrothermal alteration of iron-bearing minerals. It is also deposited directly in bogs in high latitudes and in the past has precipitated extensively under marine conditions. It is a constituent of the iron-enriched layer or iron pan of the subsoil, and under tropical weathering constitutes part of laterite. It forms the gossan or iron hat at the outcrop of sulphide mineral veins.

Use: It forms an important part of the marine sedimentary *minette* iron ores of Alsace–Lorraine, and of the Quebec–Labrador iron ores. It is a minor constituent in many ores. As yellow ochre it is used as a pigment.

Lepidocrocite $\qquad$ 4[FeO(OH)]

Orthorhombic *A mmm* $a\ 0.387$ $b\ 1.251$ $c\ 0.306$

Structure: Fe^{3+} is octahedrally co-ordinated to O^{2-} the octahedra linked by their apices to form chains ||x-axis and the chains joined by sharing octahedral edges to form corrugated sheets ||{010} (Fig. 192b). The sheets are laced together by hydrogens (H in Fig. 192a) between pairs of O^{2-}, the O^{2-} participating in this bonding being half of those in the structure. (These can equally well be regarded as $(OH)^-$ groups.)

Habit: Scaly crystals flattened ||{010}; massive, bladed or micaceous. Cleavage perfect {010} parallel to structural layers, {100} and {001}. Brittle. $H\ 5$. $D\ 4.0$. Lustre submetallic. Colour red to red-brown. Streak dull orange.

Optics: Translucent, shades of orange-red, pleochroic. $\alpha = y$ yellow to colourless 1.94, $\beta = z$ orange to yellow 2.20, $\gamma = x$ orange to yellow 2.51, $\gamma - \alpha\ 0.57$, $2V_\alpha\ 83°$. In reflected light grey-white pleochroic and anisotropic, $R\%\ 16–25$, with red internal reflections. Fairly hard in polishing, between pyrrhotite and magnetite.

H

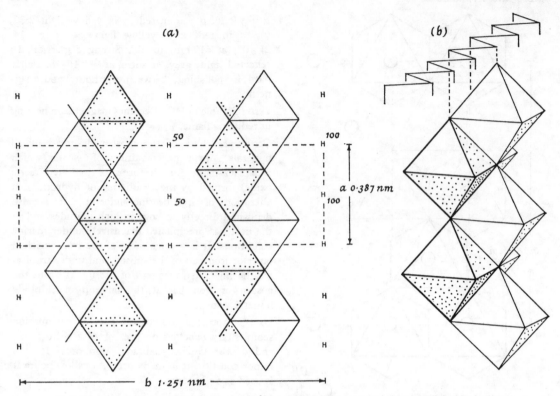

Fig. 192. Structural pattern of lepidocrocite. $Fe^{3+}O_6$ octahedra in corrugated sheets are linked by hydrogen ions at positions marked H.

Occurrence: As for goethite and associated with it. Perhaps because of the lack of attention given to iron hydroxides, it has not been widely reported. Good crystals come from Eizenzeche Mine, Seigen, Germany. It occurs in the Lake Superior iron ranges.

Boehmite   4[AlO(OH)]

Orthorhombic *A mmm*   *a* 0·370   *b* 1·223   *c* 0·287

Structure: Lepidocrocite-type.

Habit: Usually disseminated flakes or concretionary nodules.

Cleavage {010} perfect. *H* 3½–4. *D* 3·1. Colour white.

Optics: Transparent, colourless, with good relief. $\alpha = y$ or z 1·64, $\beta = y$ or z 1·65, $\gamma = x$ 1·66, $\gamma - \alpha$ 0·02, $2V_\gamma \simeq 80°$. Optical properties not well known, because of small size of crystals.

Tests: Moistened in cobalt nitrate solution and heated in the blow pipe flame it turns blue (a general test for aluminous minerals).

Occurrence: Though poorly characterized and requiring X-ray methods for certain identification, boehmite is an important mineral as a major con-

stituent of bauxite. It is hence a principal ore of aluminium. It forms with gibbsite and diaspore by prolonged chemical weathering and leaching of aluminous silicates under tropical or sub-tropical conditions. Important producers of bauxite are France (the deposits of Le Baux result from weathering of impure limestone), Hungary, U.S.A., Guyana, Jamaica, U.S.S.R., Yugoslavia, and Australia.

Manganite   8[MnO(OH)]

Monoclinic *P 2/m*   *a* 0·884   *b* 0·523   *c* 0·574 $\beta 90°$

Structure: A hexagonal close-packing of O^{2-} and $(OH)^-$ when viewed along *x*, as in diaspore. The distribution of the octahedrally co-ordinated cations is, however, different from that in diaspore.

Habit: Columnar, longitudinally-striated crystals often grouped in bundles. Stalactitic.

Twinning on {011}. Cleavage {010} perfect. Fracture uneven. Brittle. *H* 4. *D* 4·38. Lustre submetallic. Colour dark grey to black. Streak reddish brown to black.

Optics: Hardly translucent. In reflected light grey-white with a brown tone, weakly pleochroic. Anisotropic $R\%$ 14–20, with blood-red internal reflections. Rather hard in polishing; a little softer than magnetite.

Tests: In the borax bead a very small amount gives violet colour in the oxidizing flame, colourless in reducing flame. The sodium carbonate bead is bluish green in the oxidizing flame. Soluble in HCl with evolution of Cl.

Occurrence: As a low-temperature vein mineral with barytes, calcite, and siderite. Also deposited by circulating waters with pyrolusite, psilomelane and goethite, and accumulated as a residual deposit.

Psilomelane $2[(Ba, H_2O)_2Mn_5O_{10}]$

Monoclinic A $2/m$ a 0·956 b 0·288 c 1·385
                                                    $\beta 92° 30'$

Structure: Uncertain whether it is monoclinic or orthorhombic but the main features are clear. The chains of octahedra $||x$-axis in the rutile-type structure of pyrolusite (MnO_2) become double chains which link to leave square-section tubes through the structure (Fig. 193). In *cryptomelane* (KMn_8O_{16}) and *hollandite* $(BaMn_8O_{16})$, both pseudotetragonal, K and Ba lie in these tunnels. In psilomelane Ba and H_2O share the tunnel sites.

Habit: Massive, botryoidal, stalactitic or earthy. Single crystals of separable size extremely rare.

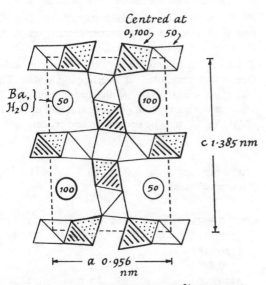

Fig. 193. Structure of psilomelane. Ba^{2+}, H_2O are in square-section tunnels through the octahedral network.

H 5–6. D 4·5. Lustre submetallic to dull. Colour dark grey to black. Streak brownish black or black.

Optics: Opaque. Grey white in reflected light. Weakly anisotropic. $R\%$ 23–24. Commonly intermixed with cryptomelane 28%, hollandite 27–34%. coronadite $(PbMn_8O_{16})$ 28–34% and pyrolusite (MnO_2) 30–42%, all of which are more reflective, and ramsdellite (MnO_2).

Occurrence: As for pyrolusite, and in association with it and the other Mn minerals just mentioned. Small amounts of Cu, Co, Ni, W, Ca, Na and K are adsorbed upon psilomelane or replace the Ba.

Use: An ore of manganese.

CARBONATES

The carbonates comprise some 69 minerals in which plane triangular $(CO_3)^{2-}$ anionic groups are linked together by intermediate cations. The principal variants arise from the entry of $(OH)^-$, F^-, Cl^-, $(PO_4)^{3-}$ or $(SO_4)^{2-}$ into the structure and from the occurrence of some of the compounds in various states of hydration, with loosely bound water molecules that may be driven off at relatively low temperatures. Most of the hydrous species are secondary products of alteration of other minerals by weathering or the action of meteoric waters.

The examples described here may be arranged in four groups.

Trigonal (Rhombohedral) Carbonates (Calcite Group)

Calcite	$CaCO_3$
Magnesite	$MgCO_3$
Siderite	$Fe^{2+}CO_3$
Rhodochrosite	$MnCO_3$
Smithsonite	$ZnCO_3$
Dolomite	$CaMg(CO_3)_2$
Ankerite	$Ca(Mg, Fe)(CO_3)_2$

Orthorhombic Carbonates (Aragonite Group)

Aragonite	$CaCO_3$
Witherite	$BaCO_3$
Strontianite	$SrCO_3$
Cerussite	$PbCO_3$

Monoclinic Carbonates with $(OH)^-$

Azurite	$Cu_3(OH)_2(CO_3)_2$
Malachite	$Cu_2(OH)_2CO_3$

Hydrated Carbonates

Thermonatrite	$Na_2CO_3.H_2O$
Trona	$Na_3H(CO_3)_2.2H_2O$

CALCITE GROUP $4[A^{2+}CO_3]$

Structure: This is the structure adopted by carbonates and nitrates of small cations (cf. Aragonite

Group). The unit cell is rhombohedral R and the point group $\bar{3}\,m$ for all except dolomite, which is $\bar{3}$. The true unit cell is a tall, thin rhombohedron containing two formula units; but the crystals cleave into a flatter rhombohedron which is highly characteristic and this cleavage form has long been used to define $\{10\bar{1}1\}$. To conform with this tradition

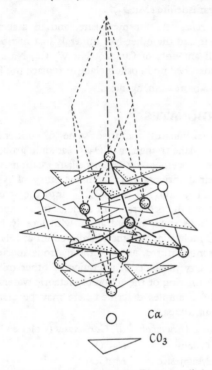

Fig. 194. The structure of calcite. The true cell shown in broken line.

a corresponding unit cell with 4 formula units is used here (Fig. 194). It may be regarded as an NaCl type arrangement of cations and $(CO_3)^{2-}$ anionic groups, in which one of the cube diagonals has been shortened. The $(CO_3)^{2-}$ groups lie with their planes normal to the triad axis. As with all

R crystals it is possible also to choose a hexagonal super-cell (a rhomb-based prism).

Complete solid solution occurs between $CaCO_3$ and $MnCO_3$ and between $MnCO_3$ and $FeCO_3$; there is a partial series from $CaCO_3$ towards $FeCO_3$ and $ZnCO_3$. There is little solid solution of Mg in $CaCO_3$ under ordinary conditions. $CaMg(CO_3)_2$ may contain Fe^{2+} up to $Fe^{2+}/Mg = 2\cdot6$; when this ratio exceeds 1 the mineral is called ankerite.

Habit: Calcite and dolomite are often well crystallized. All can occur massive in coarse- or fine-granular condition.

Calcite: Very variable. Commonly in flat rhombohedra $\{01\bar{1}2\}$ passing by rounding of faces into lensoid forms, or $\{01\bar{1}2\}$ combined with the prism to give 'nail-head spar', steep scalenohedra $\{21\bar{3}1\}$ called 'dog-tooth spar', and less commonly in prismatic forms (Fig. 22, p. 29). Stalactitic, concretionary, as spheroids a millimetre or so across (called ooliths from resemblance to fish roe), chalky.

Magnesite: Crystals usually rhombohedral $\{10\bar{1}1\}$ but rare. Commonly massive, coarse- or fine-grained, or compact porcellanous and chalk-like.

Siderite: Rhombohedral crystals, usually $\{10\bar{1}1\}$ but also $\{01\bar{1}2\}$, $\{02\bar{2}1\}$, etc. The faces are often curved and may be built up of overlapping scales. Massive, granular and cleavable, concretionary.

Smithsonite: Crystals rhombohedral $\{10\bar{1}1\}$ with rough curved faces; but usually botryoidal, reniform or stalactitic, encrusting, or as porous cavernous masses (called by the miners 'dry-bone ore').

Rhodochrosite: Crystals, which are rare, are rhombohedral $\{10\bar{1}1\}$. Usually massive, coarsely granular, or botryoidal and encrusting.

Dolomite: In rhombohedra $\{10\bar{1}1\}$ with rounded faces often composed of overlapping scales. The distortion often gives saddle-shaped crystals. In other cases rhombohedra of several slopes as $\{10\bar{1}1\}$, $\{40\bar{4}1\}$, $(02\bar{2}1\}$ may be present on one crystal.

	Rhombohedral cell $Z = 4$		Axial ratio $a:c$ (Hexagonal axes)	$0001 \wedge 10\bar{1}1$	$10\bar{1}1 \wedge \bar{1}101$
	a	α			
Calcite	0·641	101° 55′	1:0·8543	44° 37′	74° 55′
Magnesite	0·606	102° 58′	1:0·8112	43° 08′	72° 36′
Siderite	0·602	103° 05′	1:0·8184	43° 23′	73° 00′
Rhodochrosite	0·601	102° 50′	1:0·8183	43° 22′	73° 00′
Smithsonite	0·593	103° 27′	1:0·8063	42° 58′	72° 20′
Dolomite	0·618	102° 50′	1:0·8322	43° 52′	73° 45′

Twin-plane	Not recognizable optically			Recognizable in thin section		
	{0001}	{10$\bar{1}$0}	{11$\bar{2}$0}	{10$\bar{1}$1}	{01$\bar{1}$2} lamellar	{02$\bar{2}$1} lamellar
Calcite	common	not found	not found	rare	v. common	rare
Dolomite	common	common	common	rare	not found	common

In thin section dolomite often forms well-shaped rhombs. This distinguishes it from associated calcite which is generally in irregular grains.

Ankerite: Similar in habit to dolomite.

Cleavage {10$\bar{1}$1} perfect in all members except smithsonite in which it is a little less good. Calcite may have a parting parallel to the {01$\bar{1}$2} twinning, and dolomite one parallel to {02$\bar{2}$1}.

Twinning: Unrecorded or rare except in calcite and dolomite where it is extremely common. The twins of calcite and dolomite are tabulated above.

In a cleavage rhombohedron of calcite, or a rhomb-shaped section near this, {01$\bar{1}$2} twins are seen as lamellae parallel to the long diagonal of the rhomb or parallel to its edges. In a similar section of dolomite {02$\bar{2}$1} twins are seen as lamellae parallel to both the short and long diagonals of the rhomb. Herein lies a distinction between the two species. In sections with well-defined intersecting lamellae the fast vibration direction lies in the obtuse angle of intersection in calcite and in the acute angle in dolomite.

Lustre: Vitreous, inclining to pearly on cleavages, especially in dolomite. Magnesite may be porcellanous and dull in compact varieties.

Colour: Calcite: Usually colourless or white: may be brownish, red, blue, green or black.

Magnesite: Colourless, white, greyish or yellowish brown.

Siderite: Yellowish brown or brown.

Rhodochrosite: Pink of various shades.

Smithsonite: Grey-white to dark grey; also apple-green or bluish green.

Dolomite: Colourless, or white to cream and yellow brown. Sometimes pale pink. Weathers brown.

Ankerite: White, yellow, yellow-brown.

Streak: White throughout.

Fluorescence: Sometimes very marked in calcite which may also be phosphorescent after irradiation by ultraviolet rays, X-rays or cathode rays. Calcite also glows on heating and may be triboluminescent. Magnesite and smithsonite may fluoresce bluish or greenish, rhodochrosite pink or red and dolomite various shades in ultraviolet light. Dolomite is triboluminescent.

Optical and other physical properties: In conformity with the arrangement of plane anionic groups normal to z-axis the light vibrating $||z$ is less retarded than that $\perp z$, $\varepsilon < \omega$ and the minerals are optically negative with high birefringence. The rapid change in relief on rotation in plane polarized light gives the appearance called 'twinkling'. The interference colour is the blend of pale colours called the high-order white.

These properties, tabulated below, vary linearly through the solid solution series mentioned above.

Tests: All are infusible save siderite, which blackens before the blowpipe, becomes magnetic and fuses on strong blowing. Ankerite may darken and become magnetic. Calcite and dolomite glow brightly. Calcite dissolves readily with effervescence in cold dilute HCl. Dolomite, ankerite, magnesite, rhodochrosite and smithsonite effervesce in warm

	ω	ε	$\omega - \varepsilon$	D	H
Calcite	1·658	1·486	0·172	2·71	3
Magnesite	1·700	1·509	0·191	3·00	4
Siderite	1·873	1·633	0·240	3·96	4
Rhodochrosite	1·816	1·597	0·219	3·70	3½–4
Smithsonite	1·848	1·621	0·227	4·43	4–4½
Dolomite	1·679	1·500	0·179	2·85	3½–4
Ankerite	1·72	1·53	0·19	3·02	3½–4
Fe/Mg = 1					

	Calcite			Dolomite		Ankerite	Siderite
	Fe^{2+} free	Fe^{2+} poor	Fe^{2+} rich	Fe^{2+} free	Ferroan		
Soln A	pink	pink	pink	nil	nil	nil	nil
Soln B	nil	light blue	dark blue	nil	light blue	dark blue	nil (unless acid is boiling)
A + B	pink	Royal blue	purple mauve	nil	turquoise	dark turquoise	nil

dilute HCl, and siderite in the hot acid. Smithsonite heated on charcoal, moistened with cobalt nitrate solution, and reheated turns green on cooling.

Staining: A number of methods are available for differential staining of calcite and dolomite, and other carbonates. One method is to make up two solutions in 1–2 vol. % HCl (A) of 0·2 g Alizarin Red S and (B) of 2 g K ferricyanide per 100 ml. These may be applied separately or mixed 1:1. The mineral thin-section or powder is immersed in one or another of these solutions for 35–45 secs. and then washed in distilled water. The table above shows the results. (For further detail see Warne, S. St. J., 1962. *J. Sed. Petrol.*, **32**, 29–38.)

X-ray: Calcite gives a powder diffraction line at $d = 0·303$ nm, dolomite one at $d = 0·288$ nm. The relative proportions of calcite and dolomite in mixtures can be estimated from the relative intensities of these two lines, calibrated against known mixtures of the two minerals.

Occurrence:

Calcite is widespread as limestone, formed from sea shells or chemically precipitated from sea water. It precipitates also as travertine by evaporation of water in streams and lakes, and as nodules in sedimentary strata. Frequently deposited from late magmatic solutions filling vesicles and fissures. An important and widespread gangue mineral of hydrothermal veins. Limestone beds traversed by veins often act as an important precipitating agent for other minerals because of the reactive nature of calcite, and veins may expand into tabular 'flats' in limestone.

Magnesite occurs with talc and serpentine as an alteration product of peridotites, as on the Isle of Euboea and Macedonia in Greece. Also as a replacement of earlier dolomite or limestone by the agency of hydrothermal waters. Large deposits of this kind occur in Styria, Austria; at Satka, Ural

Mountains; in Manchuria and Korea; in Quebec; and in Paradise Mountains, Nevada.

Siderite: A constituent of marine oolitic iron ores as in the Mesozoic of England, Belgium, Luxembourg and N. France. As nodules and concretionary layers in shales, particularly in the Coal Measures of England where it constituted the once-important black-band iron ores. As bog iron ore in lakes and swamps of high latitudes. A widespread gangue mineral of hydrothermal veins.

Smithsonite: A secondary mineral of the oxidized zone of hydrothermal veins, derived from primary sulphides, especially sphalerite. It is often associated with malachite, azurite, cerussite, anglesite, pyromorphite and mimetite.

Rhodochrosite: Forms a gangue mineral in low to moderate temperature hydrothermal veins, with other carbonates, fluorite, barytes, quartz and sulphides. Also as a metasomatic deposit associated with pegmatite or aplite veins, with rhodonite, garnet, tephroite, braunite (Mn_2O_3) and hausmannite ($MnMn_2O_4$). As a secondary mineral in residual manganese deposits.

Dolomite occurs as bedded magnesian limestone either chemically precipitated or formed by alteration of earlier calcite limestone. A widespread gangue mineral of hydrothermal veins.

Ankerite is widespread as a gangue mineral in lead-zinc sulphide veins. It is also found in veins and as infillings of fossils in Coal Measures shales and in the cleat of the coal itself.

ARAGONITE GROUP $4[A^{2+}CO_3]$

Structure: Plane triangular $(CO_3)^{2-}$ groups lie within an octahedral arrangement of cations, the octahedra being joined face to face in columns $||z$-axis, and edge to edge with octahedra of neighbouring columns (Fig. 195). The $(CO_3)^{2-}$ groups do not lie at the centres of the octahedra, but at

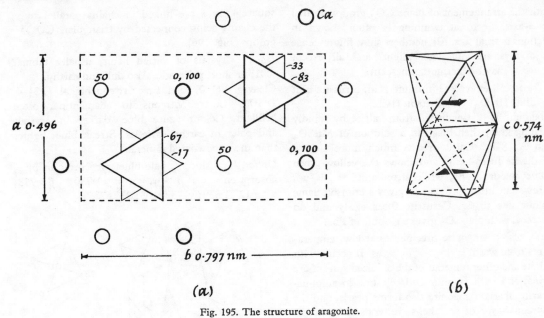

Fig. 195. The structure of aragonite.

$\frac{1}{3}$ or $\frac{2}{3}$ the height of the octahedron measured $\parallel z$ (Fig. 195b). This structure is favoured by the carbonates and nitrates of big cations: those of smaller cations adopt the calcite structure.

Habit: When crystalline short- or long-prismatic, dominated by pseudo-hexagonal forms which are often due to almost universal repeated twinning on {110} (see p. 69 and Figs 72, 73, 74). Also massive, in rounded forms.

Aragonite is usually twinned and in acicular or fibrous crystals. Also as concretions and crusts.

Witherite is always twinned with obscure forms and rough striated faces. Globular and mammillary forms common with coarse fibrous internal structure.

Strontianite forms acicular or spear-shaped crystals.

Cerussite differs in often being tabular $\parallel$ {001} or elongated along x-axis, as well as in pseudohexagonal pyramids. Often striated.

All are colourless in transmitted light, with marked variation in relief on rotation in plane polarized light. The strong negative birefringence accords

	Aragonite	Witherite	Strontianite	Cerussite
Cleavage	{010} fair	{010} fair	{110} good	{110}, {021} fair, very brittle
H	$3\frac{1}{2}$–4	3–$3\frac{1}{2}$	$3\frac{1}{2}$	3–$3\frac{1}{2}$
D	2·94	4·29	3·72	6·55
Lustre	Vitreous	Resinous	Vitreous	Adamantine
Colour	Colourless, white	Colourless to grey	Colourless, grey, yellowish or greenish	Colourless or white to grey
α	$= z$ 1·530	$= z$ 1·529	$= z$ 1·520	$= z$ 1·804
β	$= x$ 1·681	$= y$ 1·676	$= y$ 1·667	$= y$ 2·076
γ	$= y$ 1·685	$= x$ 1·677	$= x$ 1·668	$= x$ 2·078
$\gamma - \alpha$	0·155	0·148	0·148	0·274
$2V_\alpha$	18°	16°	7°	9°

with the arrangement of plane CO_3 groups normal to z-axis. Sectorial twinning is often shown in sections normal to z. All members show fluorescence in X-rays and ultraviolet light, and all except cerussite may be thermoluminescent.

Tests: All effervesce in dilute acid, cerussite best in dilute HNO_3, the others in HCl.

Aragonite is distinguished from calcite by rapidly turning grey, then black in a solution of $MnSO_4$ and Ag_2SO_4. Calcite reacts much more slowly. Witherite fuses easily and shows the yellow-green flame colour of Ba, while strontianite swells up, throws out little sprouts and gives a crimson flame colour due to Sr. Cerussite fuses easily and on charcoal with Na_2CO_3 gives a globule of Pb.

Occurrence: Aragonite precipitates at low temperatures from warm springs or in caves. It occurs with calcite and other minerals of the oxidized zone of ore veins, and with sulphur and celestine in sulphur-bearing marls. Aragonite also forms pearls, and the nacreous layer of sea shells. It tends to invert to calcite and is hence less abundant than the latter.

Witherite is rather uncommon. It occurs in low temperature veins with barytes and galena. In Durham (South Moor Colliery) and S.W. Northumberland (Settlingstones and Fallowfield) on the periphery of the lead-zinc ore-field it forms workable deposits in association with barytes. It is here an ore of barium.

Strontianite: In low temperature veins with galena and barytes at Strontian, Argyllshire and in Weardale, Co. Durham. Veins in Cretaceous marls occur in Westphalia. Also with celestine in pockets in marls as at Yate, Gloucestershire, and sporadically as nests in limestone.

Cerussite is a typical mineral of the oxidized zone of lead veins. Fine specimens come from Broken Hill, N.S.W., Australia and Tsumeb, S.W. Africa. It occurs at Leadhills, Lanarkshire, in Durham, Derbyhsire and Cornwall. The association is with anglesite, pyromorphite, malachite, smithsonite and limonite.

Azurite $2[Cu_3(OH)_2(CO_3)_2]$

Monoclinic P $2/m$ $a\ 0\cdot500$ $b\ 0\cdot585$ $c\ 1\cdot035$
$\beta\ 92°\ 20'$

Structure: Approximately square groups of two O^{2-} and two $(OH)^-$ ions surround the Cu^{2+} ions. In one type of square the $(OH)^-$ ions are along a diagonal, in the other on two adjacent corners. The

square groups are linked in chains parallel to y, the chains being connected by triangular $(CO_3)^{2-}$ groups (Fig. 196).

Habit: Crystals of varied habit, usually tabular {001} or short prismatic. Also dull and earthy.

Cleavage $||\{100\}$ good. Fracture conchoidal. $H\ 3\frac{1}{2}$–4. $D\ 3\cdot77$. Lustre vitreous to adamantine, often brilliant. Colour azure blue, darker in crystals and paler in earthy varieties. Streak blue, lighter than the unpowdered mineral.

Optics: Translucent, pale blue, pleochroic in blue, absorption $\alpha > \beta > \gamma$. $\alpha = y\ 1\cdot730$; $\beta\ 1\cdot758$; $\gamma \wedge z\ 12°\ 1\cdot838$; $\gamma - \alpha\ 0\cdot108$; $2V_\gamma 67°$.

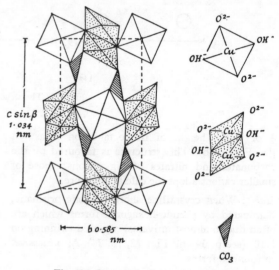

Fig. 196. Structural scheme of azurite.

Tests: As for malachite. The colour (and fine lustre of crystals) is distinctive.

Occurrence: In the oxidized zone of copper-bearing veins along with malachite and its associates. Important localities are Chessy, France (whence comes the alternative name *chessylite*), Tsumeb, S.W. Africa, Broken Hill, N.S.W., Australia, and Bisbee, Arizona.

Malachite $4[Cu_2(OH)_2CO_3]$

Monoclinic P $2/m$ $a\ 0\cdot948$ $b\ 1\cdot203$ $c\ 0\cdot321$
$\beta\ 98°$

Structure: Cu^{2+} is surrounded octahedrally by O^{2-} and $(OH)^-$, the octahedra of two types. One type has four O^{2-} and two $(OH)^-$ and the other two O^{2-} and four $(OH)^-$. Both types form chains, linked by octahedral edges, that run parallel to z. The chains are cross-linked by triangular $(CO_3)^{2-}$ groups, the corners of which are shared with octahedra (Fig. 197).

Habit: Almost always in mammillary or botryoidal form (colloform) with concentric banding and a radially fibrous structure. The very rare crystals are tufts and rosettes of twinned needles or prisms. Twins on {100}. Cleavage perfect {$\bar{2}$01}. H 3½–4. D 3·6–4·0. Lustre of fibrous aggregates silky to dull. Colour a distinctive bright green. Streak pale green.

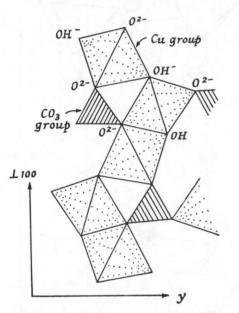

Fig. 197. Scheme of structural linkage in malachite.

Optics: Translucent green to yellow green. $\alpha \wedge z$ 23½° 1·655 very pale green, $\beta = y$ 1·875 yellowish green, γ 1·909 deep green, $\gamma - \alpha$ 0·254, $2V_\alpha$43°.

Tests: Fuses easily, colouring the flame green. Heated on charcoal gives metallic copper. Dissolves readily in dilute acids. The colour and colloform habit are distinctive.

Occurrence: Widespread as a secondary mineral in the oxidized zone of copper-bearing veins, associated with azurite, native copper, cuprite, tenorite, chrysocolla, limonite and wad. Famous localities for large masses are in the U.S.S.R. at Demidoff Mine, Nizhne-Tagilsk and at Sverdlovsk, both in the Ural Mountains.

Use: An ore of copper. Ornamental table-tops, vases, etc. cut from malachite emanated from Russia in the nineteenth century.

Thermonatrite   4[$Na_2CO_3.H_2O$]

Orthorhombic *P mmm*   *a* 1·072   *b* 0·647   *c* 0·525

Structure: Na^+ is in sixfold co-ordination with 5 O^{2-} of CO_3 groups and one $(OH)^-$. The CO_3 groups lie parallel to the *xz* plane.

Habit: As a crust or efflorescence. *H* 1–1½. *D* 2·26. Lustre vitreous. Colour grey-white or yellowish. Colourless when pure. Taste alkaline.

Optics: Transparent, colourless. $\alpha = y$ 1·420; $\beta = z$ 1·506; $\gamma = x$ 1·524; $\gamma - \alpha$ 0·104; $2V_\alpha$48°.

Tests: Soluble in water. On gentle heating loses water and falls to a powder of Na_2CO_3.

Occurrence: As a deposit from salt lakes and as an efflorescence in the soil of arid regions. It is associated with trona, halite and gypsum. It forms by dehydration at low temperature in dry air of *natron*, $Na_2CO_3.10H_2O$, which is usually found in solution in soda-lakes of arid areas, though it may occur also as crusts and efflorescences.

Trona (Urao)   4[$Na_3H(CO_3)_2.2H_2O$]

Monoclinic *C* 2/*m*   *a* 2·023   *b* 0·349   *c* 1·030
                                                 β106° 26′

Structure: Octahedrally co-ordinated Na^+ together with triangular $(CO_3)^{2-}$ anions form layers parallel to {100} separated by sheets of water molecules.

Habit: Fibrous or massive. Rarely in tablets flattened ||{001} and elongated in *y*. Cleavage perfect {100}. *H* 2½–3. *D* 2·14. Lustre vitreous, glistening. Colour grey white or yellowish. Colourless when pure. Taste alkaline.

Optics: Transparent, colourless. $\alpha = y$ 1·412; β 1·492; $\gamma \wedge z$ 83° 1·540; $\gamma - \alpha$ 0·128; $2V_\alpha$76°. γ nearly $\perp${001} cleavage.

Tests: Soluble in water. Effervesces in dilute acids.

Occurrence: As a deposit of salt lakes or an incrustation in the soil of arid regions. Associated minerals are halite, natron, and gypsum. A lava-flow composed of sodium carbonate minerals is reported from a volcano, Oldoinyo L'engai, in the East African Rift Valley.

NITRATES

Minerals of this group are structurally similar to the carbonates with plane triangular $(NO_3)^-$ groups in place of $(CO_3)^{2-}$; but because N is more electronegative than C the anionic group is less stable and the nitrate minerals are fewer in number and much more restricted in occurrence than carbonates.

They also yield oxygen readily on heating and so deflagrate (burst into flame) on heating in the presence of combustible substances.

Only one nitrate is described here.

Soda nitre   [NaNO$_3$]

Trigonal R $\bar{3}m$ a 0·507 c 1·682
                                              (hexagonal cell)

Structure: Calcite type, planar NO$_3$ groups playing the role of CO$_3$ in the carbonates.

Habit: In rhombohedra. Usually as an incrustation or impregnation in soils.

Twinning on {01$\bar{1}$2}, {0001} and {02$\bar{2}$1}. Cleavage {10$\bar{1}$1} perfect. Somewhat sectile. H 1½–2. D 2·27. Lustre vitreous. Colourless or coloured brown, yellow, etc. by impurities. Taste cooling.

Optics: Translucent, colourless: ω 1·587, ε 1·336, $\omega - \varepsilon$ 0·251, optically negative.

Tests: Deflagrates when heated on charcoal. Soluble in water.

Occurrence: As an impregnation in the soil of arid regions, along with gypsum, anhydrite, epsomite, halite and other salts. The principal occurrence is the *caliche* of the Atacama Desert along the eastern slope of the coast range of N. Chile. Formerly the main source of nitrate for fertilizers, the importance of the deposits has been diminished by the development of processes (the Haber process, cyanamide process, etc.) for direct fixation of atmospheric nitrogen.

TUNGSTATES AND MOLYBDATES

Though Mo and W both fall in Group VI B of the periodic table, have similar ionic radii, both form tetrahedral anionic groups WO$_4$ and MoO$_4$ and are both concentrated in the late fluids of granite magma, tungstates are relatively common while Mo occurs principally as the sulphide. Conversely, tungstenite WS$_2$ is an exceedingly rare mineral, recorded from only one locality.

The minerals considered here are

Scheelite	CaWO$_4$
Wolframite	(Fe, Mn)WO$_4$

Scheelite   4[CaWO$_4$]

Tetragonal I 4/m a 0·524 c 1·138

Structure: Independent WO$_4$ tetrahedral groups alternate with Ca^{+2} ions along lines parallel to

z-axis, the Ca^{2+} ion ½c from the centres of the tetrahedra above and below (Fig. 198).

Habit: Pyramidal crystals with {101} or {112}. Also massive, granular, or columnar.

Twinning common on {110}. Cleavage {101} good. Fracture uneven. Brittle. H 4½–5. D 6·1 decreasing with entry of Mo. Lustre vitreous to resinous.

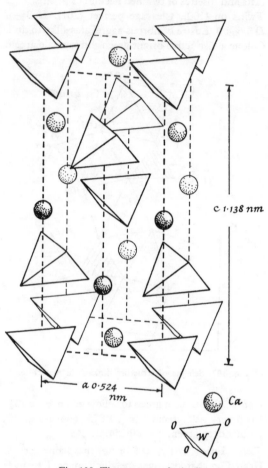

Fig. 198. The structure of scheelite.

Colour yellowish white to brownish. Streak white. Fluorescence bluish white in UV light λ254 nm. Thermoluminescent.

Optics: Translucent colourless ω 1·920, ε 1·937, $\varepsilon - \omega$ 0·017. Optically +ve.

Tests: Fused with difficulty before the blow-pipe. Decomposed by hot strong HCl or HNO$_3$, the yellow residue of tungstic oxide being soluble in ammonia. Responds to test for W with metallic tin (see under wolframite). The high density coupled with light colour is distinctive.

Occurrence: A high-temperature mineral found in quartz-veins and greisens close to granites; associated with wolframite, cassiterite, molybdenite, arsenopyrite, chalcopyrite, apatite, tourmaline, topaz, mica and fluorite (e.g. Carrock Mine, Cumberland, cassiterite veins of Cornwall). Also in contact metamorphosed limestones with garnet, diopside, vesuvianite, sphene, hornblende and epidote, as in the Humboldt Ra., Nevada. Occasionally in low-temperature veins with stibnite as at Bridge River, British Columbia (in serpentinite).

Varieties: Mo is often present, substituting for W, lowering the density and refractive indices. There is, however, no complete series to the rare mineral *powellite* $Ca(MoO_4)$.

Use: An ore of tungsten.

Related minerals: Isostructural with scheelite is the soft (H = 3) waxy orange or yellow *wulfenite*, $PbMoO_4$, a mineral of the oxidized zone of lead veins.

Wolframite $2[(Fe, Mn)WO_4]$

Huebnerite $2[MnWO_4]$

Ferberite $2[FeWO_4]$

Monoclinic P $2/m$ $MnWO_4$ a 0·484 b 0·576
(pseudo-orthorhombic) c 0·497 $\beta90°$ 53′
                $FeWO_4$ a 0·470 b 0·569
                c 0·493 $\beta90°$ 00′

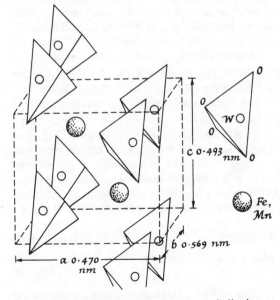

Fig. 199. A view of the wolframite structure indicating its affinity to scheelite.

Structure: Fig. 199 shows the distribution of the atoms in the pseudo-orthorhombic cell. W has been shown in a distorted tetrahedral group of O^{2-} to emphasize the relationship to scheelite, though the interatomic distances have been said to suggest a distorted octahedral co-ordination instead, with four near-neighbour O^{2-} and two more distant.

Habit: Crystals short prismatic, flattened ||{100}, which with {110} gives wedge-shaped cross-sections. Often massive and bladed. Twinning common on {100}. Cleavage {010} perfect. Fracture uneven, brittle. H 4–4½. D $MnWO_4$ 7·25 – $FeWO_4$ 7·60. Lustre submetallic, sparkling on cleavages, elsewhere dull. Colour reddish-brown, brownish black, dark greyish black, to iron-black. Streak reddish-brown to brown or brownish-black.

Optics: Translucent only in thin flakes, yellow, orange, red-brown, brown. Indices increase with Fe content.

Huebnerite $\alpha = y\,2·17,$ $\beta\,2·22,$
                $\gamma \wedge z$ 17–21° 2·32, $2V_\gamma \simeq 73°$.
Wolframite $\alpha = y\,2·26,$ $\beta\,2·32,$
                $\gamma \wedge z$ 17–21° 2·42, $2V_\gamma \simeq 78°$.

In reflected light grey-white, anisotropic $R\%$ 16–18, with red or red-brown internal reflections. Harder in polishing than magnetite, softer than hematite.

Tests: Wolframite and ferberite fuse easily, huebnerite with more difficulty, the globule being crystalline on cooling and magnetic in the case of wolframite but not in ferberite. Decomposed by aqua regia leaving yellow tungstic oxide. The mass from fusion on charcoal with sodium carbonate dissolves in HCl and gives a blue solution when boiled with a grain of metallic tin (test for W).

Occurrence: In quartz veins and greisens within or close to granites, where it is often associated with cassiterite, scheelite and their associates, and also in some high- and medium-temperature veins somewhat farther from the granite. It is not common however in the contact-metamorphosed limestones that carry scheelite. Wolframite is concentrated in eluvial (hill-wash) deposits with cassiterite and is produced as a mixed concentrate with the latter. Important deposits occur in China (Kiangsi, Kwangtung), Burma, Bolivia (La Paz District) and Portugal (Panasqueira, Beira District). Wolfram has been mined in Cornwall and at Carrock Mine, Cumberland.

Use: The principal ore of tungsten, used in lamp filaments, spark plugs, etc., and in high-speed tool steels.

PHOSPHATES, ARSENATES AND VANADATES

A large group of some 240 species many of which are of very limited occurrence. They occur over a wide range of environment from magmatic conditions to those of surface weathering, but a majority are low temperature minerals and many are hydrous. They are characterized structurally by independent $(PO_4)^{3-}$ $(AsO_4)^{3-}$ or $(VO_4)^{3-}$ anionic groups linked through intermediate cations: some have links with the sulphates by substitution of the $(SO_4)^{2-}$ group.

The following are described here:

Xenotime	YPO_4
Monazite	$CePO_4$
Vivianite	$Fe_3(PO_4)_2.8H_2O$
Erythrite	$Co_3(AsO_4)_2.8H_2O$
Amblygonite	$LiAl(OH,F)PO_4$
Apatite	$Ca_5(PO_4)_3(F,Cl,OH)$
Pyromorphite	$Pb_5(PO_4)_3Cl$
Mimetite	$Pb_5(AsO_4)_3Cl$
Vanadinite	$Pb_5(VO_4)_3Cl$
Turquoise	$CuAl_6(PO_4)_4(OH)_8.4H_2O$
Wavellite	$Al_3(PO_4)_2(OH)_3.5H_2O$
Torbernite	$Cu(UO_2)_2(PO_4)_2.8-12H_2O$
Autunite	$Ca(UO_2)_2(PO_4)_2.10-12H_2O$
Carnotite	$K_2(UO_2)_2(VO_4)_2.3H_2O$

Xenotime
$4[YPO_4]$

Tetragonal I $4/mmm$ $a\ 0.688$ $c\ 0.598$

Structure: Zircon type.

Habit: Prismatic crystals terminated by the pyramid {101}.

Cleavage perfect {100}. Fracture uneven and splintery. H 4–5. D 4.4–5.1. Lustre vitreous to resinous. Colour yellow-brown to reddish brown. Streak pale brown, yellowish or reddish.

Optics: Translucent, colourless to pale yellow or brown ω 1.720–1.721, ε 1.816–1.827, $\varepsilon - \omega$ 0.096–0.106. Optically +ve.

Tests: Much like zircon, but softer. Refractive indices lower but birefringence greater. Unattacked by acids but if the bead formed by fusion with sodium carbonate is dissolved the addition of ammonium molybdate solution gives an abundant yellow precipitate (ammonium phosphomolybdate).

Occurrence: An accessory mineral in pegmatite veins and some granites and syenites, associated with zircon, monazite, rutile and members of the pyrochlore group. Also found as detrital grains in alluvial deposits.

Varities: Other Rare Earths and Th, U and Zr may replace part of the Y. When radioactive it may become metamict.

Monazite
$4[(Ce, La)PO_4]$

Monoclinic P $2/m$ $a\ 0.676$ $b\ 0.700$ $c\ 0.644$
$\beta108°\ 38'$

Structure: Details uncertain.

Habit: Small crystals flattened || {100}, equant or wedge-shaped. Also massive and in detrital grains. Twinning on {100} common. Cleavage {100}, {010} variable in distinctness. Parting || {001}. Fracture conchoidal to uneven. Brittle. H 5–5½. D 5.0–5.2. Lustre resinous. Colour yellowish-brown, reddish-brown. Streak white. Fluorescence bright green in unfiltered UV light.

Optics: Translucent, yellow-brown to colourless. $\alpha = y$ 1.790, β 1.791, $\gamma \wedge z$ 2–6° 1.844, $\gamma - \alpha$ 0.054, $2V_y 15°$.

Tests: Infusible but turns grey before the blow-pipe. Only slowly dissolved by acids. May be tested for phosphate as under xenotime. Often metamict, requiring to be heated to produce an X-ray pattern.

Occurrence: An accessory mineral of some granites and syenites and their pegmatites. Concentrated in alluvial sands which are the chief source. Well-known localities are in S. Norway (Arendal, Risor), Madagascar, and commercially in the sands of Travancore, India, in Ceylon and in Brazil.

Varieties: The Rare Earths Nd, Pr, Dy and others substitute for Ce the ratio Ce : (La + R.E.) $\simeq$ 1 : 1. Y and Th are also present, the latter commonly to the amount of 3–9% and up to 30% ThO_2.

Use: A principal source of Ce and other Rare Earths, and of Th. Ce is used in lighter 'flints' and Th used to be used in incandescent gas-mantles, and is still employed with W in some lamp filaments. Monazite can be used for radiometric age determinations by the measurement of the Pb/Th + U ratio.

Vivianite
$2[Fe_3(PO_4)_2.8H_2O]$

Monoclinic C $2/m$ $a\ 1.008$ $b\ 1.343$ $c\ 0.470$
$\beta104°\ 30'$

Structure: Fe in octahedral spaces between H_2O oxygens and PO_4 oxygens these octahedra being

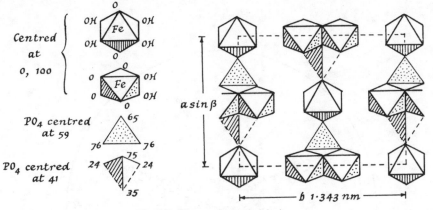

Fig. 200. The structure of vivianite.

joined into sheets parallel to {010} by the PO_4 groups (Fig. 200). A fairly open structure.

Habit: Crystals prismatic. Also reniform or globular with divergent fibrous structure or earthy and encrusting.

Cleavage {010} highly perfect, reflecting the structure. Flexible in thin flakes, sectile. $H\ 1\frac{1}{2}$–2. D 2·68. Lustre vitreous, pearly on {010} cleavage. Colourless or white when freshly exposed, becoming pale blue or greenish blue on exposure to light, due to oxidation of Fe^{2+}. Streak colourless or blue-white changing to dark blue and brown.

Optics: Transparent at first becoming translucent then opaque. Pleochroic. $\alpha = y$ deep blue 1·58, β pale blue-green 1·60, $\gamma \wedge z$ 28° yellow green 1·64, $\gamma - \alpha$ 0·06 Indices and colours vary with oxidation.

Tests: Whitens exfoliates and yields water in the closed tube. Fuses easily to a magnetic globule. Soluble in HCl.

Occurrence: Found with pyrrhotite in tin veins of Cornwall and as a secondary mineral in the gossan of sulphide veins. Also found in low temperature environments in greensands, in lignites and bog iron ores and on decayed wood and fossil bone in recent clays. The association with organic matter (the source of the phosphorus) is characteristic. Bone stained blue by vivianite is used ornamentally as *bone-turquoise* or *odontolite*. Because of the absence of Cu the HCl solution from this does not turn blue on addition of ammonia as does the solution from true turquoise.

Erythrite (Cobalt bloom)
$$2[Co_3(AsO_4)_2.8H_2O]$$
Monoclinic $\quad C\quad 2/m \qquad a$ 1·018 $\quad b$ 1·334 $\quad c$ 0·473
$$\beta 105°\ 1'$$

Structure: Vivianite type.

Habit: Usually as a crimson to pink bloom or encrustation on cobalt minerals. Crystals rare, prismatic, vertically striated.

Cleavage {010} perfect. $H\ 1\frac{1}{2}$–$2\frac{1}{2}$. D 3·18. Colour crimson to pink; crystals may be colour-banded. Streak paler than colour.

Optics: Translucent, pleochroic in red and pink. $\alpha = y$ 1·626, β 1·661, $\gamma \wedge z$ 31° 1·699, $\gamma - \alpha$ 0·073, 2V90°.

Tests: Heated gently in closed tube gives off water and turns blue. Heated on charcoal gives arsenical odour. Gives cobalt-blue borax bead.

Occurrence: A secondary mineral formed by oxidation of cobalt arsenides. Found in Cornwall and on Alston Moor, Cumberland. Abundant at Cobalt, Ontario and at La Bine Pt, Great Bear Lake, Canada, where it was a guide to the cobalt-silver-uranium veins.

Annabergite (Nickel bloom) is the corresponding nickel mineral with a green colour, less common, but found in the same setting. It occurs in mines at Creetown, Kirkcudbrightshire. A complete series exists between this and erythrite.

Amblygonite $\qquad 2[(Li, Na)Al(OH, F)PO_4]$
Triclinic $\ P\quad \bar{1} \qquad\quad a$ 0·518 $\quad b$ 0·715 $\quad c$ 0·504
$$\alpha112°\ 7'\quad \beta97°\ 48'\quad \gamma67°\ 53'$$

Structure: PO_4 tetrahedra and Al in octahedral co-ordination with (O, OH, F) form the main framework, but the positions of Li^+ ions are not accurately known.

Habit: Small crystals are equant. Often very large rough crystals and cleavable masses. Twins on {$\bar{1}\bar{1}1$} common. Cleavage {100} perfect, {110} good, {0$\bar{1}$0} distinct. $H\ 5\frac{1}{2}$–6. D 3·1. Lustre

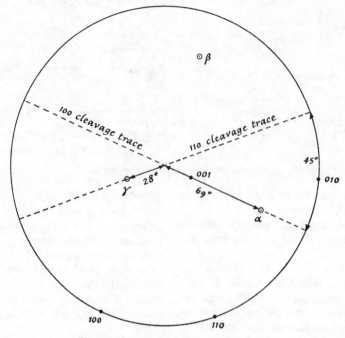

Fig. 201. Optical orientation of amblygonite.

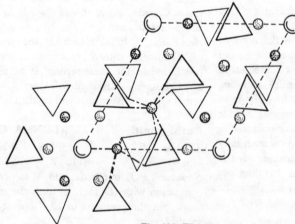
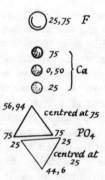

Fig. 202. The structure of apatite.

vitreous to greasy, pearly on cleavages. Colour white, milky or pale colours.

Optics: Transparent, colourless. α 1·59, β 1·60, γ 1·61, $\gamma - \alpha$ 0·02. For orientation see Fig. 201. Indices increase with substitution of OH for F, the sign changing from $-$ve (amblygonite, F > OH) to $+$ve (montebrasite, F < OH). Shows polysynthetic twinning.

Tests: Before the blowpipe, swells up and fuses easily, colouring the flame crimson (Li).

Occurrence: In granite pegmatites with spodumene, apatite, lepidolite, petalite and tourmaline, and in veins and greisens with cassiterite and its associates.

Varieties: Besides variation in the ratio F : OH, Na may substitute for Li.

Apatite $\quad\quad\quad$ 2[Ca$_5$(PO$_4$)$_3$(F, Cl, OH)]

Hexagonal $\quad P \quad 6/m \quad\quad a$ 0·937 $\quad c$ 0·688

Structure: PO$_4$ tetrahedra arranged in threes around hexad axes (screw axes 6$_3$) form tubes parallel to z, with F, Cl, or OH along their axes. 3/5 of the Ca^{2+} line the tubes internally and tie in the (F, Cl, OH) as well as holding the tubes together. The remaining Ca^{2+} lying between the tubes simply link neighbouring tubes (Fig. 202). The F, Cl, OH substitutions cause small changes in the cell edges.

Habit: Usually in long prismatic or needle-like crystals but sometimes tabular || {0001}. Also massive or concretionary and colloform.

Cleavage {0001} poor, seen as widespaced cross cracks under microscope. H 5. D 3·1–3·2 for fluorapatite and chlorapatite, 2·9–3·1 for hydroxylapatite and carbonate-apatite Ca$_{10}$(PO$_4$)$_6$(CO$_3$)H$_2$O. Lustre vitreous to sub-resinous. Colour pale green, sometimes bluish green or violet. Fluorescence common in ultraviolet light, in brown, orange or yellow. May be phosphorescent and thermoluminescent.

Optics: Transparent, colourless. ω 1·629–1·668, ε 1·624–1·667, $\omega - \varepsilon$ 0·005–0·001. Optically -ve. Indices decrease from chlorapatite through hydroxylapatite and fluorapatite to carbonate-apatite. Relief moderate, extinction straight, interference colour grey.

Tests: Distinguished from beryl and quartz by lustre and lower hardness. Soluble in HCl, carbonate-bearing varieties showing slight effervescence. The solution will precipitate CaSO$_4$ on addition of H$_2$SO$_4$. A solution in HNO$_3$ may be tested for P with ammonium molybdate.

Occurrence: A common accessory mineral of most igneous rocks. Large amounts occur in some pegmatites, as in Ontario and Quebec, and especially associated with alkali syenites, as in the Kola Peninsula, U.S.S.R., and carbonatites, as at Bukusu, Uganda. It is sometimes associated with concentrations of magnetite as at Kiruna, Sweden, and Bukusu. Apatite is common in regional metamorphic rocks, especially crystalline limestones.

Extensive sedimentary deposits of phosphate rock or phosphorite occur interbedded with limestones, sandstones and shales. These may be original chemical precipitates deriving their P from organic matter, or in some cases are rolled pebbles from erosion of phosphatized formations. On oceanic islands in the tropics the droppings of sea birds (guano) infiltrating into limestone have formed workable deposits of phosphate rock. The mineral of these sedimentary phosphates is given the sack-name *collophane*, and consists of hydroxylapatites and carbonate-apatites of variable composition. Well-crystallized carbonate-apatite may form secondarily from it, and is called *francolite* or *dahllite*. Sedimentary phosphates from Florida, Morocco, Algeria, Tunisia and Egypt are the source of much of the world's supply of phosphate for fertilizers.

Apatite is the principal material of bones and teeth. The greater resistance of fluorapatite to decay is the basis for advocating fluoridization of drinking water.

Use: Phosphorus is essential to all organic life. Large amounts of phosphate rock are used in fertilizers for agriculture.

Related minerals: Many artificial apatites have been synthesized, and have application in the electrical industry for making ceramics with specific dielectric properties.

Pyromorphite—mimetite

Hexagonal $\quad C \quad 6/m$

Pyromorphite	Pb$_5$(PO$_4$)$_3$Cl	a 0·997	c 0·732
Mimetite	Pb$_5$(As O$_4$)$_3$Cl	a 1·025	c 0·746
Vanadinite	Pb$_5$(VO$_4$)$_3$Cl	a 1·033	c 0·734

Structure: Apatite type.

Habit: Rounded barrel-shaped crystals or globular and blister-like aggregates. Also in simple prismatic crystals.

H 3$\frac{1}{2}$–4, vanadinite 3. D 7·04 (pyromorphite)–7·24 (mimetite) : 6·88 (vanadinite). Lustre sub-adamantine to resinous. Colour in pyromorphite bright green to greenish yellow or brown. Mimetite is yellow, yellow-brown, orange, sometimes white. Vanadinite is orange-red, ruby-red and brownish-red. Streak white.

Optics: Translucent, colourless.

	Pyromorphite	Mimetite	Vanadinite
ω	2·06	2·15	2·42
ε	2·05	2·13	2·35
$\omega - \varepsilon$	0·01	0·02	0·07

Optically $-$ve. Anomalously biaxial in As-rich members.

Tests: Heated on charcoal pyromorphite fuses easily giving a blue-green flame and forms a globule that is crystalline and angular. The addition of sodium carbonate is required to reduce it to lead. Mimetite reduces to lead without the flux and gives arsenical fumes. Fused with microcosmic salt vanadinite gives a bead that is yellow in the oxidizing flame, green in the reducing flame.

Occurrence: Secondary minerals of lead veins, often attractively set on manganese-oxide-coated surfaces. Associated minerals are cerussite, anglesite, smithsonite, malachite. Pyromorphite and mimetite occur in Cornwall, in the Caldbeck Fells, Cumberland, and at Leadhills, Lanarkshire, amongst many localities. Vanadinite is found at Wanlockhead, Dumfriesshire and at Leadhills. Also in Triassic sandstone at Alderley Edge, Cheshire.

Varieties: *Campylite* is a variety of mimetite in barrel-shaped yellow and orange-brown crystals. It has generally been held that complete solid solution exists between pyromorphite and mimetite, but there may be a miscibility gap in the middle of the series. There is rather better evidence that vanadinite is separated from the P and As bearing minerals by an immiscibility gap. *Svabite* $(Ca, Pb)_5 (AsO_4, PO_4)_3 (F, Cl, OH)$ is related to this group and to apatite.

Turquoise $CuAl_6(PO_4)_4(OH)_8 . 4H_2O$
Triclinic $P\ \bar{1}$ $a\ 0·748$ $b\ 0·995$ $c\ 0·768$
                                 $\alpha 111° 39'$ $\beta 115° 23'$ $\gamma 69° 26'$

Habit: Massive cryptocrystalline, in veinlets, concretionary or encrusting. Crystals rare.
Cleavage {001} perfect, {010} good. Fracture small-conchoidal. H 5–6. D 2·6–2·8. Lustre of massive material, waxy, dull. Colour blue-green of characteristic shade. Streak white or pale greenish.

Tests: In the closed tube decrepitates, yields water and turns brown. Infusible but turns brown before the blowpipe. After ignition soluble in HCl, the solution responding to the ammonium molybdate test for phosphorus. See also under Vivianite.

Occurrence: A secondary mineral formed in veins and pockets by the thorough-going alteration of trachytes and other aluminous igneous rocks, probably deriving its materials from apatite and disseminated copper sulphides. It is associated with limonite, chalcedony and kaolin. The ancient and famous deposits are in the Kuh-i-Nishapur of Iran, east of the Caspian Sea and near the border with U.S.S.R. Also found in the Sinai Peninsula (Wadi Maghara) at Jordansmühl, Silesia, Germany, and in New Mexico, S.W. of Santa Fe.

Use: A semi-precious decorative stone.

Related minerals: *Chalcosiderite* is the analogue with Fe^{3+} replacing Al.

Wavellite $4[Al_3(OH)_3(PO_4)_2 . 5H_2O]$
Orthorhombic $P\ \ mmm\ \ a\ 0·963\ b\ 1·734\ c\ 0·699$

Habit: As spherical aggregates of radiating crystals giving a rayed flower-like appearance when broken across.
Cleavage {110} perfect, {101} good. H $3\frac{1}{2}$–4. D 2·36. Lustre vitreous to pearly. Colour greenish-white, white, yellowish-brown.

Optics: Transparent, colourless. $\alpha = y\ 1·520$–1·535, $\beta = x\ 1·526$–1·543, $\gamma = z\ 1·545$–1·561, $\gamma - \alpha$ 0·025, $2V_\gamma$ 71°.

Tests: The habit is distinctive. Infusible but easily soluble in acids, the HNO_3 solution giving a yellow precipitate with ammonium molybdate solution.

Occurrence: On joints in low-grade slates (Barnstable, Devon), in phosphate deposits (Lahn District, Germany, and Florida) and in residual deposits of limonite and manganese oxides. In Bolivia in tin veins, an atypical occurrence.

Varieties: F may substitute in small amounts for (OH) and Fe^{3+} for Al.

Autunite $2[Ca(UO_2)_2(PO_4)_2 . 10$–$12H_2O]$
Torbernite $2[Cu(UO_2)_2(PO_4)_2 . 8$–$12H_2O]$
Tetragonal $I\ 4/mmm$

Autunite	$a\ 0·699$ $c\ 2·063$
Torbernite	$a\ 0·705$ $c\ 2·05$

Structure: A layer structure is postulated with UO_2PO_4 layers (PO_4 tetrahedral, UO_2 linear but the U^{6+} in a distorted octahedron of O^{2-}) parallel to {001} with Ca^{2+} and H_2O linking the layers.

Habit: Thin tabular crystals, often mica-like scales and foliated aggregates.

Cleavage {001} eminent (like mica). Flakes a little flexible in autunite, less so in torbernite. H 2–2½. D 3·1–3·22. Lustre pearly on {001}.

	Autunite	Torbernite
Colour:	lemon-yellow	emerald green
Streak:	yellowish	green, paler than the crystals
Fluorescence (UV light):	yellowish-green	not fluorescent
Optics:		
ω	yellow 1·577	sky-blue 1·592
ω'	1·575	
ε	colourless to pale yellow 1·553	green 1·582
$\omega - \varepsilon$	0·024	0·010
	Anomalously biaxial −ve $2V \sim 10° − 30°$	Optically −ve

Tests: The habit and colours are distinctive. Fused in NaF, they give a bead fluorescent in lemon-yellow under UV light λ 254 nm (a test for U).

Occurrence: Both are secondary minerals of the oxidized zone of uraninite-bearing veins, torbernite occurring where there are associated copper sulphides. Well known localities are the mines of Redruth, St Austell, St Just, etc. in Cornwall, in the Katanga mining area, Congo, and at Mt. Painter, Flinders Ra., S. Australia. The conspicuous colours make them valuable guides in uranium prospecting.

Carnotite $2[K_2(UO_2)_2(VO_4)_2 . 3H_2O]$

Monoclinic P $2/m$ a 1·047 b 0·841 c 6·91
                                      β 103° 40′

Structure: Paired square-pyramidal co-ordination groups form (V_2O_8) units joined through UO_2 groups form layers ‖ 001, which are linked by K^+ and H_2O molecules. Fig. 203 is a schematic view of part of a layer viewed ⊥ 110.

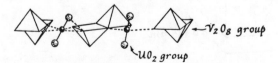

Fig. 203. Pattern of atomic linkage in carnotite.

Habit: As a powder or microcrystalline aggregate. Macroscopic crystals rare.

Cleavage {001} perfect. $H \simeq 2$. D 4·5. Lustre dull and earthy. Colour a conspicuous bright lemon-yellow. Non-fluorescent.

Optics: α 1·75–1·76, $\beta = y$ 1·90–1·93, γ 1·92–1·95, $\gamma - \alpha$ 0·017–0·019. $2V\alpha \simeq 40°$.

Tests: Radioactive. Infusible before the blowpipe.

Occurrence: Disseminated in shallow water or terrestrial sandstones or concentrated around fossil plant remains, particularly in the Mesozoic of Colorado and Utah. It is supposed to be secondary after earlier uranium and vanadium minerals.

Use: An ore of uranium.

Related mineral: *Tyuyamunite* $Ca(UO_2)_2(VO_4)_2$ nH_2O is an analogous 'uranium mica' of yellow to greenish colour found with carnotite, but distinguished from it by being easily fusible to a dark liquid.

SULPHATES

Except for barytes, the sulphates described here are low-temperature minerals of the zone of oxidation of mineral veins above the water-table, or evaporite minerals. All embody in their structure the tetrahedral $(SO_4)^{2-}$ ion which is not in itself a markedly anisotropic building block. These anionic groups are usually separate from one another, linked through intervening metal atoms or H_2O molecules, but the structure suggested for linarite, $CuPb(OH)_2SO_4$, provides an exception.

They are grouped as follows:

Anhydrous, with large cations

Barytes	$BaSO_4$
Celestine	$SrSO_4$
Anglesite	$PbSO_4$

Anhydrous with smaller cation

Anhydrite	$CaSO_4$

With $(OH)^-$

Alunite	$KAl_3(OH)_6(SO_4)_2$
Jarosite	$KFe_3(OH)_6(SO_4)_2$

Hydrates

Kieserite	$MgSO_4.H_2O$
Gypsum	$CaSO_4.2H_2O$
Polyhalite	$K_2Ca_2Mg(SO_4).2H_2O$
Epsomite	$MgSO_4.7H_2O$
Melanterite	$FeSO_4.7H_2O$
Alum	$KAl(SO_4)_2.12H_2O$

BARYTES GROUP

Barytes	$4[BaSO_4]$
Celestine	$4[SrSO_4]$
Anglesite	$4[PbSO_4]$

Orthorhombic *P mmm*

Structure: The large cation lies in 12-fold co-ordination with O^{2-} ions belonging to 7 tetrahedral SO_4 groups. In Fig. 204 heavy-lined elements are centred $\frac{1}{4}$ *b* from the front face and light-lined ones $\frac{3}{4}$ *b* back.

Habit: Members of this group are often well-crystallized, tabular ||{001}, with {210} prominent; {011} is a common form also. The crystals may be elongated || *x* or *y*. Barytes may form aggregates of tabular crystals projecting in cockscomb fashion. 'Desert roses' from Oklahoma and Kansas are rosette-like aggregates of barytes. Anglesite may be prismatic elongated || *z*. They may also be massive, compact, or in nodular or stalactitic form.

Cleavage {001} perfect {210} good {010} poor in all members.

		Barytes	Celestine	Anglesite
Unit cell	*a*	0·887	0·836	0·848
	b	0·545	0·535	0·540
	c	0·715	0·687	0·696
H		$3-3\frac{1}{2}$	$3-3\frac{1}{2}$	$2\frac{1}{2}-3$
D		4·50	3·97	6·38

Lustre	vitreous pearly on cleavage	vitreous pearly on cleavage	adamantine
Colour	white or colourless	colourless to pale blue	white or colourless
Streak	white	white	white
Optics	$\alpha = z$	1·636	1·622	1·877
	$\beta = y$	1·637	1·624	1·883
	$\gamma = x$	1·648	1·631	1·894
	$\gamma - \alpha$	0·012	0·009	0·017
	$2V_\gamma$	36°	51°	75°

Barytes and celestine sometimes fluoresce in UV light; anglesite often does so in yellow colours. The first two may be thermoluminescent and barytes may be phosphorescent.

Tests: Barytes decrepitates before the blow pipe, fuses with difficulty and colours the flame yellowish green. Celestine behaves similarly fusing to a milk-white pearl and colouring the flame crimson. Anglesite fuses readily. When heated on charcoal in the reducing flame it bubbles and yields metallic lead. All are insoluble or slowly soluble in acids.

Occurrence: *Barytes* is an abundant primary mineral in some moderate to low-temperature sulphide veins, with galena, quartz, carbonates (including witherite) and fluorite. In the North

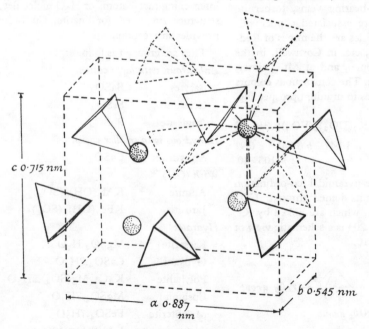

c 0·715 nm

b 0·545 nm

a 0·887 nm

Ba

Fig. 204. The structure of barytes.

Pennine Orefield, where it is widespread, it is found only outside the central, higher-temperature area of fluorite deposition. It is also found in veins, lenses and as nodules in limestones or marls. It may occur with manganese oxide nodules of lakes or the sea bed, as a hot-spring deposit and as a vesicle mineral.

Celestine occurs in marls and other sediments, associated with evaporite minerals like gypsum, anhydrite and halite, and with sulphur. Well-known occurrences are at Yate near Bristol, and in the Sicilian sulphur deposits. It is also found in the cap-rock of salt domes. It may occur in hydrothermal veins or in cavities of basic eruptive rocks.

Anglesite is a mineral of the zone of oxidation of lead-bearing veins, commonly formed from galena. Associated minerals are cerussite, pyromorphite, mimetite and the oxides and carbonates of copper. Famous localities are Broken Hill, N.S.W., Australia, and Tsumeb, S.W. Africa. The original locality is the Parys Mine, Anglesey (whence the name) and it is found at Matlock in Derbyshire, Leadhills in Lanarkshire and Wanlockhead in Dumfriesshire.

Varieties: Sr substitutes for Ba in barytes and there is probably a complete solid solution series between the two minerals. Limited amounts of Ca enter into this series. Pb may substitute for Ba to the extent of about one atom in four, leading towards anglesite.

Use: Barytes is used in the manufacture of white pigment for paint, as a filler in paper-making, and to weight the mud circulated during the drilling of oil and gas wells.

Celestine is the chief source of strontium.

Anglesite is a valuable ore of lead.

Related minerals: Isostructural with these sulphates is *crocoite*, $PbCrO_4$, found in striking purplish-red or orange-red prismatic crystals as a secondary mineral of lead veins. Well-known localities for crocoite are near Ekaterinburg in the Ural Mountains and in the mines of the Dundas District, Tasmania.

Anhydrite   4[CaSO₄]

Orthorhombic B mmm a 0·6238 b 0·6991

                                     c 0·6996

Structure: The Ca^{2+} ion is surrounded by eight nearest neighbour O^{2-} ions belonging to tetrahedral SO_4 groups (Fig. 205). This lower co-ordination of the cation commensurate with its smaller size accounts for the difference between this and the barytes structure.

Habit: Equant or thick tabular, with the three pinacoids prominent. Usually massive; also fibrous or lamellar, the lamellae sometimes contorted.

Twins on {011} either as contact twins or polysynthetic lamellae.

Cleavage good parallel to the pinacoids: in order of perfection {001}, {010}, {100}. Fracture uneven, splintery. H 3½. D 2·98.

Lustre vitreous; pearly on cleavages. Colourless or bluish in crystals. White, pink or mauve in massive material. Often discoloured by impurities. Streak white.

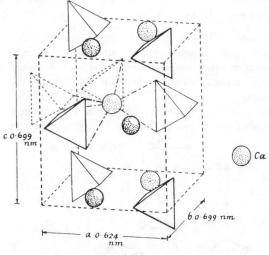

Fig. 205. The structure of anhydrite.

Optics: Transparent, colourless, usually as anhedral or subhedral grains. $\alpha = z\,1.570$; $\beta = y\,1.575$; $\gamma = x\,1.614$; $\gamma - a\,0.044$; $2V\,43°$. Extinction parallel to cleavages.

Tests: Fuses with some difficulty to a white enamel. Moistened with HCl it gives an orange-red Ca flame colour. Dissolves in boiling HCl and yields a white precipitate when $BaCl_2$ solution is added. Harder than gypsum and denser than calcite.

Occurrence: Forms stratified bodies of rock in the lower parts of evaporite sequences, typically succeeding the basal dolomite, and continues in the overlying salt beds. It is probable that the primary precipitate of the anhydrite rock was gypsum, since anhydrite does not precipitate below 42° until the salinity becomes rather high. The gypsum would be converted to anhydrite by the rise of temperature with depth of burial. The reverse change, from anhydrite to gypsum, takes place near the outcrop of evaporite beds. Anhydrite is abundant in the cap-rock of salt domes. In small amounts it may be present in hydrothermal veins and vesicles of igneous rocks.

Use: Anhydrite is treated with ammonia in the manufacture of ammonium sulphate fertilizer, the mineral acting as a source of sulphate. It is also reduced with coke to form SO_2 for sulphuric acid manufacture.

ALUNITE GROUP

Alunite $3[KAl_3(OH)_6(SO_4)_2]$
Jarosite $3[KFe^{3+}(OH)_6(SO_4)_2]$
Trigonal $R\ 3m$

Structure: The principal elements are (*i*) SO_4 tetrahedral groups pointing alternately up and down in rows parallel to *z*-axis, (*ii*) K^+ or other large cationic elements between the tetrahedral bases, (*iii*) Al or Fe^{3+} lying midway between pairs of K^+ positions, and each co-ordinated octahedrally to two O^{2-} of the SO_4 groups and to

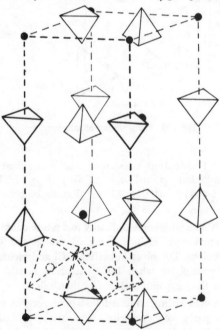

● K, Na, Ag, NH_4 etc.   ○ Al, Fe^{3+}
O^{2-} at corners of tetrahedra : OH^- at remaining corners of octahedra

Fig. 206. The structure of alunite.

four (OH^-). Only three Al positions and two octahedra are drawn in Fig. 206.

Habit: *Alunite* is usually in rock-like masses enclosing quartz and clay minerals. Crystals when they occur are {0001} tablets or flat rhombohedra.

Jarosite forms crusts, small concretionary or coralloid masses or earthy and powdery coatings. Crystals are minute rhombohedra {10$\bar{1}$2} (cubic in aspect) or {0001} tablets.

	Alunite	Jarosite
a (hexagonal axes)	0·696	0·720
c (hexagonal axes)	1·735	1·700
Cleavage	{0001}	{0001}
H	3½–4	2½–3½
D	2·6–2·9	2·91–3·26
Lustre (of crystals)	vitreous	subadamantine
Colour	white, grey, yellowish, reddish brown	yellow, ochreous to dark brown
Streak	white	pale yellow, sometimes glistening
Pyroelectric effect	strong	strong
Optics	translucent colourless	translucent pleochroic
ω	1·572	red-brown 1·817–1·820
ε	1·592	colourless 1·715
ε − ω	0·020	ω − ε 0·105
	+ve	−ve, anomalously biaxial with small 2*V*
	fast ray ∥ cleavage traces	

Tests: Alunite decrepitates in the blow-pipe flame but does not fuse. Insoluble in HCl and HNO_3. Slowly dissolved in H_2SO_4. Moistened with cobalt nitrate solution it turns blue on heating. When the bead formed by heating with sodium carbonate and powdered charcoal is moistened it will darken a silver coin (test for SO_4). After ignition dissolves readily in HNO_3.

Jarosite: Gives acid water on heating in a closed tube. Soluble in HCl. X-ray methods are required for definite identification.

Occurrence: *Alunite* forms by the hydrothermal alteration of intermediate and acid volcanic rocks, by solutions from decomposition of pyrite. Large areas of alunitized rock occur in the mining areas of Western U.S.A. Deposits at Tolfa, N.E. of Rome, have been mined for centuries.

Jarosite is a secondary mineral widespread as coatings on iron-bearing vein-stuff in the zone of oxidation, and on cracks and joints in other rocks. It is found on the cleat of coal. In general it is to be expected in environments where pyrite is oxidising. Probably often lumped together with 'limonite'.

Varieties: The different positions in the structure may be variously occupied giving *natroalunite* $NaAl_3(OH)_6(SO_4)_2$, *natrojarosite* $NaFe^{3+}(OH_6)(SO_4)_2$, *argentojarosite* $AgFe_3(OH)_6(SO_4)_2$, *ammoniojarosite* $NH_4Fe_3(OH)_6(SO_4)_2$ and *plumbojarosite* $Pb[Fe_3(OH)_6(SO_4)_2]_2$. The group links on through *svanbergite*, $SrAl_3(OH)_6SO_4PO_4$, and *hidalgoite* $PbAl_3(OH)_6AsO_4SO_4$ to a large group of analogous phosphates and arsenates, providing a remarkable example of structural flexibility.

Use: Alunite, also called alumstone, has long been used in alum manufacture and as a source of potash.

Kieserite $4[MgSO_4.H_2O]$

Monoclinic   *C*   2/m   *a* 0·689   *b* 0·761   *c* 0·763
                                                  $\beta 117° 43'$

Structure: There is some uncertainty about the atomic arrangement.

Habit: Usually massive, mingled with other salts. Twins polysynthetic, on an uncertain law. Cleavage perfect on {110} and {111}, less so on {$\bar{1}$11}, {$\bar{1}$01} and {011}. *H* 3½. *D* 2·57.

Lustre vitreous. Colourless, grey-white or yellowish.

Optics: Transparent, colourless. $\alpha 1·520$, $\beta = y 1·533$, $\gamma \wedge z 76°$ 1·584, $\gamma - \alpha$ 0·064, $2V_y$ 55–60°. Shows poly-synthetic twin lamellae, high birefringence and many cleavages in thin section.

Occurrence: In the upper horizons of marine evaporite successions associated with halite, anhydrite, polyhalite, sylvine and carnallite.

Gypsum $4[CaSO_4.2H_2O]$

Monoclinic   *C*   2/m   *a* 0·576   *b* 1·515   *c* 0·628
                                                  $\beta 113° 50'$

Structure: Consists of layers parallel to 010, each layer made up of two sheets of SO_4 tetrahedra linked by Ca^{2+} ions and faced by a sheet of H_2O groups on each surface (Fig. 207).

Angles: 100 $\wedge$ 001   66° 10'; 010 $\wedge$ $\bar{1}$11   71° 50'; 010 $\wedge$ 011 69° 16'; 010 $\wedge$ 120 55° 37'.

Habit: In simple crystals tabular ||{010} with {$\bar{1}$11} and {120}; as rosette-like aggregates; sometimes in elongated crystals twisted or curled; as fibrous masses; granular, massive.

Twinning is common on {100} or {$\bar{1}$01} giving swallow-tail twins (Fig. 72, page 68).

Cleavage {010} eminent, in conformity with the layer structure; also {100} and {011} giving rhomb-shaped fragments with angle 66° and 114°. {010} cleavage flakes flexible, inelastic. *H* 2. *D* 2·32. Lustre pearly on {010}, otherwise sub-vitreous. In massive varieties glistening. Colourless or white to grey, yellowish or pink. Streak white. Fluorescence sometimes shown in yellow and cream: also phosphorescent at times.

Optics: Transparent, colourless. $\alpha 1·520$, $\beta = y 1·523$, $\gamma \wedge z 52°$ 1·529, $\gamma - \alpha$ 0·009, $2V_y$ 58°. 2V decreases markedly with rise in temperature becoming 0° at 91° and then opening out $\perp$010. Polysynthetic twins seen in thin sections of gypsum are said to be produced by heating the section.

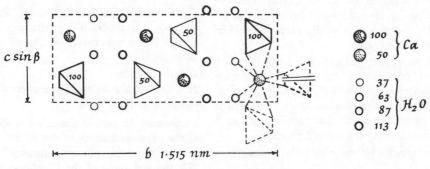

Fig. 207. The structure of gypsum.

Tests: Yields water on heating in the closed tube, a distinction from anhydrite which can be confirmed by refractive index tests. Dissolves in HCl on warming, the solution giving a white $BaSO_4$ precipitate with addition of $BaCl_2$ solution.

Occurrence: As bedded sedimentary formations interstratified with limestone, red shale, marl and clays. It is commonly associated with deposits of halite, anhydrite and magnesian limestone, as in S.E. Durham and Cumberland, but may occur under conditions that have not produced salt deposits, as in sulphur deposits of Sicilian type, and in the Tertiary deposits of the Paris basin. It is an important constituent of the cap-rock of salt domes. It may form as curled and twisted crystals on the walls of limestone caves.

Uses: Gypsum is mined, and heated at just under 200° to produce $CaSO_4 . \frac{1}{2}H_2O$, which is the principle constituent of plaster of Paris. It is used in its raw state as a filler. The fine-grained massive variety, alabaster, is used for the manufacture of carved ornaments.

Varieties: *Selenite* is clear well-crystallized gypsum. *Alabaster* is fine-grained uniform massive material. *Satin spar* is a fine-grained fibrous type. Daisy-bed ore from Longmarton, Cumberland, consists of brown rosettes of crystals in a light grey matrix.

Polyhalite $K_2Ca_2Mg(SO_4).2H_2O$

Triclinic P $\bar{1}$ $a:b:c = 0.718:1:0.466$
                                $\alpha 91° 39'$ $\beta 90° 06'$ $\gamma 91° 53'$

Structure: Not known.

Habit: Crystals small, tabular on {010} or elongated along z-axis. Massive fine- or coarse-grained, fibrous or spherulitic. Twins ubiquitous on {010} and {001}, simple or polysynthetic.
Cleavage perfect {10$\bar{1}$}. H $3\frac{1}{2}$. D 2·78. Lustre vitreous, or waxy in the mass. Colour white or grey, but nearly always stained brick-red, orange or pink by included iron oxide. Partly soluble in water, but without taste.

Optics: Transparent, colourless. α 1·547, β 1·560, γ 1·567, $2V_\gamma$60–70°. On cleavage flakes, $\gamma' \wedge$ twin composition plane {010} = 28°. Simple or polysynthetic twinning is conspicuous, the two directions giving a striped pattern (like plagioclase) or a chequer pattern, between crossed polars.

Tests: Decomposed by water with the separation of gypsum.

Occurrence: A widespread and important constituent of salt deposits, with halite, anhydrite and kieserite, as in the Stassfurt deposits of Germany, those of N.E. Yorkshire, and in the New Mexico–Texas region.

Epsomite $4[MgSO_4 . 7H_2O]$

Orthorhombic P 222 a 1·191 b 1·202 c 0·678

Structure: Built of $Mg(H_2O)$ octahedra and (SO_4) tetrahedra with a weakly bonded seventh H_2O molecule, which is readily lost in dry air to leave the hexahydrate.

Habit: Short prisms {110} terminated by the sphenoid {111} (or its enantiomorph {1$\bar{1}$1}) well seen in commercial Epsom salts. In nature acicular or hair-like crystals and fluffy efflorescences.
Cleavage {010} perfect. H 2–2$\frac{1}{2}$. D 1·68. Lustre silky. Colourless or white. Taste bitter and salty.

Optics: Transparent, colourless. $\alpha = x$ 1·433, $\beta = z$ 1·455, $\gamma = y$ 1·461, $\gamma - \alpha$ 0·028, $2V_\alpha$51°. Optic plane normal to the length. May be either length-fast or length-slow. Shows rotation of the plane of polarized light, at the rate of about 2° per mm along the optic axis.

Tests: The form, taste and solubility in water are distinctive.

Occurrence: As an efflorescence in mine-workings and limestone caves and overhangs, or on gypsiferous beds. Also as a deposit from mineral springs.

Related minerals: *Reichardtite* is the same substance found with carnallite in evaporites. *Melanterite*, $FeSO_4.7H_2O$, monoclinic, is a green to greenish-blue efflorescence found in similar environments to epsomite.

Alum $4[KAl(SO_4)_2 . 12H_2O]$

Cubic P $m3m$ a 1·216

Structure: $Al(H_2O)_6$ octahedra alternate with $K(H_2O)_6$ octahedra along the cell edges. SO_4 groups lie approximately on the diagonals of the cube.

Habit: Octahedra.
H 2. D 1·76. Colourless. Taste sweetish, astringent.
Optics: $n = 1·45$.

Occurrence: In pyritic shales upon decomposition of the pyrite and attack of the sulphuric acid upon the clay minerals as in the Jurassic Alum Shales of Whitby, and those of the Cambro-Silurian of the

Oslo district. Also around volcanic vents. Associated minerals include epsomite, melanterite and sulphur.

Use: As a mordant in dyeing, for which purpose it was obtained from the Whitby shales by roasting and extraction with water.

BORATES

Boron is rare compared to other light elements. It occurs in notable amounts in volcanic exhalations and some hot springs. Besides its presence in the complex borosilicates of granite pegmatites, a good deal of it is camouflaged away in argillaceous and iron oxide-rich sediments. The number of borates is few, and these mostly occur as evaporite deposits of desert areas on land.

Only one borate is considered here.

Borax $4[Na_2B_4O_5(OH)_4 . 8H_2O]$

Monoclinic C $2/m$ a 1·186 b 1·067 c 1·220
β106° 41′

Structure: $B_4O_5(OH)_4$ groups made up of 2 tetrahedra and 2 plane triangular elements are linked into stalks parallel to z, regularly interplanted with stalks of linked $Na(H_2O)_6$ octahedra (Fig. 208). These linear elements are joined only by weak residual bonds.

Habit: Short, squarish, prismatic crystals or tablets ‖ {100}.

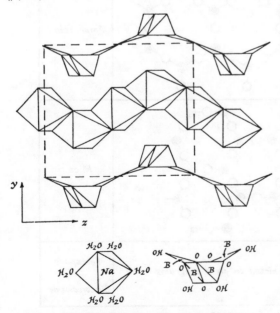

Fig. 208. The structural pattern of borax.

Cleavage {100} perfect {110} good. Fracture conchoidal. H 2–2½. D 1·71. Lustre vitreous or earthy. Colourless or white. Taste sweetish alkaline, weak.

Optics: Transparent colourless. $\alpha = y$ 1·447, β 1·469, $\gamma \wedge z$ 55° 1·472, $\gamma - \alpha$ 0·025 $2V_\alpha$ 40°. Crossed dispersion $r > v$ marked, leading to anomalous interference colours in sections normal to the optic axes.

Tests: Fuses readily with great swelling to a clear glass. Soluble in water. Turmeric paper moistened with the solution and dried at 100°C on a water bath turns a reddish brown which goes black with ammonia (test for B).

Occurrence: In evaporite deposits of ephemeral lakes in arid regions, along with halite, gypsum, calcite, trona and soda nitre. Large deposits occur, at Searles Lake, San Bernardino County, California, and in Kern County in the same State. Anciently worked sources lie in Kashmir and Tibet.

Use: Borax is used in enamelling iron, in ceramic glazes and glass-making, and as a flux. It is well-known as a water softener in laundering and has other uses in treating fabrics. Boron hydrides are amongst the hardest substances known. Borax and the related mineral *kernite*, $Na_2B_4O_6(OH)_2 . 3H_2O$, are the principal sources of borax and of boron.

SILICATES

The accessible crust of the earth and its whole body, it is believed, down to the nickel-iron core beginning at 2900 km depth, is principally composed of the great group of silicates. Of this group about eight mineral series are greatly predominant: olivines, pyroxenes, amphiboles, micas and clay minerals, feldspars, quartz, the aluminium silicates and the garnets.

All silicate structures are based on the tetrahedral $(SiO_4)^{4-}$ anionic group. These groups may be independent (island silicates) or linked together by sharing corner O^{2-} ions into couplets, rings, chains, layers or frameworks. The silicates are classified according to the type of structural element into which the tetrahedral groups are linked (Fig. 209), and this essentially $\{Si_x - O_y\}$ part of the formula is enclosed in curly brackets.

The structural elements (chains, sheets, etc.) are joined together by bonds to cations lying between them, or, in the case of frameworks, cations may be held in the spaces of the framework. There is considerable freedom of ionic substitution in the cation sites between the tetrahedral elements,

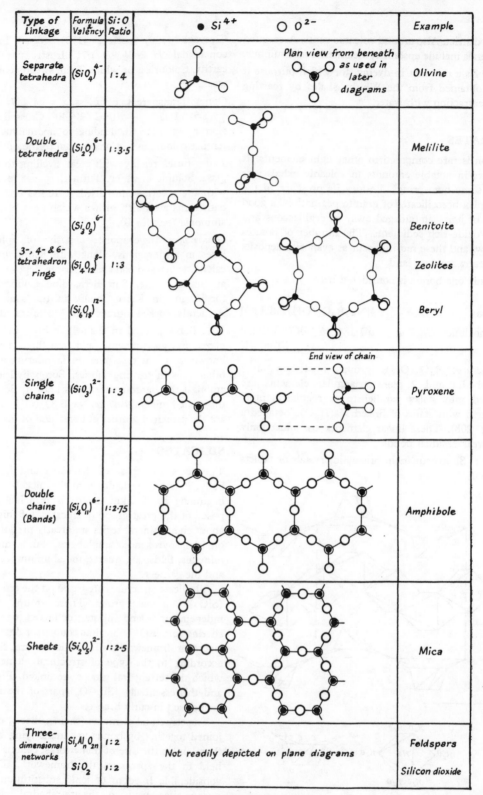

Type of Linkage	Formula & Valency	Si:O Ratio	● Si^{4+} ○ O^{2-}	Example
Separate tetrahedra	$(SiO_4)^{4-}$	1:4	Plan view from beneath as used in later diagrams	Olivine
Double tetrahedra	$(Si_2O_7)^{6-}$	1:3.5		Melilite
3-, 4- & 6-tetrahedron rings	$(Si_3O_9)^{6-}$ $(Si_4O_{12})^{8-}$ $(Si_6O_{18})^{12-}$	1:3		Benitoite Zeolites Beryl
Single chains	$(SiO_3)^{2-}$	1:3	End view of chain	Pyroxene
Double chains (Bands)	$(Si_4O_{11})^{6-}$	1:2.75		Amphibole
Sheets	$(Si_2O_5)^{2-}$	1:2.5		Mica
Three-dimensional networks	$Si_xAl_nO_{2n}$ SiO_2	1:2 1:2	Not readily depicted on plane diagrams	Feldspars Silicon dioxide

Fig. 209. Types of SiO$_4$ linkage in silicates.

governed by ionic charge and size. Within the tetrahedral groups the principal element substituting for Si^{4+} is Al^{3+} and this substitution is limited in amount, rarely exceeding $Al:Si = 1:3$.

Silicate minerals are mostly refractory and rather hard, so that blowpipe tests, scratching tests and acid solubility tests are of limited usefulness for them. On the other hand they are mostly translucent, and optical tests have special value in determining not only the species, but the composition within the range shown by one species.

Island silicates

Olivine group	$(Mg, Fe)_2\{SiO_4\}$
Zircon	$Zr\{SiO_4\}$
Sphene	$CaTi\{SiO_4\}(O, OH, F)$
Andalusite	$Al_2O\{SiO_4\}$
Sillimanite	$Al_2O\{SiO_4\}$
Kyanite	$Al_2O\{SiO_4\}$
Topaz	$Al_2\{SiO_4\}(OH, F)_2$
Staurolite	$Fe_2Al_9\{SiO_4\}_4(O, OH)_8$
Garnet	$R_3^{2+}R_2^{3+}\{SiO_4\}_3$
Vesuvianite	$CaMg_2Al_4\{SiO_4\}_5$ $\{Si_2O_7\}_2(OH, F)_4$
Chloritoid	$\{((Mg, Fe)_2Al)O_2(OH)_4$ $Al_3\{SiO_4\}_2$
Braunite	$Mn_7O_8\{SiO_4\}$

Couplet silicates

Epidote group	$Ca_2(Al, Fe)_3\{SiO_4\}$ $\{Si_2O_7\}O(OH)$
Lawsonite	$CaAl_2(OH)_2\{Si_2O_7\}H_2O$
Hemimorphite	$Zn_4\{Si_2O_7\}(OH_2)H_2O$
Melilite series	$Ca_2(Mg, Al)\{(Al, Si)_2O_7\}$

Ring silicates

Beryl	$Be_3Al_2\{Si_6O_{18}\}$
Cordierite	$Al_3(Mg, Fe)_2\{Si_5AlO_{18}\}$
Tourmaline	$NaMg_3(Al, Fe)_6(OH)_4$ $(BO_3)_3\{Si_6O_{18}\}$

Chain silicates

Pyroxene group	$X_{1-p}Y_{1+p}\{Si_2O\}_6$
Wollastonite group	
Wollastonite	$Ca_3\{Si_3O_9\}$
Pectolite	$NaHCa_2\{Si_3O_9\}$
Rhodonite	$Mn_5\{Si_5O_{15}\}$
Amphibole group	$(W, X, Y)_{7-8}\{Si_8O_{22}\}$ $(O, OH, F)_2$

Layer silicates

Kaolinite (Kandite) group	$Al_4\{Si_4O_{10}\}(OH)_8$
Serpentine	$Mg_6\{Si_4O_{10}\}(OH)_8$
Pyrophyllite	$Al_2\{Si_4O_{10}\}(OH)_2$
Talc	$Mg_3\{Si_4O_{10}\}(OH)_2$
Mica group	$K(Mg, Fe, Al)_{2-3}$ $\{AlSi_3O_{10}\}(OH)_2$
Montmorillonites (Smectites)	$0.33M(Al, Mg)_2\{Si_4O_{10}\}$ $(OH)_2 nH_2O$
Chlorite group	$(Mg, Fe, Al)_{2-3}(OH)_6$ $(Mg, Fe, Al)_{2-3}$ $\{Si_4O_{10}\}(OH)_2$

Framework silicates

Silica minerals

Quartz	SiO_2
Tridymite	SiO_2
Cristobalite	SiO_2
Opal	$SiO_2 nH_2O$

Feldspar group

Alkali feldspar series	$(K, Na)\{AlSi_3O_8\}$
Plagioclase series	$(Na, Ca)\{(Al, Si)_4O_8\}$

Feldspathoid group

Nepheline	$Na\{AlSiO_4\}$
Leucite	$K\{AlSi_2O_6\}$
Sodalite	$Na_4\{AlSiO_4\}_3Cl$

Zeolite group

Analcime	$Na\{AlSi_2O_6\}H_2O$
Chabazite	$(Ca, Na_2)\{Al_2Si_4O_{12}\}6H_2O$
Laumontite	$Ca\{Al_2Si_4O_{12}\}4H_2O$
Stilbite	$(Ca, Na_2, K_2)\{Al_2Si_7O_{18}\}$ $7H_2O$
Apophyllite	$KFCa_4\{Si_8O_{20}\}8H_2O$

Fibrous Zeolites

Natrolite	$Na_2\{Al_2Si_3O_{10}\}2H_2O$
Thomsonite	$NaCa_2\{Al_5Si_5O_{20}\}6H_2O$
Mesolite	$Na_2Ca_2\{Al_6Si_9O_{30}\}8H_2O$
Scolecite	$Ca\{Al_2Si_3O_{10}\}3H_2O$
Scapolite series	$(Na, Ca)_4\{(Al, Si)_{12}O_{24}\}$ (Cl, SO_4, CO_3)
Cancrinite	$(Na, Ca)_4\{Al_3Si_3O_{12}\}$ $(CO_3, SO_4, Cl)1-5H_2O$

ISLAND SILICATES

OLIVINE GROUP

Forsterite (Fo)	$4[Mg_2\{SiO_4\}]$
Olivine	$4[(Mg, Fe)_2\{SiO_4\}]$
Fayalite (Fa)	$4[Fe_2\{SiO_4\}]$
Monticellite	$4[CaMg\{SiO_4\}]$

Orthorhombic $P\ mmm$ Fo a 0·476 b 1·021 c 0·599
Fa a 0·482 b 1·048 c 0·611

Structure: The oxygen atoms at SiO_4 tetrahedral corners are in approximate hexagonal close-packing. Tetrahedra point alternately to $+x$ and $-x$, $+y$ and $-y$ as one follows each row $\|z$. Each of the intermediate cations is in octahedral co-ordination, their sites being of two kinds, half lying between tetrahedral bases (starred in Fig. 210) and half between an edge and a base. Mg^{2+} and Fe^{2+} replace each other in all proportions, and are randomly distributed over the cation sites. In monticellite the large Ca^{2+} ion replaces Mg^{2+} in the ratio 1:1 occupying the unstarred sites only.

Habit: Crystals equant or slightly elongated $\|z$, bounded by pinacoids {100}, {010} and {021} and the prism {110}. In some basanites forsterite may be elongated into rods $\|z$. Branching rod-like growths occur in the special peridotite called harrisite. Fayalite is sometimes almost amoeboid.

Twinning uncommon. Cleavage on {010} may be distinct but is often poorly developed. Fracture is conchoidal. Under the microscope characteristic curving cracks often traverse phenocrystic olivine. H 7(Fo), $6\frac{1}{2}$(Fa). D varies regularly through the series (Fig. 211). Lustre vitreous. Colour green to yellowish-green; monticellite grey to colourless.

Optics: Translucent, colourless, or pale yellowish in the case of fayalite. Properties vary regularly (Fig. 211) $\alpha = y$, $\beta = z$, $\gamma = x$; optic axial plane $\|\{001\}$. Shows high relief, shagreen surface, and bright interference colours of 2nd–3rd order ($\gamma - \alpha$0·035–0·050). Straight extinction. Very susceptible to alteration, the characteristic curving cracks often decorated with magnetite and bordered by green or yellow chlorite or serpentine. Pseudo-

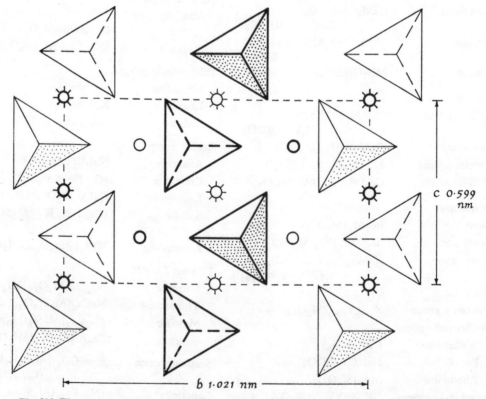

Fig. 210. The structure of olivine: plan view on 100. The units in heavy line are centred on top and bottom of the cell, those in light line at half height. Starred atoms remain Mg^{2+} in monticellite.

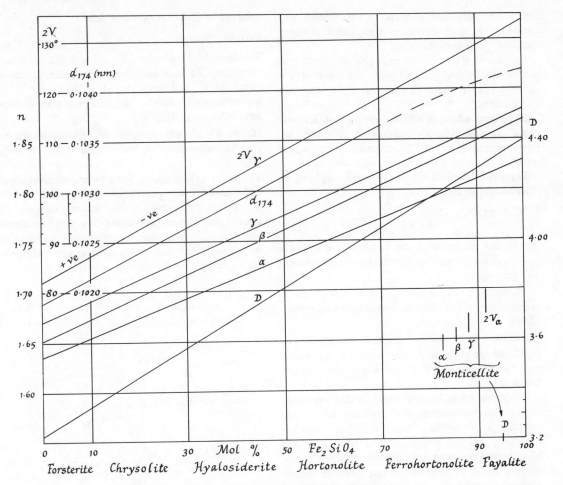

Fig. 211. Some physical properties of the olivine series.

morphs in these products are common. In unaltered crystals tiny comb-like or planar dendritic bodies of exsolved magnetite are sometimes seen.

Twinning is rare but a lamellar extinction-banding ∥ the $\alpha - \beta$ plane is sometimes shown between crossed polars in strained grains. In ordinary light no discontinuity is seen. These bands, called translation lamellae, are due to lattice distortion under tectonic stress or stresses arising during solid injection of peridotites. The optical vibration directions vary only a few degrees from one lamella to the next.

Monticellite is distinguished by its lower birefringence and smaller +ve $2V$.

Crystallization relationships. Forsterite melts at 1890°, fayalite at 1205°C, the series showing the simple liquidus and solidus curves of a perfect solid solution series. The names applied to members

of the series appear on Fig. 211. The characteristic olivine of basic igneous rocks is a forsteritic one, found associated with Ca-rich plagioclase feldspar and pyroxenes. Forsterite is not stable in the presence of free silica with which it reacts to form the pyroxene molecule $MgSiO_3$ (enstatite or pigeonite) according to the equation, $Mg_2SiO_4 + SiO_2 \rightarrow 2MgSiO_3$ and, broadly speaking, it is not found in quartz-bearing rocks. Because of the incongruent melting of this pyroxene, forsterite may precipitate at first from liquids in which it will not finally be stable, and reaction rims of $MgSiO_3$-rich pyroxene may sometimes be seen around forsteritic olivine. Forsterite is therefore characteristic of basalts and gabbros and, as accumulated masses of early-formed crystals (or perhaps in some cases incompletely melted material of the earth's mantle), it forms the principal constituent of dunite and peridotite. It is an important constituent of stony and stony-iron meteorites.

Fayalite is stable in the presence of quartz and indeed as the low-melting point member of the series is typically found in certain syenites, trachytes and granites.

In metamorphism of olivine-bearing basic igneous rocks reaction rims or coronas of orthorhombic pyroxene with spinel and hornblende develop where olivine is in contact with basic plagioclase.

Forsterite is produced in the metamorphism of siliceous dolomites:

$$2CaMg(CO_3)_2 + SiO_2 \rightarrow$$
$$Mg_2SiO_4 + 2CaCO_3 + 2CO_2$$

Subsequent alteration of the forsterite produces the green-mottled serpentine marbles known as pencatite and predazzite.

Monticellite is a high-temperature contact-metamorphic mineral of siliceous dolomitic limestones. stones.

Alteration: Rock bodies originally of dunite or peridotite are often completely altered to serpentine rock by intracrustal solutions acting at temperatures below 500°C.

Olivine in basic igneous rocks is also very susceptible to weathering; the products have been referred to serpentine, talc, carbonate, chlorite, iddingsite and bowlingite. They present a variety of appearances under the microscope from bright to pale green or yellowish-green material of low aggregate birefringence to more strongly coloured yellow-brown, brown and red-brown substances with high birefringence. It has been shown that the products called iddingsite and bowlingite are often polymineralic and may include montmorillonite-type and chlorite-type clay minerals, goethite, hematite and quartz, as well as the more obvious associated magnetite. There is often uncertainty about the relative importance of deuteric action and weathering in the alteration. Iron-rich olivines are less susceptible to alteration than magnesian ones and hortonolites do not seem to weather more easily than other silicates at outcrop. Hematite is a prominent decomposition product of fayalite. *Varieties:* Clear green forsteritic olivine of gem quality is called *peridot*.

Related minerals: Members of the *humite* group of minerals are built up of blocks of forsterite structure Mg_2SiO_4 interlayered with blocks of brucite structure $Mg(F, OH)_2$ in a stack $||z$. They are rare

minerals of limestone contact skarns.

Zircon $4[Zr\{SiO_4\}]$

Tetragonal *I* 4/*mmm*   *a* 0·662   *c* 0·602
Structure: Zr ions and SiO₄ tetrahedra alternate along lines $||z$, being centred $\frac{1}{2}c$ apart. Zr is in 8-coordination with O^{2-} at corners of surrounding tetrahedra (Fig. 212).

Habit: Square prisms {100} of variable length to breadth ratio. Also in bipyramidal {111} crystals (Fig. 62).
Twins on {112}, giving knee-twins. Cleavage not well marked. Fracture conchoidal. *H* 7½. *D* 4·65–4·70 but metamict crystals lower. Lustre vitreous to adamantine. Colour brown, red-brown; also yellowish, green, colourless (jargoon) or purplish (hyacinth). The blue stones used as gems are produced by heating the natural mineral.

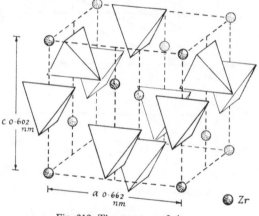

Fig. 212. The structure of zircon.

Optics: Transparent, colourless or pale yellow. ω 1·923–1·960, ε 1·968–2·015, $\varepsilon - \omega$ 0·045–0·055. Optically +ve. High relief, high double refraction and a good +ve uniaxial interference figure from square cross-sections are distinctive. Metamict crystals have lower refractive indices and birefringence than those quoted above. Often found as inclusions in biotite where it may be surrounded by a strongly pleochroic halo due to the action of its radioactive emanations on the biotite.

Atomic substitutions and radioactivity: U and Th may replace Zr in the structure making the zircon radioactive. This property may render the crystals metamict, upon which they become partly isotropic and amorphous to X-rays. Heating may restore their diffracting capacity. The altered material is called *malacon*. Hf^{4+} and the Rare Earths are also common substituents for Zr in zircon.

Occurrence: A common accessory of acid and intermediate igneous rocks, particularly the more sodic granites, syenites and felsic alkaline rocks, and their pegmatites. Particularly large concentrations occur in the alkaline rocks of the Kola Peninsula, U.S.S.R., where Zr is so concentrated in alkaline magma as to form eudialyte and a number of other rare Zr minerals in addition to zircon itself. It is abundant also in syenites and pegmatites of the Langesund and Arendal areas of S. Norway. Zircon is obtained from gem gravels in Ceylon and Thailand and beach sands in New South Wales. It is a common constituent of the heavy fraction of sandstones and its characteristics may be useful in correlation of the rocks and recognition of their provenance.

Use: An ore of Zr and a gem-stone.

Related minerals: *Thorite*, $ThSiO_4$, is isostructural with zircon, and similar in habit. It is strongly radioactive and usually metamict. *Uranothorite* contains U replacing Th, and *thorogummite* is a hydrated variety. Thorite is an accessory mineral of granites and syenites though much less widespread than zircon. Langesund and Arendal in S. Norway are well-known localities.

Sphene (titanite) $4[CaTi\{SiO_4\}(O, OH, Cl, F)]$

Monoclinic C $2m$ $a\ 0.655$ $b\ 0.870$ $c\ 0.743$
$\beta 119°\ 43'$

Structure: Besides O^{2-} forming SiO_4 tetrahedra we here have O^{2-} not bound to Si^{4+}. Along lines $\|y$ there is the repeated sequence: SiO_4 tetrahedron —O^{2-}—Ca^{2+} (Fig. 213b). Between these lines and half way between each pair of independent O^{2-} ions lies a Ti^{4+} ion. This results in an octahedral co-ordination of Ti while Ca lies within a group of $7O^{2-}$. Fig. 213a shows a plan of the arrangement on {010}.

Habit: Crystals often wedge-shaped, flattened $\|$ {001} with {111} well-developed. Also massive or in irregular grains.
Twinning on {100} rather common. Cleavage {110}; parting $\|$ {221}. $H\ 5–5\frac{1}{2}$. $D\ 3\cdot4–3\cdot56$. Lustre adamantine to resinous. Colour usually shades of brown, sometimes yellow or green.

Optics: Translucent, neutral to purplish brown. Slightly pleochroic in brown shades. $\alpha\ 1\cdot84–1\cdot95$, $\beta = y\ 1\cdot87–2\cdot03$, $\gamma \wedge z\ 35°–53°\ 1\cdot94–2\cdot11$, $\gamma - \alpha\ 0\cdot10–0\cdot16$, $2V_y\ 20°–50°$. Though the optical constants vary considerably, the typically double-wedge-shaped sections, neutral to purplish colour, high relief, and extreme birefringence giving pale pinkish high-order interference tints make sphene crystals easily recognizable under the microscope. When it occurs as part of leucoxene, a whitish fine-granular alteration product of ilmenite or titanomagnetite, positive identification is more difficult.

Tests: Soluble in H_2SO_4 the solution giving a yellow colour with H_2O_2 (test for Ti).

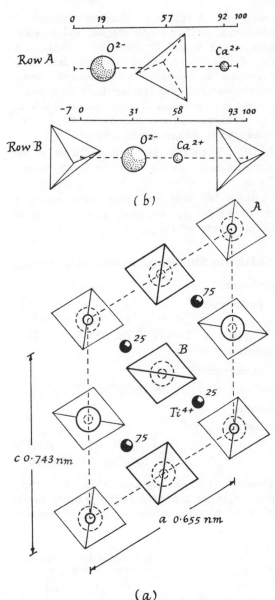

Fig. 213. The structure of sphene.

Occurrence: A common accessory mineral of granites, granodiorites and syenites and in some gneisses and schists. It occurs in abundance in the Zr–Ti–Nb-rich alkali syenites and their pegmatites in association with eudialyte, zircon and various rare Zr and Ti minerals, as, for example, in the Lovozero massif, Kola Peninsula, U.S.S.R., and at Langesund and Arendal in S. Norway. Sometimes found as a detrital mineral.

Varieties: Y, Ce and other Rare Earths or Na and K may replace Ca, while Zr, Nb or Ta may replace Ti.

Varieties: *Leucoxene* (see above under optics) is probably partly sphene, partly anatase or one of the other forms of TiO_2.

Andalusite, Sillimanite, Kyanite $4[Al_2O\{SiO_4\}]$

Andalusite

Orthorhombic				Cell Vol.
	a	b	c	(nm^3)
P mmm	0·779	0·790	0·556	0·342

Sillimanite

Orthorhombic				
P mmm	0·749	0·767	0·577	0·332

Kyanite

Triclinic				
P $\bar{1}$	0·712	0·785	0·557	0·311
	$\alpha 89° 59'$, $\beta 107° 07'$, $\gamma 106° 00'$			

Structure: The cell dimensions of these three polymorphs are controlled by the presence, in all, of chains of AlO_6 octahedra joined by common

edges, running parallel to z (Fig. 214). The chains are cross-linked by SiO_4 tetrahedra and by additional Al co-ordination polyhedra which may be octahedra (kyanite), tetrahedra (sillimanite) or a unique kind of triangular bipyramid shape giving 5-fold co-ordination (andalusite).

Distinguishing features: *Andalusite* may form anhedral crystals when small but soon grows to square section prisms often with a cross of dark inclusions along the diagonals (*chiastolite*) (Fig. 215). Square pattern {110} cleavage seen in cross-sections. Cordierite is a common associate.

Sillimanite may build well-formed prisms that stand proud on weathered rock surfaces. In other cases it is fibrous in habit (fibrolite) and may be intergrown with quartz to form dead-white knots with a silky lustre (Faserkiesel) that stand out on the weathered surface. Yet again a patchy yellowish-stained appearance of some gneisses may indicate its presence as microscopic crystals. Under the microscope the rather strong birefringence, length-slow orientation and cross-fracture are distinctive. Garnets, micas and potash feldspar are its associates.

Kyanite is distinguished in hand specimen by its blue colour and bladed form. The high relief, well-cleaved appearance, extinction angle of 30°, and the biaxial negative interference figure with large $2V$ given by the broad sections ||{100}, or by cleavage flakes, are distinctive. Bx_a is nearly $\perp 100$.

Occurrence: These three polymorphs are all found in metamorphosed aluminous sediments. Andalusite is most commonly found in contact hornfelses but appears also in regional metamorphism where stress has been deficient as in the Banff and Fraserburgh

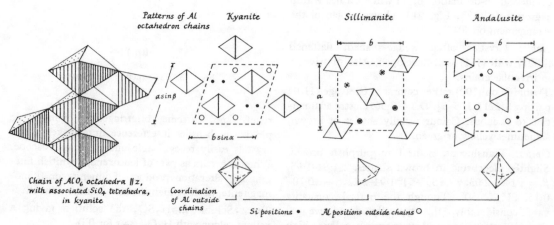

Fig. 214. Comparison of structural patterns in the aluminium silicates.

	Andalusite	Sillimanite	Kyanite
Habit	prismatic crystals of square cross-section $110 \wedge 1\bar{1}0\ 89°\ 12'$	long slender prisms or a felted mass of fibres	bladed crystals: may show lamellar twinning on {100} and {001}
Cleavage	{110} distinct, {100} poor	{010} perfect, also a cross-fracture $\perp z$	{100} perfect {010} good {001} parting
H	$7\frac{1}{2}$	6–7	5–7 varying with direction: least on {100} $\|z$, greatest on {010}.
D	3·16–3·20	3·23–3·27	3·56–3·67
Lustre	vitreous, weak	vitreous to almost adamantine	vitreous, pearly
Colour	white, pink, violet-grey	greyish-white, yellow, hair-brown, greyish-green	blue, sometimes white at margins
Optics:	transparent colourless or pink	transparent colourless	transparent colourless or pale blue. Pleochroism weak.
	$\alpha = z\ 1·629–1·649$	$= x\ 1·654–1·661$	1·715 (nearly $\perp$ 100)
	$\beta = y\ 1·633–1·653$	$= y\ 1·658–1·670$	1·722
	$\gamma = x\ 1·638–1·660$	$= z\ 1·673–1·684$	$\gamma \wedge z$ in 100 $\simeq 30°\ 1·730$
	$\gamma - \alpha\ 0·011$	0·020	0·015
Int. colour	1st order yellow $2V_\alpha\ 75°–85°$	2nd order blue $2V_\gamma\ 20°–30°$	1st order red $2V_\alpha\ 82°$

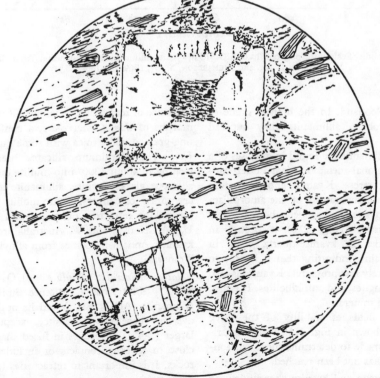

Fig. 215. Andalusite, variety chiastolite, in thin section normal to z-axis, showing regularly arranged inclusions. One section shows the cleavage.

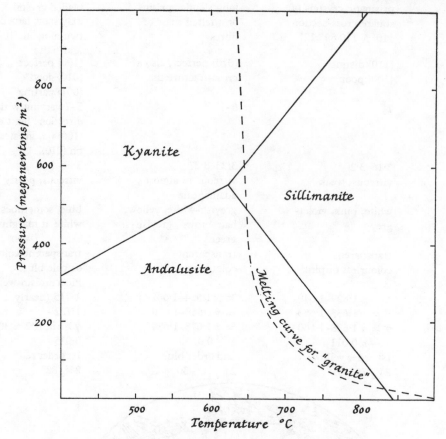

Fig. 216. Stability fields of the polymorphs of AlO{SiO₄}, kyanite, sillimanite and andalusite. (After GILBERT, M. C., BELL, P. M., and RICHARDSON, S. W., 1969, pp. 135–136.)

district of N.E. Scotland. In the highest grade of contact metamorphism sillimanite may succeed andalusite.

Kyanite and sillimanite are found in gneisses of a high grade of regional metamorphism under conditions of strong stress. Kyanite appears before sillimanite as the temperature of metamorphism increases. Though kyanite has been held to be a high-pressure mineral (Fig. 216) it occurs in unorientated blades in kyanite-quartz veins in metamorphic terrains, indicating that it is able to form under relatively low pressures. Kyanite is also found in some eclogites, and amphibolites. It is also found as a detrital mineral.

The supposed fields of stability of the three polymorphs are shown in Fig. 216. Many experiments have been made to determine these relationships but finality has not been reached.

Alteration: Andalusite and kyanite alter under the influence of retrograde metamorphism into pseudomorphs of *shimmer-aggregate*, a confused mass of fine-grained white mica with some quartz.

Use: The aluminium silicates break down on heating to about 1300° into mullite (see below) and siliceous glass. Kyanite, sillimanite and andalusite are mined for conversion into mullite which is used for making refractory porcelains and bricks. India, U.S.A., Kenya, and W. Australia are suppliers of kyanite. Andalusite comes from alluvial deposits in Transvaal.

Related minerals: *Mullite* $Al_6O_5\{SiO_4\}_2$, or $3Al_2O_3.2SiO_2$, has a structure similar to that of sillimanite $(Al_2O_3.SiO_2)$ and its optical constants overlap those of sillimanite, except that $2V$ is larger $(>45°)$. It occurs in fused shales (buchites) close to igneous contacts or included in igneous rocks. It is important in refractories (see preceding paragraph).

Topaz $4[Al_2\{SiO_4\}(OH, F)_2]$

Orthorhombic $P\ mmm$ $a\ 0.465$ $b\ 0.880$ $c\ 0.839$

Structure: Independent SiO_4 tetrahedra are linked by octahedrally co-ordinated Al. Four of the apices of an Al octahedron are O^{2-} shared with SiO_4 tetrahedra and the two remaining apices, shared between two Al octahedra, are occupied by F^- or $(OH)^-$. Fig. 217 shows the arrangement.

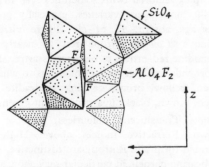

Fig. 217. Scheme of atomic linkage in topaz.

Habit: Prismatic crystals with {110} and {120}, vertically striated, often with a variety of terminal pinacoids and pyramids. $110 \wedge 1\bar{1}0 = 55° 40'$, $120 \wedge \bar{1}20 = 86° 50'$. Also columnar or granular. Cleavage {001} perfect. Fracture sub-conchoidal. Brittle. $H\ 8$ (the standard for this value). $D\ 3.49–3.6$ falling with increasing (OH). Lustre vitreous. Colour amber or paler yellow, colourless; also pale blue or green. Streak uncoloured. Sometimes fluorescent in yellow, cream or pale green. Sometimes slightly pyroelectric, but this is an anomaly in view of the centrosymmetrical structure.

Optics: Transparent, colourless in section. $\alpha = x\ 1.607–1.629$, $\beta = y\ 1.610–1.631$, $\gamma = z\ 1.617–1.638$; $\gamma - \alpha\ 0.009–0.010$; $2V_\gamma\ 48°–68°$. The indices rise and the 2V falls with increase in (OH). Dispersion $r > v$ distinct. The interference colour is similar to that of quartz, but the relief is moderately high and the crystals generally euhedral. Basal sections or cleavage flakes give a +ve biaxial interference figure with a fairly large 2V. It may alter to white mica.

Tests: Its crystal form, hardness and cleavage are distinctive. The optical properties of a cleavage flake should be decisive.

Occurrence: Found in cavities in granites, in granite pegmatites and high-temperature veins, and in a few rhyolites. Its usual associates are cassiterite, tourmaline, fluorite and muscovite. The crystals in pegmatites occasionally attain a size measured in metres. Large crystals come from Transbaikalia and the Ural Mountains in the U.S.S.R. and from Minas Gerais province in Brazil. It is found in the tin veins of Cornwall and in cavities in the Mourne Mountains granite, N. Ireland.

Use: As a gem stone.

Staurolite $2[Fe_2Al_9\{SiO_4\}_4(O, OH)_8]$

Monoclinic $C\ 2/m$ $a\ 0.782$ $b\ 1.652$ $c\ 0.563$ (pseudo-orthorhombic) $\beta\ 90°$

Structure: Based on the same pattern of chains of AlO_6 octahedra and linking SiO_4 tetrahedra as kyanite. Between each pair of cells (a little offset)

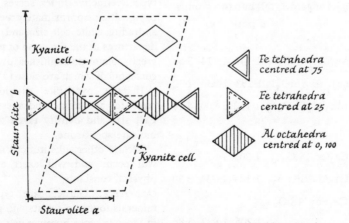

Fig. 218. Structural plan of staurolite, showing its relationship to kyanite.

of kyanite structure is interposed, parallel to 100 of kyanite, a sheet of atoms $Fe_2Al(O, OH)_4$. These pick up bonds to the kyanite elements on each side to form the polyhedra shown in Fig. 218. Ideally the formula contains $O_6(OH)_2$ but this is invariably exceeded (with omission of Si and Fe^{2+}).

Habit: Prismatic crystals with {110}, {010} and {001}. $110 \wedge 1\bar{1}0 = 50° 26'$.

Twins. Famous for its interpenetrant twinning on {031} giving a rectangular cross (Fig. 72) and on {231} giving a 60° cross. Also repeated twins. Cleavage {010} fair. H 7–7½. D 3·75. Lustre vitreous to resinous. Colour yellowish to dark reddish brown.

Optics: Translucent, pleochroic in yellow tints. $\alpha = y$ 1·739–1·747 colourless; $\beta = x$ 1·745–1·753 pale yellow; $\gamma = z$ 1·752–1·761 golden yellow; $\gamma - \alpha$ 0·014; $2V_\gamma$ 80°–90°. Usually in euhedral prismatic crystals, with quartz inclusions. Extinction is straight in most sections, or symmetrical in cross-sections. Interference colour yellow to orange. Dispersion $r > v$.

Occurrence: In metamorphosed argillaceous sediments in which sufficient iron is present. Its occurrence in a sequence of increasing metamorphism is somewhat erratic on account of this requirement. It is associated with kyanite and often intergrown with it {010} staurolite being parallel to {100} kyanite. Garnet, micas, quartz and plagioclase are other associates. It shows as yellow-brown prisms standing out on the weathered surface of the rock. Also found as a detrital mineral.

GARNET GROUP $8[R_3^{2+}R_2^{3+}\{SiO_4\}_3]$

Cubic I $m3m$ Cell sizes below

The principal species of garnets fall into two groups:

		a(nm)	n	D
Pyralspite				
Pyrope	$Mg_3Al_2\{SiO_4\}_3$	1·146	1·714	3·58
Almandine	$Fe_3^{2+}Al_2\{SiO_4\}_3$	1·153	1·830	4·32
Spessartine	$Mn_3Al_2\{SiO_4\}_3$	1·162	1·800	4·19
Ugrandite				
Uvarovite	$Ca_3Cr_2\{SiO_4\}_3$	1·201	1·87	3·85
Grossular	$Ca_3Al_2\{SiO_4\}_3$	1·185	1·734	3·59
Andradite	$Ca_3Fe_2^{3+}\{SiO_4\}_3$	1·205	1·887	3·86

Structure: Imagine two kinds of totem poles (Fig. 219a and b). Type (b) may be either a right-handed or left-handed screw axis. These are interplanted alternately along diagonals of the 001 face of the cubic unit cell (Fig. 219c).

Habit: The basic crystal forms are the rhombic dodecahedron {110} and the trapezohedron {211}, or {110} with its edges bevelled by {211}. {321} occurs sometimes. Also granular massive.

Complex sectorial twins sometimes seen in anomalously birefringent varieties, especially grossular. Cleavage absent. H 6½–7½. Lustre vitreous to resinous. *Colour*: pyrope red to nearly black; almandine red-brown to black; spessartine dark red to brownish red; grossular colourless to white, pale green, yellow, brown or pink; andradite yellow, green, brown, black; uvarovite bright green.

Optics: Translucent, colourless, pink or yellow brown. Refractive indices above. High relief, rounded form and isotropy are distinctive. But the Ca bearing garnets in particular may show appreciable birefringence. The grey interference colours may define pyramidal sectors, with their apices towards the centre and their bases on the {110} faces, in which optical behaviour differs. Sectors may be measurably biaxial.

In metamorphic rocks inclusions are common, which may be zonally or spirally arranged.

Distinguishing features. The three parameters D, n and a may be used to obtain a rough idea of garnet composition on the assumption that the change in properties is linear between the end-members. A break in compositional variation separates pyralspite from ugrandite. We may therefore make an arbitrary division at $a = 1·170$ nm into Ca-rich and Ca-poor garnets (Fig. 220). Amongst Ca-rich types, refractive index serves to locate the garnet composition approximately between grossular and andradite while cell size and density will indicate departures in the direction of pyralspite. The bright green colour distinguishes uvarovite, the physical constants of which are close to those of andradite.

In the pyralspite division refractive index separates almandine and spessartine from pyrope. Cell size and density taken together will help to indicate spessartine.

Clearly other substitutions than those mentioned will occur and will exert an influence on the physical constants.

Occurrence: Garnets are typically metamorphic minerals found in schists and gneisses. Some occur in igneous rocks however as set out below.

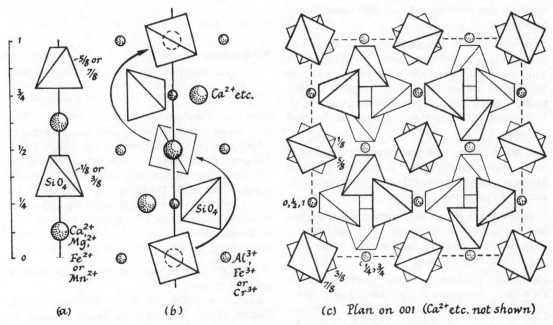

Fig. 219. The structure of garnet (based on STRUNZ, H., 1957).

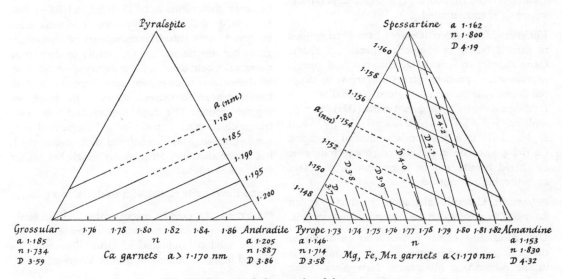

Fig. 220. Some physical properties of the garnet group.

Pyrope is found in ultrabasic igneous rocks particularly kimberlite, and related serpentinites, and in the dense rock eclogite, which is equivalent to basalt in composition but has crystallized under high pressures.

Almandine is the common garnet of middle and high grades of regional metamorphism, being found in a wide range of schists and gneisses formed from mudstones, impure sandstones, or basic igneous rocks. It also occurs in contact metamorphic rocks. It is found, rather rarely, in certain granites and rhyolites.

Spessartine is found in granite pegmatites and in metasomatic skarn deposits. The garnet of the lower grades of regional metamorphism is usually relatively rich in the spessartine molecule.

Grossular is the garnet of metamorphosed impure limestones, whether regionally or contact altered. It is associated with calcite, diopside, wollastonite and vesuvianite (idocrase).

Andradite is produced by contact metamorphism of limestones accompanied by metasomatic introduction of iron. It is common in the limestone skarn iron-ores of Sweden and elsewhere. Its titaniferous analogue *melanite* is found in nepheline syenites as at Loch Borolan, Sutherland, and elsewhere in carbonatites associated with alkali syenites.

Uvarovite is uncommon, being found in serpentinites where Cr is concentrated.

Varieties: Many different names have been applied to garnets falling in the series mentioned above. Gem varieties are *rhodolite*, rose-red and purple (pyrope-rich); precious garnet or *carbuncle*, deep-red (almandine); *demantoid*, green (andradite). Hydrogrossular is a Ca garnet with $(OH)_4$ groups replacing some SiO_4, which has n 1·711–1·724 and D 3·39–3·47. It is a secondary mineral replacing Ca-rich plagioclase in the altered gabbro rodingite, in altered basalts and in some lime-rich contact metamorphic assemblages.

Alteration: Chloritic alteration products, called *kelyphite*, may form coronas around garnet crystals.

Use: Almandine garnet is useful as an abrasive: the principal source of material for this purpose is in the Adirondack Mountains, New York. Gem quality garnets come from Czechoslovakia (a pyrope), Ceylon, India and Brazil (almandines), and Nizhne-Tagilsk in the Ural Mountains (demantoid).

The name garnet comes from pomegranate, the red seeds of which suggest the typical occurrence of the mineral.

Vesuvianite (Idocrase)

$$4[Ca_{10}(Mg, Fe)_2Al_4\{SiO_4\}_5\{Si_2O_7\}_2(OH, F)_4]$$

Tetragonal P 4/*mmm* a 1·552 c 1·179

Structure: Two quadrants of the plan view (Fig. 221) are of garnet structure, with the two types of totem poles (a) and (b) shown in Fig. 219, in which SiO_4 is in independent tetrahedra. The other quadrants have Al omitted, the SiO_4 tetrahedra being joined in Si_2O_7 groups, and strings of Ca ions run parallel to z-axis. In linking position is a chain of octahedrally co-ordinated Fe with two octahedral apices formed by $(OH)^-$, the other apices being shared with SiO_4 groups.

Habit: Prismatic or pyramidal crystals; columnar or granular massive.
Cleavage or parting on {001}. Fracture uneven. Brittle. H 6½. D 3·33–3·45. Lustre vitreous to resinous. Colour brown to yellow-brown or green. Streak white.

Optics: Transparent, colourless to pale green or brown, sometimes with colour zoning. There may be weak dichroism. ω 1·703–1·752, ε 1·700–1·746, $\omega - \varepsilon$ 0·003–0·006. Optically −ve. The weak birefringence gives interference colours of first-order grey; but anomalous brown, purple or deep blue colours not belonging to the ordinary spectrum may be shown, due to dispersion superimposed on weak birefringence. Sometimes shows a biaxial interference figure. The high indices and low birefringence coupled with the crystal form, if shown, are distinctive. May be confused with zoisite which has better cleavage, or grossular which has higher indices and equant form.

Tests: Fuses quite easily, with swelling, to a glass.

Occurrence: In contact-metamorphic impure limestones along with lime-garnet, diopside, wollastonite, epidote and calcite. Originally found in ejected limestone blocks at Mount Vesuvius.

Varieties: *Wiluite* is positive biaxial material containing boron, from Siberia. *Colophonite* comes from Arendal, S. Norway, and *cyprine*, which is blue, from Telemarken also in S. Norway.

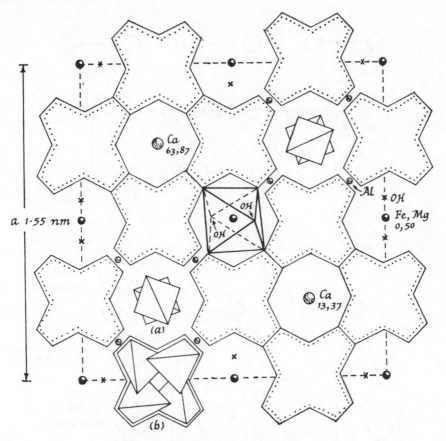

Fig. 221. The structure of vesuvianite (cf. Fig. 219).

Chloritoid

$$4[\{((Fe, Mg, Mn)_2, Al)O_2 (OH)_4\}Al_3\{SiO_4\}_2]$$

Monoclinic C $2/m$ a 0·952 b 0·547 c 1·819
(also Triclinic) β 101° 39′

Structure: A layer structure related to micas but differing in that the SiO_4 tetrahedra are not linked to each other in a continuous layer. Layers of (Mg etc.)$O_2(OH)_4$ octahedra and of AlO_6 octahedra alternate along z joined by planes of independent SiO_4 tetrahedra. All O^{2-} are joined on one side to Si^{4+} (Fig. 222).

Habit: Scales or thin plates, disseminated or grouped in rosettes; foliated masses.

Twins on mica-type laws with {001} composition plane. Cleavage on {001} good (less so than micas) into brittle flakes; also two poor prismatic cleavages intersecting at about 60°. H 6½. D 3·5–3·8. Lustre rather weak, pearly. Colour dark green.

Optics: Transparent; pleochroic, grey-green to colourless. $\alpha = y$ grey-green 1·71–1·73; β grey-blue, indigo 1·72–1·73; $\gamma \wedge z = 3$–30° 1·72–1·74; $\gamma - \alpha$ 0·006–0·02; $2V_\gamma$ variable, 36° to 70°, also larger, becoming optically negative. Orientation variable. Dispersion strong, $r > v$. Relief high, greater than that of chlorites. Weak birefringence coupled with strong dispersion may lead to anomalous interference colours. Extinction angle variable, not straight (cf. mica and chlorite). Interference figure shows moderate to large $2V$. Often full of inclusions which may be arranged in hourglass pattern.

Occurrence: In phyllites and mica schists of low-grade regional metamorphism of Al- and Fe-rich rocks.

Varieties: Mg-rich material is called *sismondine*, the Mn-rich variety is *ottrelite*.

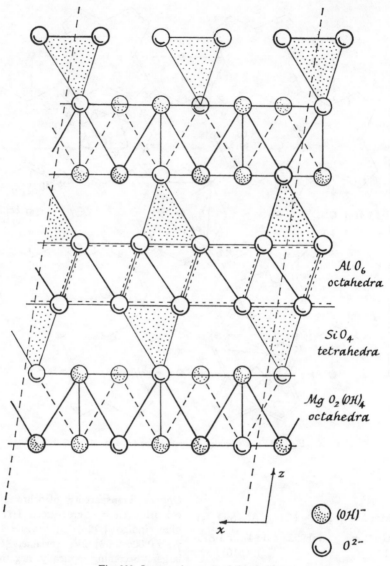

Fig. 222. Structural pattern of chloritoid.

Braunite $8[Mn^{2+}Mn_6^{3+}O_8\{SiO_4\}]$

Tetragonal I 4/m $\quad$ $a\,0 \cdot 938$ $\quad$ $c\,1 \cdot 867$

Structure: An island silicate, 8 separate SiO_4 tetrahedra forming an essential part of the structure. There are 48 Mn in octahedral, and 8 in 12-fold co-ordination.

Habit: In pseudo-octahedra {101}, or massive. Contact twins on {112}. Cleavage {112} perfect. Brittle, with uneven fracture. $H\,6–6\frac{1}{2}$. $D\,4 \cdot 75–4 \cdot 82$. Lustre sub-metallic. Colour and streak dark brown-black to dark grey.

Optics: Opaque, greyish-white with a brown tinge, $R\%$ 18–20, polishing hardness near that of pyrite.

Tests: Fused in borax, gives an amethyst-coloured bead in the oxidizing flame, indicating Mn. The sodium carbonate bead is green. Soluble in HCl with a residue of gelatinous silica.

Occurrence: With pyrolusite and psilomelane in secondary manganese ores formed by weathering. Also in veins where it may be primary. Found in many manganese deposits including Ilfield, Harz Mts; Långban, Sweden; Nagpur, India; and Batesville, Arkansas.

COUPLET SILICATES

EPIDOTE GROUP

Zoisite $4[Ca_2AlAl_2\{SiO_4\}\{Si_2O_7\}O(OH)]$

Orthorhombic P mmm a 1·619 b 0·545 c 1·013

Clinozoisite–Epidote series

$2[Ca_2(Al, Fe^{3+})Al_2\{SiO_4\}\{Si_2O_7\}O(OH)]$

Monoclinic P $2/m$ a 0·896 b 0·563 c 1·030
$\beta 115° 24'$

Allanite (Orthite)

$2[(Ca, Ce)_2(Fe^{2+}, Fe^{3+})Al_2\{SiO_4\}\{Si_2O_7\}O(OH)]$

Monoclinic P $2/m$ a 0·898 b 0·575 c 1·023
$\beta 115° 00'$

Structure: In epidote (Fig. 223), chains of AlO_6 alternating with chains of $AlO_4(OH)_2$ octahedra joined edge to edge run along y (cf. the chains along z in sillimanite). The chains are linked together by independent SiO_4 groups and Si_2O_7 groups. In the spaces of this framework are Ca^{2+} and (Al, Fe^{3+}) ions. Zoisite has $a \simeq 2a \sin \beta$ of epidote, the relation between monoclinic and orthorhombic structures being similar to that between ortho- and clinopyroxene.

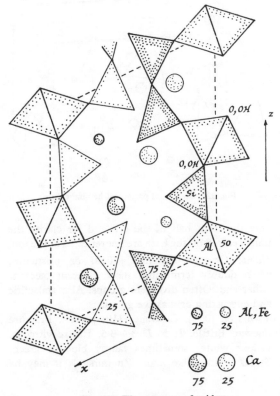

Fig. 223. The structure of epidote.

Habit: Crystals elongated and striated along y, with {001}, {100} and {$\bar{1}$01} prominent; 001 $\wedge$ 100 64° 36'. Columnar, or coarse- or fine-granular. Twins on {100}, sometimes lamellar, may be shown by clinozoisite-epidote. Cleavage {100} perfect in zoisite, {001} perfect in the others. Fracture uneven. Brittle. H 6½–7, 5½–6 in allanite. D 3·25–3·6, increasing with Fe; up to 4·2 in allanite. Lustre vitreous, to resinous or submetallic in allanite. Colour: zoisite grey to white or yellow-brown, also rose-pink in variety *thulite*; epidote a characteristic yellow-green (pistachio green) to greenish black; piemontite, Mn-bearing epidote, a violet-red to red-brown; allanite brown to black. Streak uncoloured or grey-brown.

Optics: All members have straight extinction relative to the length of the crystals and (except for α-zoisite) show $+$ve/$-$ve elongation.

Zoisite has two forms, α- and β-zoisite. Colourless in thin section. Thulite is pink and pleochroic. In α-zoisite $\alpha = y$, $\beta = x$, $\gamma = z$ Opt. ax. pl. 100 ($=$ cleavage). In β-zoisite $\alpha = x$, $\beta = y$, $\gamma = z$ Opt. ax. pl. 010.

α 1·685–1·705, β 1·688–1·710, γ 1·697–1·725, $\gamma - \alpha$ 0·004–0·008, $2V_y$ 0–60°. Dispersion marked $r > v$, $r < v$ respectively in the two types. Salient features are anomalous blue to normal grey interference colours (the former due to dispersion $+$ low double refraction) and straight extinction referred to cleavage in cross-sections.

Clinozoisite-epidote is colourless to yellowish in section, the colour often patchily distributed and increasing with Fe. When coloured shows distinct pleochroism. $\alpha \wedge z$ 0–15° 1·67–1·75, $\beta = y$ 1·67–1·78, γ 1·69–1·80; $\gamma - \alpha$ 0·005–0·05; $2V_\alpha$ 14°–90° (clinozoisite) $2V_\alpha$ 60°–90° epidote. Extinction angle $\gamma \wedge 001$ cleavage 33°–41° in cross-sections. The indices increase with Fe, but more noticeable is the increase in birefringence.

Clinozoisite, the iron-poor end-member has weak grey to yellow interference colours, epidote, richer in iron, has bright 3rd order orange-reds and greens of distinctive quality. They are often unevenly distributed over the crystal. The division of the series at $2V$ 90° corresponds to a ratio $Fe^{3+}:(Al + Fe^{3+})$ of about 1:7.

Piemontite with Mn from 1 in 10 to 1 in 4 of the interstitial cations has indices at the upper end of, and even higher than, the range quoted above, and $2V_y$ 60–90°. It is violet-red or pink and pleochroic in red and yellow; it is detectable by its colour in hand specimens even when finely disseminated.

Allanite is yellow-brown to brown and pleochroic. $\alpha \wedge z$ up to $40°$ $1·69$–$1·79$, $\beta = y$ $1·70$–$1·82$, γ $1·71$–$1·83$; $\gamma - \alpha$ $0·013$–$0·036$; $2V_\alpha$ large, variable. Also metamict due to Th content, when it may be isotropic. Radial cracking of surrounding minerals is then observed.

Tests: Epidote fuses with swelling to a black magnetic glass. Allanite is often somewhat radioactive.

Occurrence: Zoisite and clinozoisite-epidote form by low-temperature alteration of lime-plagioclase in basic igneous rocks and form part of the aggregate called *saussurite* replacing plagioclase. Epidote forms in vesicles and as veins in basic lavas, often with calcite and chlorite. The group is of major importance in schists and amphibolites of the low and medium grades of metamorphism, and epidote is common in metamorphosed limestones with lime-garnet, actinolite, vesuvianite and calcite.

Piemontite is found in schists in which Mn is present.

Allanite is found in granites, syenites and their pegmatites with other Rare Earth minerals. It may contain Ce, Y, La, Th, Ti and Mn.

Related minerals: *Withamite* from Glencoe, Argyllshire, Scotland, is a pink epidote with some Mn, found in vesicles and veins in andesite. *Pumpellyite* is a hydrous epidote-like mineral of wide distribution found in the vesicles of ancient lavas (as in the original locality in the Keweenawan lavas of the Canadian Shield) and in low grade metamorphosed tuffs. It forms small crystals elongated along y and often interwoven, that show anomalous blue interference colours. It is distinguished from epidote and zoisite by the characteristic blue-green colour of the β vibration direction. α and γ are colourless to pale yellow-green or yellow-brown.

Lawsonite $4[CaAl_2(OH)_2\{Si_2O_7\}H_2O]$

Orthorhombic B 222 a $0·89$ b $1·33$ c $0·58$

Structure: Chains of Al(O, OH) octahedra along z linked by Si_2O_7 groups, rather like epidote, with Ca and H_2O in the interstices. Two orientations with y and z interchanged, have been used.

Habit: Usually as disseminated porphyroblasts with rhombic or rectangular cross-sections, flattened parallel to {010}, elongated ||α. Twins on {101} may be lamellar. Cleavage {010}, {100} perfect, {101} fair. $H7$. $D3·09$. Lustre vitreous to greasy. Colour white to blue-grey.

Optics: Transparent, colourless, blue-green, yellow, pleochroic. $\alpha = z$ $1·665$, $\beta = x$ $1·675$, $\gamma = y$ $1·685$; $\gamma - \alpha$ $0·02$; $2V_\gamma$ $80°$; $\alpha \wedge$ {101} $34°$. Interference colour up to 2nd order blue, and the cleavages distinguish it from zoisite. Prehnite has higher birefringence. Twinning is commonly seen. May be intergrown with pumpellyite.

Occurrence: In glaucophane schists and associated rocks along with chlorite, epidote, pumpellyite, mica, garnet, sphene and quartz. Initially described from California it is being increasingly widely recognized in rocks metamorphosed under low temperature and high pressure.

Hemimorphite $2[Zn_4\{Si_2O_7\}(OH)_2.H_2O]$

Orthorhombic I mm a $0·838$ b $1·072$ c $0·512$

Structure: Pairs of SiO_4 tetrahedra linked by one corner to form Si_2O_7 groups are arranged with their bases parallel to 001 and their apices all pointing in the same sense along z. This confers a polar character on the whole structure, the two ends of z being structurally different. Zn and OH ions, and H_2O molecules are arranged between the groups as shown in Fig. 224. The figure shows

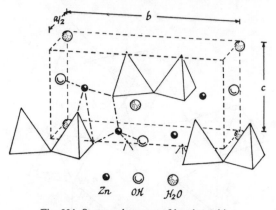

Fig. 224. Structural pattern of hemimorphite.

only the front half of the cell and for clarity the units centred on the base are not repeated at the top.

Habit: Crystals flattened ||{010} or prismatic; when doubly terminated show different faces at either end. Often divergent groups. Also stalactitic and encrusting or massive granular.

Twins on {001}. Cleavage {110} perfect. Fracture uneven. Brittle. H 5. D $3·4$–$3·5$. Lustre vitreous. Colour white; sometimes faintly blue or green. Streak white. Like many Zn minerals it may be fluorescent in UV light, the colours being orange, bluish, white or yellow. Phosphorescent when rubbed. Strongly pyroelectric and piezoelectric.

Optics: Transparent, colourless. $\alpha = y\,1.614$, $\beta = x\,1.617$, $\gamma = z\,1.636$; $2V_\gamma\,46°$.

Tests: Decrepitates and yields water on heating in the closed tube. Heated with sodium carbonate on charcoal it gives a coating that is yellow hot, white cold, which moistened with cobalt nitrate solution and reheated turns green. Gelatinizes with acids.

Occurrence: A mineral of the zone of oxidation of zinc-bearing veins. It is said to invert to *willemite* Zn_2SiO_4 at 240°C and must therefore form below this temperature. It is found with smithsonite, sphalerite, galena and their associates. Occurrences in England include Roughten Gill, and the Alston district in Cumberland and Matlock, Derbyshire.

Use: An ore of zinc.

The old name *calamine* should be discarded as it has led to confusion with smithsonite $ZnCO_3$ to which it was also formerly applied.

MELILITE SERIES

| **Åkermanite** | $2[Ca_2Mg\{Si_2O_7\}]$ |
| **Gehlenite** | $2[Ca_2Al\{AlSiO_7\}]$ |

Tetragonal $P\ \bar{4}2m$ $a\ 0.78$ $c\ 0.50$

Structure: SiO_4 tetrahedra joined in pairs to form Si_2O_7 groups are linked by octahedrally co-ordinated Mg^{2+} or Al^{3+}. Sheets of these elements are united through bonds to Ca^{2+} in 8-co-ordination.

Habit: Short square prisms or square tablets.

Cleavage {001} fair. H 5. D 2.95–3.05. Lustre vitreous. Colour white or yellowish.

Optics: Transparent, colourless to pale yellow. ω 1.632 Åk.–1.669 Geh., ε 1.640 Åk.–1.658 Geh. $\omega - \varepsilon$ 0.00–0.013, usually about 0.005. Optically +ve or −ve usually the latter. Characterized by oblong cross-sections of fairly high relief and low birefringence giving grey interference colours or an anomalous blue-grey. Extinction is straight and the interference figure usually negative. Lines normal to the length of the sections constitute the so-called 'peg-structure' which may help in recognition.

Occurrence: In silica-deficient lavas with nepheline or leucite and in melilite-basalts and alnoite. More rarely in coarse-grained alkali gabbro associated with incorporation of limestone. Gehlenite is found in metamorphic limestones, and åkermanite in metamorphosed dolomites.

RING SILICATES

| **Beryl** | $Be_3Al_2\{Si_6O_{18}\}$ |
| Hexagonal C $6/mmm$ | a 0.919 c 0.919 |

Structure: Six-membered rings of SiO_4 tetrahedra, giving Si_6O_{18} groups, are centred at heights 0, 50, 100 one above the other along the c dimension. Sheets of Be and Al ions lie at heights 25, 75, between the layers of rings (Fig. 225). Four bonds from Be and six from Al tie the rings together horizontally and vertically. Large cations (e.g. Na^+) may reside in the tunnels along the stacks of rings.

Habit: Simple hexagonal prisms; sometimes columnar groups; massive.

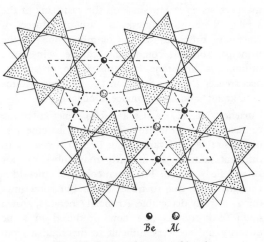

Be Al

Fig. 225. Structural pattern of beryl.

Cleavage {0001} poor. Fracture conchoidal to uneven. Brittle. H 8. D 2.66–2.80 increasing in a general way with entry of foreign ions and fall in Be. Because of the variety of substituents D, is only a rough guide to Be content. Lustre vitreous. Colour pale green to bright green (emerald), bluish green (aquamarine), yellow and white.

Optics: Translucent, colourless in thin section. In thicker sections may show weak pleochroism in greens. ω 1.568–1.601, ε 1.564–1.594, $\omega - \varepsilon$ 0.004–0.008. Optically negative, length-fast. Interference colour grey to white with a hint of yellow. The refractive indices, like the D, increase with the entry of foreign ions into the structure and give a rough guide to the content of the latter, and to the Be content.

Tests: Distinguished from apatite by hardness and lower refractive indices. Insoluble in acids. From quartz, which it much resembles when white and massive, by its hardness and generally slightly greater D (test against quartz in a tube of heavy liquid). A better test is to fuse briefly in KOH, dissolve in water, and add a little of a weak solution of quinalizarin and KOH in water. A cornflower blue colour indicates Be.

Occurrence: In cavities in granites, in granite pegmatites and greisens. The crystals may grow to great size, many metres in length and tons in weight. Its associates are quartz, feldspar, micas, topaz, tourmaline, cassiterite, columbite and pyrochlore. Aquamarine from Nerchinsk Mountains, Transbaikalia, U.S.S.R. has this paragenesis. Fine aquamarines come also from Brazil. Also in some mica schists as at Mursinsk on the east flank of the Urals, N. of Sverdlovsk, a famous source of emeralds. Emeralds from Colombia, one of the principal sources, come from calcite veins in a bituminous limestone but this is an anomalous occurrence. Beryl is found in the Mourne Mountains granite and near Dungloe, Donegal, Eire.

Varieties: Na, K, Ca, Cs and H_2O may enter the tunnels in the structure with some balancing loss of Be, so that BeO varies from 13·8 to 10·5 wt% in different samples. Li and Mg may substitute for Be and Fe^{3+} for Al. The green colour of emerald is thought to be due to traces of Cr but sometimes this is scarcely detectable. Artificial emeralds, grown upon seed crystals, are now produced in some quantity and are very difficult to distinguish from the natural stone.

Use: Beryllium metal, of which beryl is the ore, is finding increasing use in industry in alloys with copper, in electrical porcelains, for the windows of X-ray tubes, and (formerly) in fluorescent lighting tubes. It is dangerous to people working with it, the oxide giving rise to an inflammatory lung disease or an inflamed skin condition.

Cordierite $\quad$ $Al_3(Mg, Fe^{2+})_2\{Si_5AlO_{18}\}$

Orthorhombic $\quad C$ $\quad mmm$ $\quad a\ 0·969$ $\quad b\ 1·706$ $\quad c\ 0·937$

Structure: Closely allied to that of beryl, with stacks of Si_5AlO_{18} rings running parallel to z, Al occupying the Be sites and Mg the Al sites of beryl. Additional Al replaces 1 in 6 of the Si in the rings. The y-axis corresponds to the long diagonal of the beryl cell-plan. H_2O may reside in the tunnels of the structure. Indialite, a dimorph of cordierite, is isostructural with beryl.

Habit: Pseudohexagonal short prisms, $110 \wedge \bar{1}10 =$

60° 50′; usually as anhedral or enwrapping crystals. Twins on {110} and {130} yield pseudohexagonal forms. The twins may be sectorial or interpenetrant. Cleavage {010} poor; parting on {001} seen in altered grains. $H\ 7$. $D\ 2·55$–$2·75$ increasing with Fe. Colour pale blue to violet. Lustre vitreous.

Optics: Transparent, colourless in thin section but pleochroic in thick sections with α pale yellow, β pale blue, γ dark blue or violet. $\alpha = z\ 1·530$–$1·560$, $\beta = x\ 1·535$–$1·574$, $\gamma = y\ 1·538$–$1·578$; $\gamma - \alpha\ 0·008$–$0·011$; $2V_\alpha 42°$–$90°$, sometimes positive with $2V_\gamma 76°$–$90°$. The refractive indices increase with Fe. Optical properties may also be influenced by the degree of approach to the hexagonal symmetry shown by indialite, $2V_\alpha$ decreasing and n increasing. Often in anhedral crystals enclosing other minerals and with dusty opaque inclusions. Sectorial twinning seen in basal sections. Pleochroic haloes in otherwise colourless material may surround radioactive inclusions (commonly zircon). Frequently altered to fine-grained mica, the aggregate being known as pinite. Exceedingly like quartz on casual examination. Sector twinning and enwrapping habit help to distinguish it and the biaxial character clinches the identification.

Occurrence: Widespread in hornfelses produced by thermal metamorphism of argillaceous rocks. Also in argillaceous rocks subjected to regional metamorphism in which stress has been deficient as in the Banff district of N.E. Scotland. Often indicated as oval bluish spots or oval cavities from which it has weathered away. In these occurrences associated with biotite, andalusite or sillimanite, spinel or garnet. Again in norites resulting from the incorporation of argillaceous rocks by gabbroic magma, as at Haddo House, Aberdeenshire. It also occurs in some granites (e.g. Dartmoor) from inclusions of metamorphosed shale or crystallized from contaminated magma. Muscovite and garnet accompany it. It may crystallize from fused xenoliths of shale (buchites) in some cases: this high-temperature material is likely to be indialite.

Varieties: *Iolite* is a name applied to violet-coloured material, and *dichroite* referring to the colour change in polarized light is another name. A progressive change of cordierite to indialite with rise of temperature is reported, probably due to progressive disordering of (Si$_5$Al) and $Al_3(MgFe^{2+})_2$ over available lattice sites. The departure from hexagonal symmetry is measurable from the degree of separation of the three X-ray powder lines into which the $21\bar{3}1$ reflection of indialite is resolved.

Use: Blue crystals from alluvium in Ceylon are used as gem stones (saphir d'eau).

Tourmaline

$$(Na, Ca)(Li, Mg, Al)_3(Al, Fe, Mn)_6(OH)_4$$
$$(BO_3)_3\{Si_6O_{18}\}$$

Trigonal *R* 3*m* *a* 1·60 *c* 0·717 (hexagonal axes)

Structure: The rather daunting formula reduces readily to order. A string of alternating Na^+ and $(OH)^-$ runs up the centre of a stack of Si_6O_{18} rings, in which the individual tetrahedral bases lie in a plane and the apices all point in the same direction through the stack (Fig. 226). Interlayered with the rings are sheets of triangular BO_3 groups,

three above each ring. The Li bracket of atoms, lining the tube, hold the stack together; and a ring of (Al, Fe, Mn) plus a triangle of (OH), outside the stack, link it to adjacent stacks.

Habit: Long prismatic crystals of triangular cross-section, often terminated by the rhombohedron $\{10\bar{1}1\}$, or $\{02\bar{2}1\}$ as in Fig. 21, and vertically striated. Also columnar masses and radiating groups.

Cleavage not well marked. *H* 7½. *D* 3·0–3·2, increasing with Fe (and Mn), the upper limit corresponding to about 15 wt% of total iron oxide, largely FeO. Lustre vitreous. Colour commonly black, also brown, blue, green, red. Colour-zoning is common and the colour may vary along

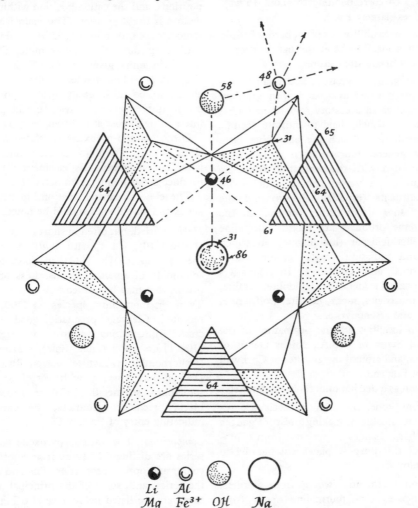

Fig. 226. The structure of tourmaline.

the crystal, e.g. from green to blue or black. Rarely it is colourless. Strongly pyroelectric and piezo-electric, also electrified by friction.

Optics: Translucent, the depth of colour varying from blue, neutral tint, olive-green or yellow-brown to colourless. Iron-rich types (schorl) have the deeper colours, magnesium-rich types (dravite) are yellow, and lithium-rich types (elbaite) are colourless except in thick section, when they may be pink. Coloured varieties are strongly dichroic with absorption $\omega > \varepsilon$ so that light transmission is much less for the ray vibrating across the length of the crystal. $\omega\, 1\cdot635{-}1\cdot675$, $\varepsilon\, 1\cdot610{-}1\cdot650$, $\omega - \varepsilon\, 0\cdot02{-}0\cdot035$. Optically $-$ve. The indices increase regularly with Fe (and Mn) content, a value of $\varepsilon = 1\cdot64$ corresponding to about 15 wt% total iron oxides (largely FeO).

Tests: Variable in fusibility but often hardly fusible. Unattacked by acids. The habit and optical properties are the best diagnostic features.

Occurrence: In granites, greisens and granite peg-matites or quartz-veins associated with granites. Also widespread as an accessory mineral in schists and gneisses, and usually held to indicate granitic magma at depth from which the boron has been supplied. In general tourmaline is a late-stage mineral, associated with the separation from the granite magma of a vapour phase enriched in volatile constituents (especially boron) and the tourmaline often replaces earlier minerals of the granite. Extreme products are quartz-tourmaline rock with relict feldspar (luxullianite), in which may be found radiating crystal groups called 'tourmaline suns', and schorl-rock. In pegmatites and veins associated minerals are lepidolite, petalite, spodumene, muscovite, topaz, fluorite, wolframite, molybdenite, and arsenopyrite.

Well-known localities are the pegmatites of the New England States of U.S.A., Minas Gerais of Brazil, Ceylon, and around the granites of Cornwall and Devon in England.

It is common as a detrital mineral in sandstones.

Use: As a gem stone. Its strong piezoelectric response makes it valuable in making pressure gauges to measure sharp changes in pressure from explosions etc. For this purpose plates cut parallel to 0001 are employed.

Varieties: Iron tourmaline (*schorl*) from granites, schists and gneisses; alkali tourmaline (*elbaite*) from pegmatites; magnesium tourmaline (*dravite*) from limestones and dolomites. *Rubellite* is a well-known pink alkali tourmaline found embedded in lepidolite.

CHAIN SILICATES

PYROXENE GROUP

The pyroxenes, though crystallizing in two systems, the orthorhombic and monoclinic, have a close community of structure and chemistry.

Structure: All are built on chains of SiO_4 tetrahedra (Fig. 227) running parallel to z cross-linked by bonds to intermediate cations distributed in two kinds of space, one set between the bases of tetra-hedra in adjacent chains and the other between apices. These two sites are designated X and Y positions, and the cation position within each tetra-hedron is the Z position. The main features of the structure are shown in Fig. 227b, c, using the mono-clinic diopside as the type example. The repeat of the chain units gives $\{Si_2O_6\}^{4-}$ in the formula (Fig. 227a). The tendency for bonds between chains to be most easily broken confers on all members two sets of cleavage cracks parallel to z, intersecting almost at right-angles (Fig. 227d).

Orthorhombic pyroxene is related to the mono-clinic form by a doubling of the a dimension of the latter, coupled with a displacement of the chains to produce orthorhombic symmetry (Fig. 227e).

The close structural relationship between pyro-xenes and amphiboles should be noted.

Habit: Varied, but often in short prisms bounded by $\{100\}$, $\{010\}$, $\{110\}$, with $\{0kl\}$ in orthorhombic species and $\{001\}$, $\{\bar{1}11\}$ in monoclinic. Most commonly in anhedral grains. Less often in long prisms that may be curved.

Twins in monoclinic species on $\{100\}$ sometimes repeated. Cleavage prismatic, good, two sets of cracks intersecting almost at right angles: $110 \wedge 1\bar{1}0 = 93°$ (monoclinic), $210 \wedge 2\bar{1}0 = 92°$ (orthorhombic). Fracture uneven. Brittle. H 5–6. D 2·96–3·55 varying with Fe. Lustre vitreous, sometimes submetallic. Colour brown, bronze, black, green or colourless, the darker colours indicating entry of Fe and Ti.

Composition: The various pyroxene solid solution series are distinguished one from another by their colour and optical properties. The end members of the series, and, some of the principal intermediate members are listed below (see also Fig. 228). The optics and occurrence of each will be described separately.

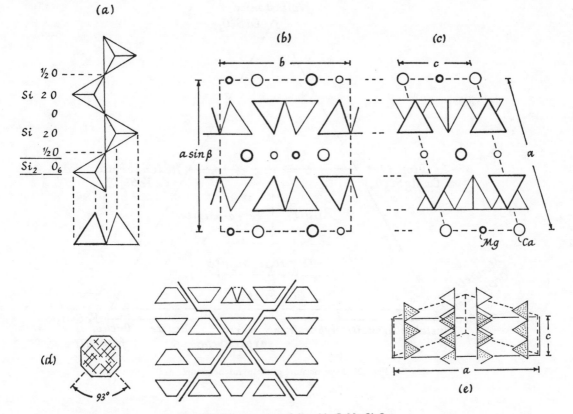

Fig. 227. The structure of diopside CaMg{Si$_2$O$_6$}.

(a) A single chain of tetrahedra in side view and end view.

(b) The unit cell viewed along z-axis. Heavy outlines denote units centred at 0, 100; light outlines, those at 50.

(c) The cell viewed along y. Only the upper half of the cell contents is shown. Heavy outlines denote units above; light outlines, those below $\frac{1}{4}b$.

(d) Relation of cleavage to pattern of chains.

(e) Doubling the a dimension and adjusting the chains vertically gives the orthopyroxene cell.

Orthopyroxenes	X, Y		Z
Enstatite	Mg		{SiO$_3$}
Hypersthene	(Mg, Fe)		{SiO$_3$}
Ferrosilite (not naturally occurring)	Fe		{SiO$_3$}
Clinopyroxenes	X	Y	Z
Diopside	Ca	Mg	{Si$_2$O$_6$}
Hedenbergite	Ca	Fe^{2+}	{Si$_2$O$_6$}
Augite	Ca	(Mg, Fe^{2+})(Al, Fe^{3+}, Ti)	{Si$_2$O$_6$}
Pigeonite	(Mg, Fe^{2+}, Ca)	(Mg, Fe^{2+})(Al, Fe^{3+})	{Si$_2$O$_6$}
Aegirine	Na	Fe^{3+}	{Si$_2$O$_6$}
Jadeite	Na	Al	{Si$_2$O$_6$}
Spodumene		(Li, Al)	{Si$_2$O$_6$}

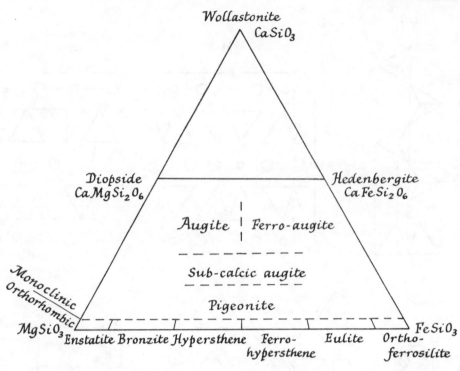

Fig. 228. Nomenclature of the pyroxene group.

Enstatite–Ferrosilite

$$16[MgSiO_3]–16[FeSiO_3]$$

Orthorhombic $P\ mmm$ a 1·823–1·843 b 0·880– 0·906 c 0·518–0·526

The series is divided as shown in Fig. 228 into sub-species on the basis of the mg ratio (i.e. Mg/Mg + Fe + Mn); of these bronzite mg 0·90–0·70 and hypersthene mg 0·70–0·50 are the commonest. Ferrosilite is unstable and does not occur in nature. The content of Ca is small, Ca/(Ca + Mg + Fe) not more than about $3\frac{1}{2}$%.

Structure: The cell dimensions increase with Fe content. A high temperature polymorph of $MgSiO_3$, protoenstatite, with the a dimension halved is found in experimental melts.

Optics: Transparent; enstatite colourless, bronzite and hypersthene more or less pleochroic. $\alpha = y$ colourless to pink 1·652–1·768, $\beta = x$ colourless to pale yellowish or greenish 1·653–1·770, $\gamma = z$ colourless to pale green 1·661–1·788, $\gamma - \alpha$ 0·009– 0·020. $2V_\alpha$ 52°–127° (see Fig. 229). The optic axial plane is 100 (Fig. 230).

The pleochroism, low birefringence, straight extinction in sections in the [001] zone and (in common members) the large $2V$ distinguish ortho-pyroxenes. Also a fine striation parallel to {100} often confined to the central area of the crystal, and due to exsolution of extremely fine lamellae of clinopyroxene is characteristic. Platy, brown, regularly-arranged inclusions, of uncertain nature (? TiO_2) are often seen in plutonic or metamorphic examples.

Occurrence: Widely distributed in ultrabasic rocks, norites, gabbros and some basalts. It is found in some rhyolites and pumices also. Also in thermally metamorphosed shales and in regional metamorphic rocks of a high grade of metamorphism belonging to the granulite and charnockitic groups.

Alteration: To serpentinous products or fibrous amphibole. Platy serpentine pseudomorphs in serpentinites are called *bastite*. Usually alters more readily than associated augite.

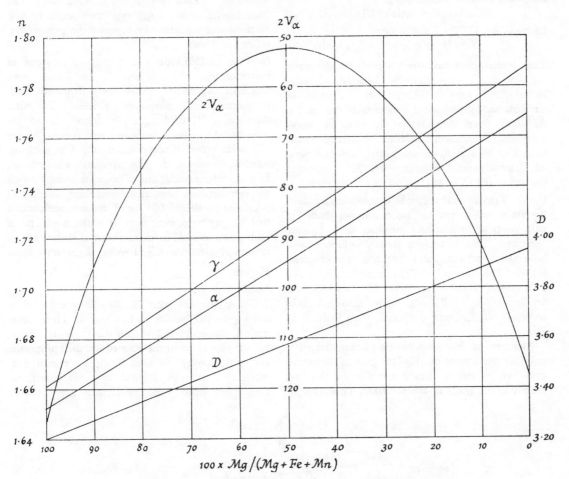

Fig. 229. Optical properties of orthopyroxenes. (After LEAKE, B. E., 1968, p. 746.)

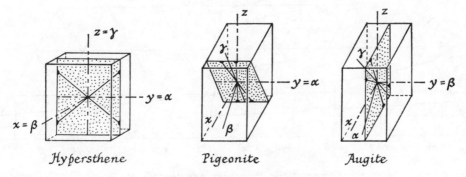

Fig. 230. Optic orientation of pyroxenes.

Diopside–Augite–Hedenbergite

$$4[Ca(Mg, Fe^{2+})AlFe^{3+}Ti\{(Si, Al)_2O_6\}]$$

Monoclinic C $2/m$ Diopside:

$$a\ 0{\cdot}971\quad b\ 0{\cdot}889\quad c\ 0{\cdot}524\quad \beta 105°\ 50'$$

The transition from the diopside-hedenbergite series to augite takes place by the entry of tetrahedral Al^{3+} ions, permitting the inclusion of trivalent and quadrivalent ions outside the chains. Although pure diopside is common in metamorphosed magnesian limestone, elsewhere transitions abound and so all are treated together here. Al^{3+} usually only replaces 3–4% Si^{4+} ions but this may rise to 7 or 8%.

Optics: Transparent; diopside is colourless, hedenbergite brown-green, and augite varying shades of pale green to brown, and purplish in Ti-bearing varieties. These last often zonally coloured with hour-glass pattern, and marked pleochroism. $\alpha\ 1{\cdot}66$–$1{\cdot}735$, $\beta = y\ 1{\cdot}67$–$1{\cdot}74$, $\gamma \wedge z\ 36°$–$54°\ 1{\cdot}69$–$1{\cdot}76$, $\gamma - \alpha\ 0{\cdot}018$–$0{\cdot}031$. $2V_\gamma 40°$–$62°$ (lower values are found in sub-calcic augites, see below). Indices increase, birefringence decreases, with increase in Fe.

Members of this series have a parting along 100 and fragments lying on this face give an uncentred optic axis interference figure that allows β to be set readily in the plane of the polarizer. This with $2V$ allows an estimate of composition (Fig. 231). The high birefringence, large extinction angle in 010 sections and moderate $2V$ separate the group from other pyroxenes.

Occurrence: Diopside and hedenbergite occur in metamorphic rocks. Diopside is found with forsterite and calcite, sometimes with monticellite, in thermally-metamorphosed, siliceous, dolomitic limestones. Hedenbergite is found in metamorphosed iron-rich sediments and skarns. Augite of a wide range of composition occurs in gabbros, basalts, diorites and some syenites and granites. In magmatic associations it becomes more Fe-rich as crystallization proceeds. Exsolution lamellae of pigeonite parallel to 001, or of hypersthene parallel to 100, may be present in it. It is also a mineral of high grade thermal metamorphism of shales, and of regionally-metamorphosed rocks of granulite type.

Varieties:

Subcalcic augite

Clinopyroxenes with $2V$ $20°$–$40°$ are found in quickly-cooled basalts and andesites. Their other optical properties fall within the range of augite, but they are often strongly zoned and their properties are consequently variable. They are regarded as metastable minerals, formed by rapid crystallization, that bridge the miscibility gap between

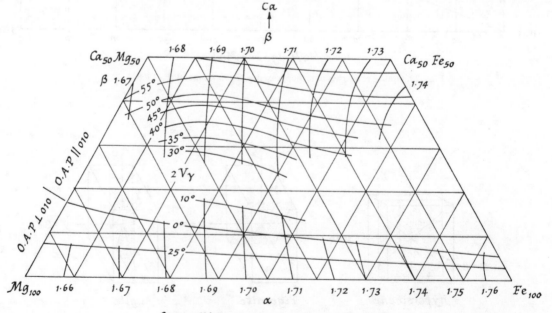

Fig. 231. Refractive index $\parallel$ y-axis and $2V_\gamma$ of common clinopyroxenes. Part of the grid drawn by MUIR, I. D., 1951, p. 713.)

augite and pigeonite. They may in some cases consist of more than one phase in extremely fine intergrowth.

Titanaugite

Titaniferous augite is distinguished by its purple-brown or violet colour, pleochroism in shades of violet and buff, and its high dispersion, $r > v$, which often leads to incomplete extinction in white light. It is characteristic of alkali olivine-basalt and alkali-gabbro, and is often zoned outwards to green aegirine-augite.

Omphacite

Omphacite is an unusual bright green variety rich in the $NaAlSi_2O_6$ molecule, occurring in eclogites. It is characterized by a large $2V_\gamma 58°-83°$.

Alteration: Augite is often altered to an aggregate of pale fibrous amphibole called uralite. It may also alter to chlorite and calcite.

Pigeonite $4[Ca(Mg, Fe)\{Si_2O\}_6]$

Monoclinic P $2m$ $a\ 0.973$ $b\ 0.895$ $c\ 0.526$
$\beta 108°\ 33'$

Structure: The shape and relative positions of the Si-O chains are slightly different from those of diopside, and the structure is intermediate between those of diopside and clinoenstatite (which is the pyroxene found in artificial melts).

Composition: Pigeonite is Ca-poor clinopyroxene with only about 1 in 12 Ca^{2+} amongst the interstitial cations, compared with 1 in 2 in diopside and about 2 in 5 in augites. Although sub-calcic augites (see above) partly bridge the compositional gap between augites and pigeonite these are metastable products. There is a real structural and compositional break between the two series. Pigeonite contains about three times as much Ca as orthopyroxene (Fig. 228).

Optics: Transparent, colourless or faintly yellow-green; less strongly coloured than associated augite. $\alpha = y$ (usually) $1.68-1.72$, β (sometimes $= y$) $1.68-1.72$, $\gamma \wedge z\ 37-44°$, $1.70-1.75$, $\gamma - \alpha\ 0.02-0.03$, $2V_\gamma 0-30°$ (Figs. 230, 231). The indices lie in the upper part of the range for augite and the extinction angle is at the lower end of the augite range. The optic axial plane is usually $\perp 010$, rarely $\|010$. The distinctive feature is the small $2V$, usually not much over 20°. The lack of colour, and tendency to readier alteration also help to distinguish pigeonite from augite. Pigeonite may contain exsolution lamellae of augite on 001, representing Ca-rich

material exsolved on slow cooling. If the pigeonite later inverts to orthopyroxene, these lamellae lie on an irrational plane in the orthopyroxene and disclose that it is an inversion product. Clinopyroxene exsolved from crystals already possessing orthopyroxene structure forms lamellae on the orthopyroxene 100 plane.

Occurrence: In quickly-cooled basalts, dolerites and andesites, usually associated with augite. According to experimental studies the first Ca-poor pyroxene to form in simple melts is clinopyroxene (clinoenstatite in the crucible, pigeonite in nature) which inverts to orthopyroxene on slow cooling. In natural melts crystallization may begin below the inversion temperature in Mg-rich compositions and orthopyroxene may be precipitated from the outset. In more Fe-rich compositions the inversion temperature is lower than the temperature of initial crystallization and pigeonite forms first. In plutonic rocks it will later invert to orthopyroxene exsolving Ca-rich pyroxene.

Aegirine (Acmite) $4[NaFe^{3+}\{Si_2O_6\}]$

Aegirine-augite
$4[(Na, Ca)(Fe^{3+}, Mg, Fe^{2+})\{Si_2O_6\}]$

The entry into augite of Na^+ in X position coupled with Fe^{3+} in Y position leads to an irregular increase in density (up to 3.59) and to a progressive change in optical properties which is continuous and, to a first approximation, linear from augite through aegirine-augite to aegirine. The habit changes from short prismatic crystals in aegirine-augite to long prismatic with acute terminations, or needle-like, in aegirine. The assumption of pleochroism in greens, and a fall in extinction angle $\alpha \wedge z$ to below 40° marks the change from augite to aegirine-augite.

Optics: Transparent, pleochroic in green and yellow. $\alpha \wedge z\ 0-32°$ light to dark green $1.71-1.78$, $\beta = y$ yellow-green to light green $1.73-1.82$, γ yellow $1.75-1.84$, $\gamma - \alpha\ 0.03-0.06$. $2V_\alpha 105°-60°$. Indices and birefringence increase with Fe, as does the depth of colour. Because of the strong absorption, the interference colour is often an anomalous yellow-orange to brown. The small extinction angle distinguishes this series from other pyroxenes. α is in the acute angle β' in aegirine-augite, with a maximum $\alpha \wedge z$ of 32°; this falls somewhat irregularly to 0°, at $Fe^{3+} \simeq 0.75$ atoms per formula unit (about 80 wt% $NaFe^{3+}Si_2O_6$), and then opens out in the obtuse angle β to 10° in aegirine. Aegirine-augite is optically +ve; $2V$ passes through 90° at about $0.35 Fe^{3+}$ per formula unit and more Na, Fe^{3+} rich specimens are $-ve$.

Occurrence: Aegirine-augite crystallizes as a late phase in alkali olivine-basalts and alkali-gabbros, often forming an outer zone on titanaugite phenocrysts and constituting the entire groundmass pyroxene. More sodic members occur in alkali-syenites and trachytes and alkali-granites. It is the principal dark silicate of the strongly alkaline ijolite rock series. It occurs in fenites, the metasomatically altered wall-rocks of alkali intrusions and in some schists where soda metasomatism may have operated.

Spodumene $4[LiAl\{Si_2O_6\}]$

Habit: Prismatic crystals flattened on {100}. In addition to pyroxene cleavages there is a good {100} parting.

Colour: White, greenish or greyish-white, emerald-green (hiddenite) or purple (kunzite). Sometimes fluorescent in orange, yellow or white, kunzite showing persistent phosphorescence after irradiation by UV light or X-rays.

Optics: Transparent, colourless except as noted. α (green in hiddenite, purple in kunzite) $1.65-1.66$, $\beta = y\ 1.66-1.67$, $\gamma \wedge z\ 26°\ 1.67-1.68$, $\gamma - \alpha\ 0.015-0.025$, $2V_\gamma 55°-77°$. The extinction angle is small compared with other colourless pyroxenes.

Tests: Before the blow-pipe whitens, swells, and fuses after long blowing, colouring the flame purple red (Li flame).

Occurrence: In granite pegmatites associated with lepidolite, beryl, tourmaline, albite and quartz. Crystals tens of metres long and a metre or so across have been recorded from Etta Mine, Keystone, South Dakota. Hiddenite is found in N. Carolina, kunzite in California, and Madagascar. Spodumene is recorded from Killiney on the coast between Dublin and Bray in Eire, and Peterhead in Scotland.

Use: An ore of lithium, the chief producer being the Etta Mine, S. Dakota.

Jadeite $4[NaAl\{Si_2O_6\}]$

Habit: Usually as a compact tough mass of fine fibres.

Colour: Green, rarely white. $D = 3.33-3.35$, often useful in identification of ornamental pieces.

Optics: Transparent, colourless. $\alpha\ 1.658$, $\beta = y\ 1.663$, $\gamma \wedge z\ 40°\ 1.673$, $\gamma - \alpha\ 0.015$, $2V_\gamma 70°$.

Occurrence: As stream-worn boulders in Burma, the source of oriental jade. Now known to occur in metamorphic rocks associated with serpentinite bodies in Burma, Guatemala and elsewhere, and

also in rocks of glaucophane schist type. It is known from experiment to form only under high pressures, representing under these conditions compositions that would normally crystallize as the commoner minerals nepheline and albite.

$$NaAlSiO_4 + NaAlSi_3O_8 \leftrightharpoons 2NaAlSi_2O_6$$
$$\text{nepheline} \quad \text{albite} \quad \text{jadeite}$$

Use: Ornamental jade is partly jadeite, partly the compact amphibole, nephrite.

WOLLASTONITE GROUP

Wollastonite	$Ca_3\{Si_3O_9\}$
Pectolite	$NaHCa_2\{Si_3O_9\}$
Rhodonite	$(Mn, Ca)_5\{Si_5O_{15}\}$

Structure: Members of this group are built of chains of SiO_4 tetrahedra that differ from those of diopside in that the tetrahedra do not have their bases approximately in a plane, but are twisted (Fig. 232) so as to repeat after every third tetrahedron (wollastonite, pectolite) or after every fifth (rhodonite). In wollastonite and pectolite the chains run along y, in rhodonite along z.

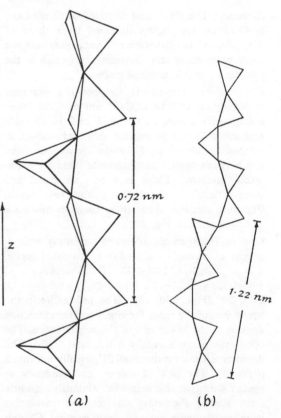

Fig. 232. Chains of SiO_4 tetrahedra in (*a*) wollastonite and pectolite, and (*b*) rhodonite.

Wollastonite $2[Ca_3\{Si_3O_9\}]$

Triclinic $P\ \bar{1}$ $a\ 0.794$ $b\ 0.732$ $c\ 0.707$
$\alpha 90°\ 02'$ $\beta 95°\ 22'$ $\gamma 103°\ 20'$

Habit: Crystals elongated along y, in columnar or fibrous masses. Twins on {100}.

Cleavage {100} perfect; {001}, {$\bar{1}02$} good: $100 \wedge 001 = 84\frac{1}{2}°$, $100 \wedge \bar{1}02 = 70°$. $H\ 4\frac{1}{2}$–5. $D\ 2.9$. Lustre vitreous. Colour white or greyish. Sometimes fluorescent in orange-red or yellow: also phosphorescent.

Optics: Transparent, colourless. $\alpha\ 1.618$, $\beta \simeq y\ 1.630$, $\gamma\ 1.632$, $\gamma - \alpha\ 0.014$. $2V_\alpha 40°$. O.A.P. approx. $\|\{010\}$ $\alpha \wedge 100$ cleavage in sections normal to $\beta \simeq 36°$. Sections normal to β equant, with three cleavages. Crystals both length-slow and length-fast.

Tests: Dissolves in HCl with separation of silica.

Occurrence: In thermally metamorphosed limestones, at igneous contacts, often associated with lime-garnet and diopside. In some deposits the silica may have been introduced from the magma.

Pectolite $2[NaHCa_2\{Si_3O_9\}]$

Triclinic $P\ \bar{1}$ $a\ 0.799$ $b\ 0.704$ $c\ 0.702$
$\alpha 90°\ 03'$ $\beta 95°\ 17'$ $\gamma 102°\ 28'$

Habit: Aggregates of needles elongated along y, sometimes radiating.

Cleavage {100}, {001} perfect. $H\ 5$. $D\ 2.9$. Lustre vitreous or silky. Colour white or grey.

Optics: Transparent, colourless. $\alpha\ 1.60$, $\beta\ 1.61$, $\gamma \simeq y\ 1.635$, $\gamma - \alpha\ 0.035$, $2V_y 50°$. $\alpha \wedge 100$ cleavage = $15°$ in sections normal to γ. Length-slow.

Tests: Yields water on heating in a closed tube. It is distinguished from zeolites by higher refractive indices.

Occurrence: A secondary mineral found in vesicles or veinlets in basic eruptive rocks. It is usually one of the earliest of the amygdale minerals to form and occurs with apophyllite, natrolite and prehnite.

Rhodonite $2[(Mn, Ca)_5\{Si_5O_{15}\}]$

Triclinic $P\ \bar{1}$ $a\ 0.668$ $b\ 0.766$ $c\ 1.220$
$\alpha 111°\ 01'$ $\beta 86°\ 00'$ $\gamma 93°\ 02'$

Habit: Crystals uncommon, tabular on {001}. Usually massive aggregates.

Cleavage {110}, {$1\bar{1}0$} perfect, almost at right angles, {001} good. Fracture conchoidal. Tough. $H\ 5\frac{1}{2}$–$6\frac{1}{2}$. $D\ 3.4$–3.7. Lustre vitreous. Colour flesh pink or rose. Blackened on surface or fractures by manganese oxides. Streak white. Sometimes fluoresces red.

Optics: Transparent, colourless. $\alpha\ 1.71$–1.74, $\beta\ 1.72$–1.74, $\gamma\ 1.73$–1.75, $\gamma - \alpha\ 0.011$–0.014. $2V_y 60°$–80°. $\gamma' \wedge z$ on {110}, {$1\bar{1}0$} cleavage pieces $\simeq 10°$. On {001} cleavage pieces α' approx. bisects the angle between the prismatic cleavages and lies nearly in the 010 plane.

Tests: The colour is distinctive. Compared with rhodochrosite it is harder and fuses before the blowpipe with slight swelling. Also it does not effervesce with HCl, but as it often occurs with rhodochrosite, or other carbonate, care is needed.

Occurrence: In hydrothermal and contact-metamorphic veins and segregations and in metamorphosed sediments rich in manganese. The occurrence at Sverdlovsk in the Ural Mountains is famous and rhodonite from there has been used as an ornamental stone. It occurs at Långban, Sweden in iron ore, at Broken Hill, N.S.W., Australia and at Franklin and Sterling Hill, New Jersey.

AMPHIBOLE GROUP

The group includes both orthorhombic and monoclinic members but all are closely related structurally and chemically.

Structure: Double chains of SiO_4 tetrahedra (Fig. 233a) run parallel to z and are linked together by intermediate cations. There are four types of cation position. (*i*) Large spaces between the rings of opposed tetrahedral bases in the chains. These are designated W positions. (*ii*) Between tetrahedral bases but not opposite the ring-shaped spaces of the chains. These are X positions (cf. pyroxenes). (*iii*) Between opposed tetrahedral apices. These are Y positions. (*iv*) The Z positions are those within tetrahedra. Finally $(OH)^-$ or F^- lies in the rings of the SiO_4 double chains, at the level of the tetrahedral apices (Fig. 233b).

Cleavage takes place by breaking of bonds between the chains giving two sets of cleavage cracks intersecting at 56° and 124°. (Fig. 233c).

The main outlines of the structure can be derived from the pyroxene arrangement by introducing 010 mirror planes along the single pyroxene chains, and the b dimension of amphibole is approximately double that of pyroxene. Orthorhombic amphibole is related to monoclinic amphibole by a doubling of the a cell dimension, in the same way as in pyroxenes. These relations between cell dimensions are shown in Fig. 233d.

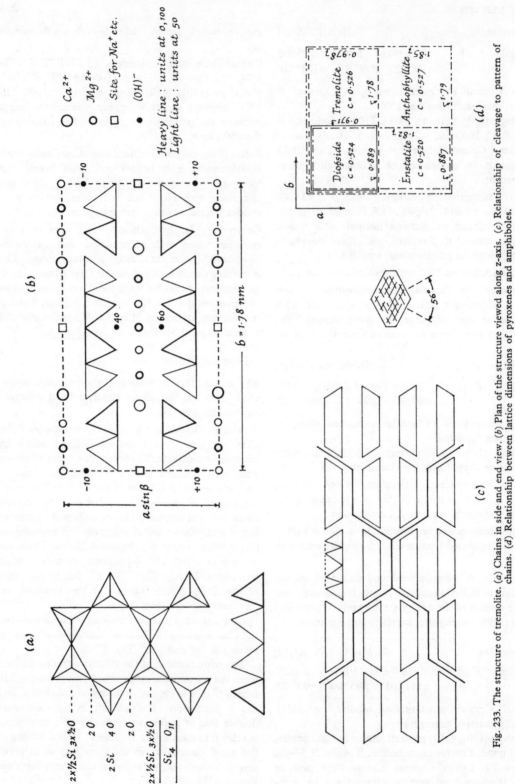

Fig. 233. The structure of tremolite. (a) Chains in side and end view. (b) Plan of the structure viewed along z-axis. (c) Relationship of cleavage to pattern of chains. (d) Relationship between lattice dimensions of pyroxenes and amphiboles.

Orthoamphibole

<table>
<tr><td></td><td>(X, Y)</td><td>Z</td></tr>
</table>

Anthophyllite series $(Mg, Fe^{2+})_7\{Si_8O_{22}\}(OH, F)_2$

Clinoamphiboles

Cummingtonite series $(Fe^{2+}, Mg)_7\{Si_8O_{22}\}(OH, F)_2$

<table>
<tr><td>X</td><td>Y</td><td>Z</td></tr>
</table>

Tremolite-actinolite series $Ca_2(Mg, Fe^{2+})_5\{Si_8O_{22}\}(OH, F)_2$

<table>
<tr><td>W</td><td>X</td><td>Y</td><td>Z</td></tr>
</table>

Hornblende series $(Na, K)_{0-1}Ca_2(Mg, Fe^{2+}, Al, Fe^{3+}\ etc.)_5\{(Al, Si)_8O_{22}\}(OH, F)_2$

<table>
<tr><td>W</td><td>(X,Y)</td><td>Z</td></tr>
</table>

Glaucophane series $Na_2(Mg, Fe^{2+})_3(AlFe^{3+})_2\{Si_8O_{22}\}(OH)_2$

Habit: Generally in long prisms, sometimes columnar, in needles, or in fibres. Also in anhedral grains usually with some elongation along z.

Twins on {100} are common in monoclinic members. Cleavage {110} ($=$ {210} orthorhombic) perfect, giving two sets of cracks intersecting at 56° and 124°. Fracture uneven. Brittle when in stout crystals; fibres pliant, capable of being woven. H 5–6. D 2·9–3·5. Lustre vitreous, splendent on cleavages. Colour usually black or green, also brownish black, cinnamon, more rarely colourless, blue, amber.

Optics: Most members have $2V$ −ve, large, and a small extinction angle.

Composition: Extensive solid solution takes place, the different series listed below being cross-connected by intermediates. They are distinguished by colour and optical properties but optical estimations of composition are far from reliable.

There is an unfortunate tendency to proliferation of names, in the amphibole group, for minerals of overlapping physical properties. These names can only be used after complete chemical analysis of the mineral concerned, and even then there is lack of agreement on boundaries between the types. A simplified classification is adopted here.

Anthophyllite

$$4[(Mg, Fe^{2+})_7\{Si_8O_{22}\}(OH, F)_2]$$

Orthorhombic $P\ mmm$ a 1·85 b 1·79 c 0·527

Structure: The unit cell has a approximately twice that of monoclinic species.

Habit: Usually in aggregates of slender prisms: sometimes asbestiform. Colour grey-brown, cinnamon-brown, green.

Optics: Transparent, colourless to pale brown, pale yellow-green. $\alpha = x$ pale brown 1·59–1·68, $\beta = y$ pale brown 1·60–1·69, $\gamma = z$ yellow-green 1·62–1·70, $\gamma - \alpha$ 0·015–0·025. $2V_\gamma$ 80–110°. Optically +ve or −ve. (Indices increase with Fe.) The general amphibole properties coupled with straight extinction in the [001] zone are distinctive.

Occurrence: A metamorphic mineral of magnesium-rich sediments, associated with cordierite. Some introduction or local concentration of Mg may be necessary to its formation. Also formed during metamorphism of ultramafic rocks, when it is associated with talc.

Use: The asbestiform variety (*amosite*) is used for insulation, but the fibres are brittle and unsuitable for woven products.

Related minerals: Aluminous anthophyllite is called *gedrite*. The entry of Al may be associated with higher Fe content extending the indices to higher values than those quoted; but Fe-rich compositions generally crystallize as monoclinic cummingtonite.

Cummingtonite $2[(Fe, Mg)_7\{Si_8O_{22}\}(OH)_2]$

Monoclinic $C\ 2/m$

Usually in fibrous crystals, often radiating, of pale- to dark-brown colour.

Optics: Transparent, colourless or neutral tint. Fe-rich types may show slight pleochroism. α 1·64–1·68, $\beta = y$ 1·65–1·71, $\gamma \wedge z$ 10–20° 1·66–1·73, $\gamma - \alpha$ 0·02–0·04, $2V_\gamma$ 70–95°, mostly +ve. The indices and birefringence increase with Fe, the extinction angle decreases, and the optic sign changes from +ve to −ve. The refractive indices are higher than those of tremolite-actinolite and the inclined extinction separates it from anthophyllite.

Occurrence: Mainly a metamorphic mineral in regionally metamorphosed basic igneous rocks and some hornfelses. The iron-rich members occur in metamorphosed iron-rich sediments, notably the iron ores of the Lake Superior region.

Variety: *Grunerite* is the name given to the iron-rich optically −ve members of the series.

Tremolite–Actinolite

$$2[Ca_2(Mg, Fe^{2+})_5\{Si_8O_{22}\}(OH, F)_2]$$

Monoclinic *C* 2/*m* Tremolite: *a* 0·984 *b* 1·805
                                                      c 0·528   *β*104° 42′

In colourless to dark green fibrous to columnar crystals. *D* increases regularly with Fe from 3·0–3·45.

Optics: Transparent, colourless to pleochroic in pale greens. α 1·60–1·67, β = *y* 1·61–1·68, γ ∧ *z* 10–20° 1·62–1·69, γ − α 0·020–0·027. $2V_\alpha$65–85°. The indices increase regularly with Fe, and the extinction angle decreases. The lack of colour distinguishes magnesian members from hornblende and the extinction angle of the more Fe-rich green actinolites is less than that of hornblende.

Occurrence: This series is widespread particularly in metamorphic rocks. Tremolite occurs in thermally metamorphosed magnesian limestones, associated with dolomite, calcite, diopside, forsterite and quartz. Tremolite and actinolite are found in altered ultrabasic rocks associated with talc. Actinolite is common in a variety of schists associated with albite, chlorite and epidote, especially in metamorphosed basic igneous rocks. Actinolite probably also constitutes the uralitic amphibole that often fringes pyroxene in plutonic igneous rocks.

Varieties: Nephrite, one of the materials known as jade, is finely-fibrous compact actinolite. It is tough, homogeneous, of a variety of attractive green shades, and takes a high polish. Found in association with metamorphosed ultrabasic rocks, it is used as an ornamental stone. The Maoris of New Zealand fashioned implements and ceremonial objects of great beauty from it during their neolithic cultural period.

Hornblende

$$(Na, K)_{0-1}Ca_2(Mg, Fe^{2+}, Al, Fe^{3+}Mn, Ti)_5$$
$$\{(Al, Si)_8O_{22}\}(O, OH, F)_2$$

With such a formula, it is clear that hornblendes can be subdivided into a large number of subsidiary

molecules that might theoretically exist independently. In nature such end-members exist only rarely, but the dominance of one or another of them may be suggested by the properties of any particular hornblende.

Taking the formula of tremolite Ca_2Mg_5 $\{Si_8O_{22}\}(OH)_2$ as a starting point, on the one hand there may be entry of Na (and K in small amounts) into the *W* position, balanced by substitution of Al^{3+} for Si^{4+} in *Z*. On the other hand, trivalent or quadrivalent ions may enter the *Y* position with charge compensation again by Al substitution for Si in *Z*. The limit of this substitution appears to be (Al_2Si_6). If Na, as well as occupying *W*, replaces some of the Ca^{2+} in *X*, charge compensation may be achieved by entry of trivalent ions in *Y* without the need for substitution in *Z*. The charge balance is also affected by replacement of $(OH)^-$ by O^{2-} in oxyhornblendes.

Optics: Transparent, pleochroic.

α 1·61–1·70	β = *y* 1·62–1·72
pale green	green
yellow-green	olive-green
yellow-green	yellow
yellow-brown	red-brown

γ ∧ *z* 12–34° 1·63–1·73
bluish green
dark green
brown
dark brown

Absorption usually γ > β > α but axes of maximum and minimum absorption do not always coincide with those of refraction. γ − α 0·014–0·03. $2V_\alpha$ usually 60–90°, but pargasite, $NaCa_2Mg_4$-$Al\{Al_2Si_6O_{22}\}(OH)_2$, is +ve with $2V_\gamma$75–90°, and hastingsite, $NaCa_2Fe_4^{2+}Fe^{3+}\{Al_2Si_6O_{22}\}(OH)_2$, has a small −ve $2V_\alpha$10–50°. The indices increase, and 2V decreases, with increase in Fe.

Occurrence: Common hornblende is widely distributed in igneous rocks, most prominently in intermediate and acid types. It is also a major mineral of metamorphic rocks, particularly metamorphosed basic igneous rocks. Because of the freedom of atomic substitution possible in hornblende a wide range of compositions can be represented in metamorphism by the two minerals hornblende and plagioclase.

Hornblendes rich in the pargasite molecule (above) are found in metamorphosed dolomites, and the hastingsite type in metamorphosed iron-rich rocks. Both are uncommon.

Oxyhornblende

$NaCa_2(Mg, Fe^{2+}, Fe^{3+}, Al)_5\{(Al,Si)_8O_{22}\}O_2$

Optics: Pleochroic in yellow to dark brown; absorption $\gamma > \beta > \alpha$, α 1·67–1·69, $\beta = y$ 1·68–1·73, $\gamma \wedge z$ 0–15° 1·69–1·76, $\gamma - \alpha$ 0·02–0·07, $2V_\alpha$ 60–80°. The dark colour, high indices and often very high birefringence are characteristic. These properties overlap, however, with those of brown alkali amphiboles, kaersutite and barkevikite, and the environment must be relied on to separate them.

Occurrence: Common hornblende can be oxidized by heating in air so that its hydroxyl is replaced by O^{2-} and a high proportion of its Fe^{2+} becomes Fe^{3+}. Its colour changes to dark brown and the indices and birefringence rise. Natural hornblendes low in $(OH)^-$ and with these optical properties occur in volcanic rocks of basaltic to trachytic type, and have probably been oxidized during and after eruption. They may be decorated with magnetite granules and show other signs of resorption.

Barkevikite

$(Na, K)Ca_2(Mg, Fe^{2+}, Fe^{3+}, Mn)_5$
$\{(Al, Si)_8O_{22}\}(OH)_2$

Kaersutite

$(Na, K)Ca_2(Mg, Fe^{2+}, Fe^{3+}, Ti)_5$
$\{(Al, Si)_8O_{22}\}(OH, F)_2$

Composition: Barkevikite has a higher total Fe relative to Mg than kaersutite, and has rather high Mn. Kaersutite has a rather low degree of oxidation of Fe and is high in Ti.

Optics: Translucent, strongly coloured. α yellow-brown 1·67–1·69, $\beta = y$ reddish brown 1·69–1·74, $\gamma \wedge z$ 0–19° dark brown 1·70–1·77, $\gamma - \alpha$ 0·014–0·018 (barkevikite), 0·030–0·032 (kaersutite). $2V_\alpha$ 40–80°. Though their properties overlap, kaersutite tends to have higher indices and has definitely higher birefringence though the absorption may mask the interference colours. Very high $\gamma - \alpha$ (0·083) found in some kaersutite is not typical and usually indicates oxyhornblende.

Occurrence: These two dark brown amphiboles are characteristic of undersaturated alkaline rocks, kaersutite in analcime-bearing rocks and nepheline-bearing basic rocks while barkevikite occurs rather in nepheline-syenites.

Glaucophane $Na_2Mg_3Al_2\{Si_8O_{22}\}(OH)_2$
Riebeckite $Na_2Fe_3^{2+}Fe_2^{3+}\{Si_8O_{22}\}(OH)_2$
Arfvedsonite $Na_3Fe_4^{2+}Fe^{3+}\{Si_8O_{22}\}(OH)_2$

Optics: These are the blue alkali amphiboles. All are strongly pleochroic.

The optical properties quoted for these minerals show considerable variability even in material from adjacent localities. Dispersion in many of them is large, leading to anomalous extinction and interference colours, while the latter may further be masked by absorption.

Occurrence: Glaucophane is a metamorphic mineral found in schists derived from greywackes and argillites under conditions of low-temperature, possibly high-pressure, metamorphism.

Riebeckite forms a series with glaucophane. The intermediate member of the series, crossite, occurs with glaucophane. Riebeckite itself occurs in soda-trachytes, rhyolites and granites (e.g. the riebeckite-microgranite of Ailsa Craig, Firth of Clyde).

Arfvedsonite is found in highly alkaline syenites and their pegmatites.

Varieties: *Crocidolite* (blue asbestos) is asbestiform riebeckite interbanded with jaspers, that is mined in Cape Province, S. Africa. When silicified it forms a compacted ornamental stone that takes a high polish. This is known as *cat's-eye*, or when partly oxidized to a golden brown colour as *tiger's-eye*.

	Glaucophane	Riebeckite	Arfvedsonite
α	colourless	dark blue	blue-green
	1·61–1·66	1·65–1·70	1·66–1·70
β	lavender	indigo	blue-violet
	1·62–1·67	1·66–1·71	1·67–1·71
γ	sky-blue	yellow-green	greenish-yellow
	1·63–1·67	1·67–1·72	1·67–1·71
$\gamma - \alpha$	0·008–0·022	0·003–0·016	0·005–0·018
	$\beta = y$, $\gamma \wedge z$ 4–14°	$\gamma = y$, $\alpha \wedge z$ 3–21°	$\gamma = y$, $\alpha \wedge z$ 0–35°
$2V_\alpha$	0–70°	40–90°	varies widely
Elongation	+ve	−ve	−ve

LAYER SILICATES

The structures of this group are based upon layers of SiO_4 tetrahedra with their bases all in one plane, sharing the three corners of the base with neighbours, and with their apices all pointing in the same direction (Fig. 234). The shared O^{2-} ions at the base are valency-satisfied: only the apical oxygens have spare negative charges for the attachment of ions external to the *tetrahedral layer*. The tetrahedral layer comprises three sheets of atoms, the basal O^{2-}, the central Si^{4+} and the apical O^{2-}. Each sheet and the layer as a whole has 6-fold symmetry.

When external ions are attached to the apical oxygens each is bonded to two apical oxygens and to an $(OH)^-$ ion which lies in the centre of the hexagonal ring of apices. Fig. 234 shows the arrangement in plan and side view.* The bonding of the external ions will be completed by attachment to a sheet of anions on the other side which might be another set of opposed tetrahedral apices (Fig. 239b) or a sheet of $(OH)^-$ ions (Fig. 234). In either case, the external ions are now in a set of octahedral co-ordination polyhedra as shown in Fig. 234. This set of polyhedra constitutes the *octahedral layer* (sometimes called the 'brucite layer').

The layer silicates are built up of different combinations of these two types of layer (Table 22). The cations populating the layers vary in different species.

* The side view provides a little mnemonic diagram (due to Pauling) used on later pages, and nicknamed the crown and the corset.

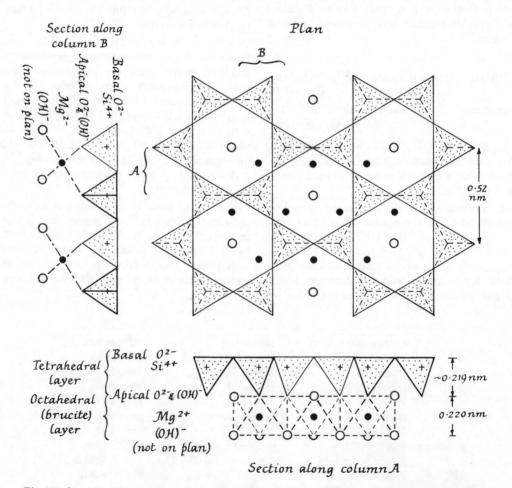

Fig. 234. Structure of a layer silicate of serpentine type, $Mg_6\{Si_4O_{10}\}(OH)_8$, to show basic structural units.

Composition of Layers	Group Name	Population of Octahedral Sheet	Mineral Name	Formula
Diphormic 1:1 (0.7 nm) (0.7 nm basal spacing)	Kandites	Dioctahedral	KAOLINITE / Nacrite / Dickite / HALLOYSITE	$Al_4[Si_4 O_{10}](OH)_8$ / $Al_4[Si_4 O_{10}](OH)_8 . 4H_2O$
	Serpentines (Septechlorites)	Trioctahedral	ANTIGORITE / CHRYSOTILE	$Mg_6[Si_4 O_{10}](OH)_8$
Triphormic 2:1 (1.0 nm)	Pyrophyllite	Dioctahedral	PYROPHYLLITE	$Al_2[Si_4 O_{10}](OH)_2$
	Talc	Trioctahedral	TALC	$Mg_3[Si_4 O_{10}](OH)_2$
	Micas	Dioctahedral	MUSCOVITE → Illite / Paragonite / Glauconite	$KAl_2[AlSi_3 O_{10}](OH)_2 \rightarrow (K,H_3O)Al_2[(Si,Al)_4 O_{10}](OH)_2$ / $(K,H_3O)(Al,Fe)_2[(Al,Si)_4 O_{10}](OH)_2$
	Micas	Trioctahedral	BIOTITE	$KMg_3[AlSi_3 O_{10}](OH)_2$ / $K(Mg,Fe)_3[AlSi_3 O_{10}](OH)_2$ / $KFe_3[AlSi_3 O_{10}](OH)_2$
	Brittle Micas	Dioctahedral	MARGARITE	$CaAl_2[Al_2 Si_2 O_{10}](OH)_2$
	Brittle Micas	Trioctahedral	CLINTONITE	$CaMg_3[Al_2 Si_2 O_{10}](OH)_2$
	Vermiculites	Dioctahedral or Trioctahedral	VERMICULITE	$0.33 M^+(Mg,Fe\ etc.)_3[(Al,Si)_4 O_{10}](OH)_2 . 4H_2O$
	Smectites (Swelling clays)	Dioctahedral	MONTMORILLONITE / Beidellite / Nontronite	$0.33 M^+(Al,Mg)_2[(Al,Si)_4 O_{10}](OH)_2 . nH_2O$
	Smectites (Swelling clays)	Trioctahedral	Saponite / Hectorite	$0.33 M^+ Mg_3[(Al,Si)_4 O_{10}](OH)_2$
Tetraphormic 2:1:1 (1.4 nm)	Chlorites / Leptochlorites	Trioctahedral-dioctahedral	Various	
	Chlorites / Orthochlorites	Trioctahedral-trioctahedral	PENNINE / CLINOCHLORE etc	$(Mg,Fe)_5 Al[AlSi_3 O_{10}](OH)_8$

Table 22. Classification of layer silicates.

The cations in the octahedral layer may be of either divalent or trivalent elements. If divalent, then charge balance is achieved when all the octahedral sites are occupied, i.e. three sites per tetrahedral ring. This gives a *trioctahedral* layer silicate. If the cations are trivalent, charge balance is maintained when only two out of the three possible octahedral sites are occupied. These are *dioctahedral* layer silicates. The hydroxide minerals brucite and gibbsite exemplify this kind of difference (Fig. 190).

Because variations in the union and stacking of standard layers produce the different species, great importance attaches to the repeat distance up the stack in identifying them. This is the *basal spacing* $d_{(001)}$ determined by X-ray powder photography.

The union of tetrahedral and octahedral layers in different sequences provides a classification into diphormic, triphormic and tetraphormic groups with the ratios of tetrahedral to octahedral layers of 1:1, 2:1 and 2:1:1 respectively (see Table 22).

The eminent {001} cleavage shown by all but the fibrous members of the group is parallel to the layer structure.

KANDITE GROUP

Kaolinite $1[Al_4\{Si_4O_{10}\}(OH)_8]$
Nacrite $2[Al_4\{Si_4O_{10}\}(OH)_8]$
Dickite $6[Al_4\{Si_4O_{10}\}(OH)_8]$

Kaolinite
Triclinic P 1 a 0·514 b 0·893 c 0·737
                         $\alpha 91° 48'$ $\beta 104° 42'$ $\gamma 90°$

Structure: Kaolinite is a stack of units each made of one tetrahedral layer and one dioctahedral layer joined together. Each pair has the tetrahedral layer on the same side so that the stack is polar (upper and lower surfaces different). The pairs of layers are valency-satisfied so that only weak residual bonding forces hold the layers together (Fig. 235★). Variations in the stacking arrangement made possible by the vacant sites in the dioctahedral layer cause the differences between kaolinite, nacrite and dickite.

Habit: Clayey masses. Hexagonal platelets usually seen only with the electron microscope. Sometimes stacks of these flakes make vermiform aggregates seen under the light microscope.

Cleavage {001} perfect. Flakes flexible, not elastic. *H* 2–2½. *D* 2·6. Lustre pearly on rare large flakes,

★$d_{(001)}$ in the figure is an older value than the one quoted above.

usually dull and earthy. Colour white but often stained yellow or brown.

Optics: Transparent, colourless. α 1·553–1·565, $\beta \wedge x$ 1°–3½°   1·559–1·569, $\gamma = y$ 1·560–1·567, $\gamma - \alpha$ 0·005. $2V_\alpha$ 25°–50°.

Tests: X-ray study is required to identify clay minerals. The kandites are recognized by showing a strong X-ray powder line with $d_{(001)} \simeq 0.715$ nm as the largest inter-planar spacing. This is not changed by saturating the clay with glycerol. Heating to 1000°C destroys this reflection and prolonged heating may lead to the appearance of lines characteristic of γ-Al_2O_3 or mullite.

Differential thermal analysis or weight loss curves during progressive heating show that the 'bound water' (OH groups in the structure) is lost between 400° and 525°C.

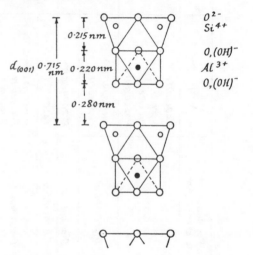

Fig. 235. Pauling diagram of layer sequence in kaolinite.

Occurrence: A widespread mineral of soils and of many shales, formed by weathering, principally of feldspar. Also as a decomposition product more or less *in situ* of feldspathic rocks, especially granites, by hydrothermal action, as in the Cornish kaolin deposits. Well-crystallized material is sometimes found inside concretions in shales.

Use: For china clay when pure. In the brick-making industry when less pure.

Related mineral: *Halloysite* $Al_4\{Si_4O_{10}\}(OH)_8.4$–$8H_2O$ has a variable number of water molecules (often 4·4 to 4·6) between the kaolinite type layers. It occurs in massive form and though soft, *H* 1–2, it is hardly at all plastic, but breaks with a conchoidal fracture. The blocks have a waxy lustre

Optically isotropic, n 1·47–1·55 depending on water content. $d_{(001)}$ 1·0 nm rising to 1·1 nm on saturation with glycerol. Under the electron microscope halloysite crystals are seen to be rolled tubes (cf. chrysotile serpentine).

Serpentine $4[Mg_6\{Si_4O_{10}\}(OH)_8]$

Chrysotile (fibrous serpentine)

a 0·532 b 0·92 c 1·464

Monoclinic or orthorhombic

β93° 20′ or 90° 00′

Lizardite (fine platy serpentine)

a 0·531 b 0·92 c 0·731

β90° 00′

Antigorite (platy serpentine)

a 4·35 b 0·925 c 0·726

β91° 23′

Structure: The basic structure is the trioctahedral analogue of kaolinite. Because the dimensions of the tetrahedral layer are about 9% smaller than corresponding lengths on the octahedral (brucite) layer, the sheets in chrysotile roll up into minute cylinders, with the tetrahedral layer on the inside of the roll. The radius of the roll is about 12·5 nm. These rolls, or bundles of them, are the fibres of chrysotile serpentine. The axis of the fibre is usually x, though it may sometimes be y.

Lizardite and antigorite are platy, but antigorite layers possibly run in a series of waves with y as the direction of wave crest. The tetrahedral-octahedral layer changes its direction of facing at each node of the wave form, so that the tetrahedral layer is on the inside of the curve (Fig. 236). The radius of curvature is about 7·5 nm and the wavelength $= a$ (about 8 × a of the other members of the group).

Habit: Chrysotile forms veinlets with fibres normal to the walls (cross-fibre serpentine). Subordinate amounts of slip-fibre material have fibres in the plane of the veinlets.

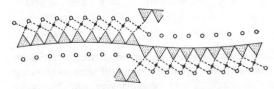

Fig. 236. Alternating curvature of serpentine-type layers that can account for the diffraction pattern of antigorite. Basic units as in Fig. 234. (Based on ZUSSMAN, J., 1954, p. 498.)

Lizardite is compact but seen to be flaky when highly magnified.

Antigorite is in rather coarser flakes.

In thin section, veinlets of cross-fibre in regular pattern separate fields of platy serpentine, giving mesh-structure (Fig. 237).

Cleavage {001} perfect in flaky kinds. Fracture conchoidal or splintery. H 2½–3½, 5½ in some polished material. D 2·55–2·60. Lustre silky in fibrous kinds to waxy or dull. Colour pale to dark green, greenish black or stained dull red.

Optics: Transparent, colourless to pale green or yellowish green. α 1·53–1·57, γ 1·54–1·575, $\gamma - \alpha$ 0·004–0·017. In antigorite $2V_\alpha$35–60° with $\alpha \simeq z$ and approximately normal to the {001} cleavage. Chrysotile usually has +ve elongation. Low birefringence with grey or white interference colours is common.

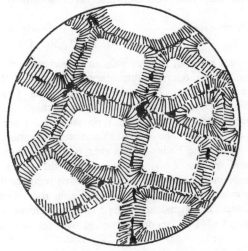

Fig. 237. Chrysotile in cross-fibre veinlets in antigorite, giving mesh-structure as viewed microscopically.

Tests: The habit separates chrysotile from the other two. Apart from this, careful study of the X-ray diffraction pattern is required to distinguish between the different types. Data for this is provided by Whittaker and Zussman (*Min. Mag.*, **31**, 107–126, 1956, particularly pp. 117, 121). From chlorite it is distinguished by lower refractive indices.

Occurrence: Chrysotile forms veins cutting rock made of other serpentine minerals. Serpentine in general forms large bodies of rock derived from hydrothermal alteration of the ultramafic rocks dunite and pyroxenite. It may form borders to unaltered cores of these rocks or replace them entirely. The rock bodies often have the form of pipes or of lenses emplaced along tectonic zones. Associated minerals are talc, magnesite, dolomite and chromite. Pseudomorphs of orthopyroxene made of serpentine may stand out as lustrous plates in serpentinite and are called *bastite*.

Serpentine also replaces olivine in basalts and gabbros.

Serpentine marbles result from serpentinization of olivine formed in the thermal metamorphism of dolomitic limestone, giving an ornamental stone called ophicalcite.

Varieties: *Bowenite* and *tangiwai* are names given to clear green serpentine of exceptional hardness (5½) resembling jade. The former comes from Rhode Island, U.S.A., and the latter from New Zealand where it was used by the Maoris for polished *greenstone* implements of great beauty.

Use: Chrysotile is the chief mineral of asbestos.

Chamosite $(Fe^{2+}, Fe^{3+}Mg, Al)_6$
$\{(Al, Si)_4O_{10}\}(O, OH)_{3-4}$

This mineral, which forms an important part of the marine oolitic sedimentary iron ores of W. Europe, occurs in ferrous (green) and ferric (brown) forms of orthogonal or monoclinic symmetry, with a serpentine type structure having $d_{(001)}$ 0·71nm. The original chamosite of Chamoson in the Rhône Valley is, however, a chlorite with a basal spacing $d_{(001)}$ of 1·4 nm, and this poses a problem of names which can be solved temporarily by using the term chamosite-chlorite for the 1·4 nm mineral, and chamosite for the 0·7 nm one. Chamosite has n 1·61–1·66 (no doubt varying with the state of oxidation), $\gamma - \alpha$ 0·00–0·008. $2V_\alpha$ small.

It forms ragged flaky crystals with mica-like cleavage within and sometimes between the ooliths of the ironstone. Associated minerals are siderite, limonite, calcite and collophane.

Pyrophyllite $4[Al_2\{Si_4O_{10}\}(OH)_2]$
Monoclinic C $2/m$ a 0·515 b 0·890 c 1·855
β99° 55′
Talc $4[Mg_3\{Si_4O_{10}\}(OH)_2]$
Monoclinic C $2/m$ a 0·526 b 0·910 c 1·881
β100° 00′

Structure: Two tetrahedral layers with inward pointing apices are united by a dioctahedral (pyrophyllite) or trioctahedral (talc) layer (Fig. 238). These sandwiches are held together by weak residual bonding forces only, there being no cations between them.

Habit: Foliated masses, sometimes in radiating groups; also granular, compact or crypto-crystalline. Cleavage {001} perfect, into flexible inelastic flakes. Sectile. H 1–2 (talc is the standard for 1) with greasy feel. D 2·7–2·9. Lustre pearly. Colour white, greyish, apple-green (pyrophyllite); apple-green, white, silvery white, greenish grey (talc).

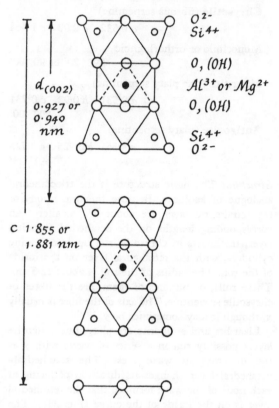

Fig. 238. Pauling diagram of layer sequence in pyro-phyllite and talc.

Optics: Transparent, colourless.

		Pyrophyllite		Talc
α	⊥001	1·534–1·556	⊥001	1·539–1·550
β		1·586–1·589		1·589–1·594
γ	$= y$	1·596–1·601	$= y$	1·589–1·600
$\gamma - \alpha$		0·05		0·05
$2V_\alpha$		57°		0–30°

Both are notable for a marked change in relief on rotation in plane polarized light, of sections showing the cleavage. These sections give soft 3rd order interference colours, while sections parallel to the cleavage give whites and greys. Extinction is approximately parallel to the cleavage. The smaller optic axial angle of talc separates it from pyrophyllite and muscovite. It may be difficult to separate pyrophyllite from muscovite optically.

Tests: If pyrophyllite is heated in the blowpipe flame, moistened with cobalt nitrate solution and reheated it turns a deep blue (indicating Al). Talc will turn pale red. Both yield water only on intense heating in the closed tube.

Occurrence: Talc is commonly found in association with serpentine often in veins with magnesite or quartz, or along faults. It also forms by the low grade metamorphism of siliceous dolomites, and talc schists are not uncommon. Extensive beds of compact talc called *steatite* or *soapstone* occur in some metamorphic areas.

Pyrophyllite is found associated with kyanite, andalusite and sillimanite as a product of their alteration. It also forms by hydrothermal alteration of feldspar.

Use: Talc is the first mineral with which the student comes into contact. Besides its lubricant property it is widely used for electrical and thermal insulation, as refractory slabs, and as a filler for paints and papers. Pyrophyllite has a similar range of uses.

MICA GROUP

Muscovite (White mica) $4[KAl_2\{AlSi_3O_{10}\}(OH)_2]$

Monoclinic C $2/m$ a 0·519 b 0·900 c 2·010
                                      β95° 11′

Structure: The basic mica sandwich has two tetrahedral layers joined by an octahedral layer as described under pyrophyllite. In micas these sandwiches form a stack, with sheets of K^+ between them, the large K^+ ion lying between the ring-shaped holes of the tetrahedral layers above and below (Fig. 239a). The valency requirement of the K^+ is met by the replacement of one-quarter of the Si^{4+} ions in the tetrahedra by Al^{3+}.

Each mica sandwich has a built-in offset, or stagger, of the upper tetrahedral layer relative to the lower, equal to one-third the diameter of one of the rings of tetrahedra. This stagger arises because of the way the octahedral cations are placed in relation to the tetrahedral apices, and is the reason why the hexagonal symmetry of the individual layers is reduced to monoclinic in the whole structure (Fig. 239b).

However, the upper and lower face of each sandwich has a hexagonal atom array, and each new sandwich could be fitted on to the last with its internal direction of stagger parallel to that of the sandwich below, or turned 60°, 120°, 180° or 240° with respect to it. Regular variations in the stacking orientation can give a varying repeat distance along z before the stagger directions coincide again. Most muscovites repeat after two sandwiches (2 M micas), but some repeat exactly in each successive layer (1 M type) and others after three sandwiches (3 T). M and T stand for monoclinic and trigonal, the resulting symmetry system. Other possibilities exist such as 2O (orthorhombic) and 6H (hexagonal) but have not been found in nature. Fig. 239c shows these staggering possibilities. Also, disordered arrangements are possible and are found in low-temperature micas in sediments. It is, indeed, an intriguing question how the mica remembers its stacking plan as it grows. In general it does so pretty well, but it sometimes makes a mistake which, if the proper pattern is at once resumed, will result in a twinned crystal. J. V. Smith and Yoder (*Min. Mag.*, **31**, 209–235, 1956) deduced these relationships. Data to distinguish the polymorphs is given in *The X-ray identification and crystal structures of clay minerals* (ed. G. Brown, Mineralogical Soc., London, 1961).

Habit: Pseudohexagonal crystals tabular on {001}, or in large anhedral plates and 'books' that may be decimetres or metres across. Also in small scales and flakes.

Twins on [310] as twin axis, with {001} as composition plane (Mica law). Cleavage {001} eminent, into flexible elastic laminae. H 2–2¼. D 2·8–3·0. Lustre vitreous to pearly. Colourless or pale grey, green or brown.

Optics: Transparent, colourless. α nearly normal to {001} 1·55–1·57. β 1·58–1·61, $\gamma = y$ 1·59–1·62, $\gamma - \alpha$ 0·036–0·05. $2V_\alpha$30–50°. Straight extinction in sections normal to the well-marked close-spaced cleavage. Bright 2nd order interference colours, in these sections. At the moment of extinction a watered silk or moirée effect appears in the colours.

Occurrence: Widespread in low temperature granites, usually appearing after biotite in the last crystallized fraction. In greisens and, above all, in pegmatites where the large crystals are found. Also abundant in schists and gneisses formed from sedimentary rocks.

Being resistant to weathering and easily borne along by water, because of its lightness and flaky habit, it is a common detrital mineral, the flakes lying in the bedding planes of shales and sandstones.

Varieties: *Sericite* is fine-grained white mica

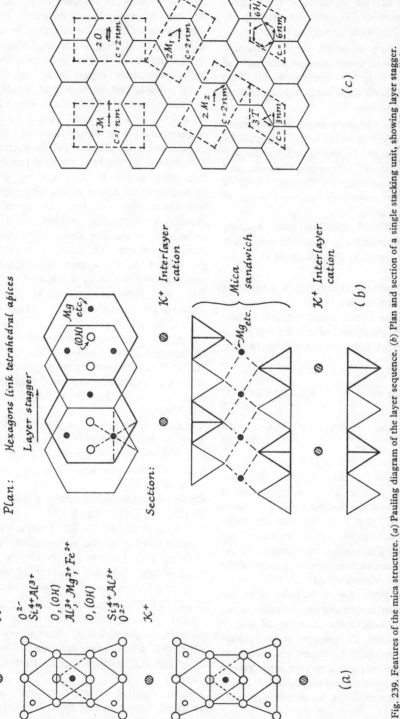

Fig. 239. Features of the mica structure. (a) Pauling diagram of the layer sequence. (b) Plan and section of a single stacking unit, showing layer stagger. (c) Hexagonal net, with unit cells outlined, showing sequences of stagger movements which, oft-repeated, would produce a lattice. 1M = monoclinic, one stacking unit per cell; 2M = monoclinic, two units per cell; 2O = orthorhombic, two units per cell; 3T = trigonal, three units per cell; 6H = hexagonal, six units per cell. (Based on SMITH, J. V. and YODER, H. S., 1956, p. 209.)

formed by the alteration of feldspar, and often crowding the centres of feldspar crystals in thin sections. *Pinite* is used for similar material formed from cordierite. *Illite* is a degraded white mica, partly leached of its K^+, found in soils and unconsolidated sediments. The interlayer K^+ is partly replaced by weakly held water molecules or the hydroxonium ion. On diagenesis it appears to regain K^+ from the environment and to form true muscovite. The general word hydromuscovite may be used to cover this and *damourite, gilbertite, pinite, phengite* etc. *Paragonite*, $NaAl_2\{AlSi_3O_{10}\}$ $(OH)_2$, the sodium analogue of muscovite, is much rarer. It is reported from some phyllites and schists. Its properties differ little from muscovite and there seems to be little data on its X-ray pattern.

Use: Muscovite is widely used for insulation and as a dielectric, as well as for heat resistant supports and windows. The chief production of sheet mica is from Bihar and Madras, India, but a great deal of mica is produced also in Canada, S. Africa, etc.

Glauconite (Table 22) occurs as granules or pellets which are microcrystalline aggregates of flaky crystals, shown by X-rays to have a mica-type structure. Rare larger flakes have perfect {001} cleavage. It is dark green in hand-specimen, pleochroic in green to yellow-green in thin-section. $\alpha\perp\{001\}$ cleavage $1\cdot59-1\cdot61$, $\beta=\gamma$ $1\cdot61-1\cdot64$, $2V_\alpha<20°$. $\gamma-\alpha$ $0\cdot02-0\cdot03$, but confused aggregate polarization is a common feature.

Glauconite is found in sedimentary rocks laid down during slow sedimentation in shallow seas. It may initially be a colloidal precipitate that later becomes crystalline, but is probably partly of diagenetic origin. It is the characteristic mineral of greensands, and is found also in some limestones, especially in their basal layers, where it may form infillings of small shells, or be associated with layers of phosphatic nodules.

Greensands have been worked on a small scale as a source of potash.

Biotite $2[K(Mg, Fe)_3\{AlSi_3O_{10}\}(OH, F)_2]$
Phlogopite $2[KMg_3\{AlSi_3O_{10}\}(OH, F)_2]$
Monoclinic C $2/m$ a $0\cdot53$ b $0\cdot92$ c $1\cdot02$
                                                    $\beta 100°$

Structure: As for muscovite, with the difference that the structure is trioctahedral with all the octahedral sites occupied by Mg^{2+}, or (Mg, Fe^{2+}). The 1 M polymorph predominates amongst phlogopites and biotites.

Habit: Twinning and cleavage as for muscovite. H $2\frac{1}{2}-3$. D $2\cdot7-3\cdot4$ increasing with Fe. Lustre vitreous or pearly in phlogopite, submetallic splendent in biotite. Colour of phlogopite golden brown; of biotite black.

Optics: Translucent, pleochroic in yellow and yellow-brown, greenish brown, grey-brown, dark brown, red-brown. Absorption $\alpha < \beta = \gamma$. $\alpha\perp\{001\}$ cleavage $1\cdot53-1\cdot63$, $\beta = \gamma$ $1\cdot56-1\cdot70$, γ $1\cdot56-1\cdot70$, $\gamma-\alpha$ $0\cdot03-0\cdot08$. $2V_\alpha 0-15°$, nearly uniaxial. A watered silk or moirée appearance at the moment of extinction is characteristic of biotite. Round, highly pleochroic spots (pleochroic haloes) may be present around inclusions of radioactive minerals, usually zircon. Phlogopite may show asterism (a rayed pattern of light, due to minute inclusions). This is seen when a candle flame is viewed through a cleavage flake.

Tests: The habit, colour, cleavage and softness are distinctive.

Occurrence: Phlogopite occurs in marbles produced from the metamorphism of siliceous dolomitic limestones. Mg-rich biotites occur in mica peridotites and kimberlite, and as the igneous rocks are traversed from basic to acid the contained biotites become more Fe^{2+}-rich, those from granite pegmatites being the most extreme. On the whole, biotite does not become abundant until diorites and granodiorites are reached. It is more likely to occur in plutonic rocks, where the water vapour pressure remains high, than in extrusives; nevertheless many biotite-rhyolites and dacites occur, and in them the iron is Fe^{3+} rather than Fe^{2+}.

Biotite is abundant in metamorphic rocks and occurs over a wide range of conditions in thermal and regional metamorphism. The biotites of schists and gneisses have intermediate values of the Fe^{2+}/Mg ratio.

Uses: Phlogopite has the same uses as muscovite and is superior in some applications, having higher heat resistance.

Lepidolite $K(Li, Al)_3\{AlSi_3O_{10}\}(OH, F)_2$

This mica is a silvery pink to pale purple colour, colourless in thin section. α $1\cdot52-1\cdot55$, β $1\cdot55-1\cdot58$, γ $1\cdot55-1\cdot58$, $\gamma-\alpha$ $0\cdot02-0\cdot04$. $2V_\alpha 0-58°$. The indices are somewhat lower than those of muscovite.

Lepidolite fuses easily, with swelling, and colours the flame carmine red. This provides a simple distinction from muscovite.

Occurrence: In granite pegmatites with minerals like spodumene and pink tourmaline. It is mined as a source of lithium. Important localities are Elba, Madagascar, S.W. Maine, and San Diego County, California.

Zinnwaldite

$$K_2(Fe, Li, Al)_{2-3}\{AlSi_3O_{10}\}(OH, F)_2$$

A rare lithium-iron mica of grey-brown, yellow-brown or violet colour. Its optical properties are like those of biotite. It is found in cassiterite and topaz-bearing pegmatites, with tourmaline and perhaps other Li minerals. From a small number of localities including the Erzgebirge of Saxony, and St. Just, Cornwall.

Brittle Micas

Margarite $CaAl_2\{Al_2Si_2O_{10}\}(OH)_2$ and **clintonite** $CaMg_3\{Al_2Si_2O_{10}\}OH_2$ are uncommon minerals. Their chief point of interest is that their cleavage flakes are brittle, unlike those of other micas which are flexible and elastic, and this reflects the higher charge of the interlayer cation and stronger interlayer bonding. The charge balance is maintained by an increase in the substitution of Al for Si in the tetrahedral layers.

Margarite occurs as pink or yellowish-white brittle flakes, formed as an alteration product of corundum. More rarely it has been found in chlorite schists. Clintonite in reddish brown, yellowish or copper-coloured flakes in limestone skarns with serpentine, spinel, lime-garnet, vesuvianite and aluminous diopside.

Vermiculite

$$0{\cdot}33M^+(Mg, Fe\ etc.)_3$$
$$\{(Al, Si)_4O_{10}\}(OH)_2 . 4{\cdot}5H_2O$$

Structure: This may be regarded essentially as a leached biotite mica in which some of the interlayer cations are replaced by water molecules.

The principal distinctive property is that on rapid heating to about 400°C the sudden release of interlayer water causes the flakes to expand rapidly to twenty or thirty times their original thickness normal to the cleavage, like an opening concertina. They writhe about as they do so, and form long twisted cylinders made up of thin laminae—hence the name, from Latin *vermiculare* = worm-like.

There is probably a continuous series from biotite, through hydrobiotite to vermiculite. Distinction between soil vermiculites and smectites is not sharp, but vermiculite does not rehydrate readily or at all fully after heating.

Habit: Scales and flakes like those of biotite, from which it differs in being golden-brown to buff in colour and of duller lustre.

Occurrence: Commercial deposits of vermiculite have formed from biotite by hydrothermal alteration. They are generally associated with basic or ultrabasic igneous rocks or with carbonatites (e.g. Palabora, Transvaal). Vermiculite is also found in soils.

Use: In expanded form as an insulating material, a constituent of light-weight plasters, and when finely ground, as a lubricant.

Montmorillonites (Smectites) $0{\cdot}33M^+(Al, Mg)_2$ $\{Si_4O_{10}\}(OH)_2 . nH_2O$

Monoclinic $a\ 0{\cdot}515{-}0{\cdot}535$ $b\ 0{\cdot}895{-}0{\cdot}922$ $c\ 1{\cdot}0{-}1{\cdot}8$

Structure: Basically of pyrophyllite type with substitution chiefly in the octahedral layer allowing the weak bonding of exchangeable cations and water molecules in interlayer position. The water is readily lost and regained, as are the interlayer cations, with accompanying large changes in the c-dimension from a mica-like figure of $1{\cdot}0$ nm to values sometimes more than $2{\cdot}0$ nm. These are therefore the *swelling clays*.

Habit: Generally in extremely small particles.

Identification: Clays containing smectites may sometimes be recognized in the field by their large volume changes on wetting and drying and consequent liability to extensive slumping. The outcrop surface may show a centimetre-scale pattern of polygonal cracking.

In the laboratory identification depends on X-ray and chemical methods (*The X-ray identification and crystal structures of clay minerals*, ed. G. Brown, Min. Soc. London, 1961). Briefly, smectites are distinguished from most other clay minerals (except a few chlorites) by the fact that, on moistening with glycerol or ethylene glycol, $d_{(001)}$ expands to about $1{\cdot}78$ nm. (The clay is first immersed in a solution of a Na, Ca or Mg salt to saturate it with one of these ions and then air-dried.) On heating an untreated sample of the clay to 500°C, montmorillonite will collapse to $d_{(001)} \simeq 1{\cdot}0$ nm. Chorite, by contrast, will retain $d_{(001)} \simeq 1{\cdot}4$ nm.

Occurrence: Smectite clays result from contemporaneous weathering of volcanic ash-beds in the geological succession giving rise to a rock-type called *bentonite*. The presence of this material in the geological profile often gives rise to slumping of overlying strata on a very large scale.

Minerals of the group also occur in soils.

Varieties: Of the dioctahedral types *beidellite* is an aluminous variety, *nontronite* is rich in Fe^{3+}. In the trioctahedral group *saponite* is Mg^{2+}-rich and shows replacement of Si^{4+} by Al^{3+} to account for interlayer bonding. *Hectorite* has Li^+ substituting for Mg^{2+}.

In these clay minerals interstratification of one structure type with another (mixed-layer structures) adds to the problems of identification.

Use: The capacity to adsorb cations between the layers, which can subsequently be washed out again, leads to the use of smectites in clearing oils and other liquids, and in catalysis.

Bentonite is used in the mud that is circulated during rock-drilling to seal pervious formations penetrated by the drill and prevent loss of the fluid which is essential to the cutting operation.

CHLORITE GROUP $2[(Fe, Mg, Al)_{2-3}(OH)_6$
$(Mg, Fe, Al)_{2-3}\{(Al, Si)_4O_{10}\}(OH)_2]$

Monoclinic* C $2/m$ a 0·53 b 0·92 c 1·43
$\beta 97° 6'$

Structure: Double layers of pyrophyllite or talc type are separated by independent gibbsite or brucite type octahedral layers. This leads to a repeat distance $d_{(001)}$ of 1·4 nm (Fig. 240).

Composition: The possibilities of atomic substitution are very wide, as both types of layer may assume either dioctahedral or trioctahedral configuration. Al may also substitute for Si in the tetrahedral layers presumably with corresponding replacement of R^{2+} by R^{3+} in the octahedral part of the talc layer or in the brucite layer.

As far as Fe^{2+} and Fe^{3+} is concerned, chlorites fall into two groups, orthochlorites with Fe^{2+} and leptochlorites with $Fe^{2+} + Fe^{3+}$. The leptochlorites are the oxidized equivalents of orthochlorites, and the latter can be changed into the former by careful heating in air.

Habit: Crystals six-sided, tabular on {001}. Usually foliated masses or scales, often in fan-shaped or radiating groups.

Twins on {001} with {001} as composition plane (the Pennine law). Also on the Mica law. Cleavage {001} highly perfect, into flexible laminae with little elasticity. H 2–2½. D 2·6–3·3, increasing with Fe. Lustre pearly. Colour green of various shades.

* Four types of unit cell have been recognized. The one quoted is typical.

K

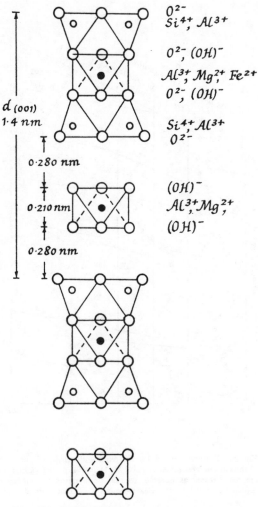

O^{2-}
Si^{4+}, Al^{3+}

$O^{2-}, (OH)^-$

$Al^{3+}, Mg^{2+}, Fe^{2+}$
$O^{2-}, (OH)^-$

Si^{4+}, Al^{3+}
O^{2-}

$(OH)^-$
$Al^{3+}, Mg^{2+},$
$(OH)^-$

$d_{(001)}$
1·4 nm

0·280 nm

0·210 nm

0·280 nm

Fig. 240. Pauling diagram of the layer sequence in chlorites.

Optics: Transparent, colourless, pale or bright green. Pleochroic though sometimes only weakly so. Though monoclinic or triclinic it is usually, for all practical purposes, uniaxial. α 1·56–1·67, β 1·57–1·69, γ 1·57–1·69, $\gamma - \alpha$ 0·0–0·01 rising to 0·02 in leptochlorites. $2V$ usually small, optically +ve (Fe-poor) or −ve (Fe-rich). Bx_a (either α or γ) ⊥{001}. Extinction straight with respect to close-spaced {001} cleavage traces. Optic axial plane 010. Some chlorites show anomalous interference colours in dark blue or violet, a feature said to be characteristic of the variety pennine.

The physical properties of orthochlorites are related to their composition in Fig. 241 (after Hey, M. H., *Min. Mag.*, **30**, 277, 1954). The correspon-

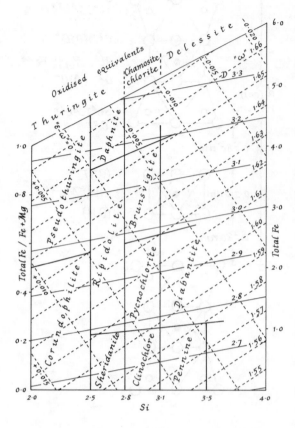

Fig. 241. Refractive index, birefringence and density of members of the chlorite group. (After HEY, M. H., 1954, p. 284.) Chlorites are here treated as uniaxial, so that "ω" means the vibration direction parallel to the cleavage, and "ε", that normal to it.

ding leptochlorites have higher values of γ, $\gamma - \alpha$ and D than orthochlorites.

Occurrence: Chlorites are widespread as late-stage minerals or secondary products in vesicles and veins in igneous rocks; as products of alteration of ferromagnesian minerals and volcanic glass; along joint and fault planes; in soils and sediments as products of weathering and diagenesis, especially as a major constituent of greywacke-type sandstones; in low-grade phyllites and schists of early regional metamorphism; and as a product of retrograde metamorphism of high-grade metamorphic rocks.

In short, chlorite is the stable low-temperature silicate form of (Mg, Fe, Al). The range of substitution is wide, and there is a chlorite for every low-temperature occasion.

Varieties: The principal varieties are shown in Fig. 241. The term chamosite-chlorite has been used for a leptochlorite of intermediate Si content, instead of chamosite which is used by Hey, because chamosite is widely applied to the serpentine-type mineral of the W. European marine oolitic iron ores (see p. 272).

Prehnite $2[Ca_2Al\{AlSi_3O_{10}\}(OH)_2]$
Orthorhombic $P\ mmm\ a\,0{\cdot}464\ b\,0{\cdot}548\ c\,1{\cdot}848$

Structure: Tetrahedral layers $\{AlSi_3O_{10}\}$ lie $\parallel$ $\{001\}$ and are linked by chains of Al atoms, coordinated octahedrally to 4O and 2(OH), running $\parallel b$, instead of by octahedral sheets as in the micas. Additional voids are thus available to accommodate Ca^{2+}.

Habit: Crystals are thick tablets elongated $\parallel y$ with $\{001\}$, $\{110\}$ and $\{010\}$ prominent; but usually in rounded massive, stalactitic or foliated aggregates with granular crystalline surfaces.
Cleavage $\{001\}$, good. $H\,6\frac{1}{2}$. $D\,2{\cdot}90$–$2{\cdot}95$. Lustre vitreous. Colour usually pale apple-green, sometimes white or grey.

Optics: Translucent, colourless. $\alpha = x\,1{\cdot}611$–$1{\cdot}630$, $\beta = y\,1{\cdot}617$–$1{\cdot}641$, $\gamma = z\,1{\cdot}632$–$1{\cdot}669$, $\gamma - a\,0{\cdot}021$–$0{\cdot}039$. $2V_\gamma\,65$–$69°$. Indices increase with entry of Fe. Anomalous interference colours and diminished $2V$ (to $0°$) have been interpreted as due to submicroscopic twinning.

Tests: Fuses with swelling to a bubbly enamel. Colour and habit are characteristic.

Occurrence: A hydrothermal mineral of cavities in basic igneous rocks often associated with zeolites. Also in low grade metamorphic rocks (zeolite facies) associated with pumpellyite. May replace calcic plagioclase in altered basic plutonic rocks.

FRAMEWORK SILICATES

The structures in this group are based on a three-dimensional framework of SiO_4 tetrahedra in which all the corner O^{2-} ions are shared between two tetrahedra. Quartz provides the type example of this arrangement. When all tetrahedra have Si^{4+} at their centres, the O^{2-} ions are all valency-satisfied, and there are no bonds for the attachment of interstitial cations. It is only by virtue of substitution of Al^{3+} for Si^{4+} in some of the tetrahedra that spare units of negative charge are made available for this purpose. Some interstitial cations appear to be necessary to stabilize the more open frameworks of the group and there is a good deal of flexibility in some of the frameworks to accommodate different cationic candidates for this function.

Apophyllite, described here amongst the zeolites because of its affinity with them in physical properties and occurrence, does not conform with these generalizations, and might rather be considered a layer silicate.

SILICA MINERALS

The crystalline forms of SiO_2 were used in Chapter 1, as an example of polymorphism. Their crystal structures and stability fields are illustrated there (Figs. 16–20) and some of their physical and optical properties are given in Table 8 (p. 23) and Fig. 15.

Quartz $3[SiO_2]$

Trigonal (low-quartz) P 32 a 0·4913 c 0·5405
Hexagonal (high-quartz) P 62 a 0·499 c 0·545

Structure: Chains of SiO_4 tetrahedra spiralling up the z-axis, the outer apices of the tetrahedra shared with adjacent chains (see further, pp. 23-4 and Fig. 17).

Habit: Commonly in prismatic crystals with unequally developed rhombohedral terminations (Fig. 18). The prism faces are horizontally striated. Right- and left-handedness can only be determined macroscopically when $\{11\bar{2}1\}$ is present: this face lies to the right of $10\bar{1}0$ which is below the larger rhombohedron $\{10\bar{1}1\}$ in a right-handed crystal and to the left of $10\bar{1}0$ in a left-handed crystal. With $\{11\bar{2}1\}$ may be associated the right or left form $\{51\bar{6}1\}$ (Fig. 70, p. 66). Striations on $11\bar{2}1$ are parallel to the edge $10\bar{1}1/01\bar{1}0$. Quartz is also often in anhedral granular masses, or cryptocrystalline or in rolled grains.

Twinning is common but often not recognizable externally. The Dauphiné twin is an interpenetrant twin about z as twin axis, all axes in the two parts being parallel (see p. 72). The Brazil law is an interpenetrant twin with twin plane $11\bar{2}0$. Contact twins with twin plane $11\bar{2}2$ constitute the Japanese law. There may also be twinning on the plane $10\bar{1}1$. Cleavage absent. Fracture conchoidal. H 7. D 2·65. Lustre vitreous. Colourless, white, or with a wide range of tints (see under varieties). Piezoelectric and pyroelectric. Fluid inclusions, either aqueous solutions or gases, sometimes arranged in regular planes, are common in quartz. Where a liquid inclusion has a bubble in the cavity also, heating until the bubble disappears gives an indication of the temperature of original formation. Geochemical data may be obtained from the analysis of the contents of these inclusions.

Optics: Transparent, colourless. $\omega 1\cdot544$, $\varepsilon 1\cdot553$, $\varepsilon - \omega\, 0\cdot009$. Optically positive. Recognized microscopically by its absence of relief, freedom from cleavage or alteration, refractive index a little above the mounting cement, and clear white maximum interference colour perhaps faintly yellowish. Strained quartz grains may be not extinguished uniformly between crossed polars, a shadow sweeping across the grain through a small angle of rotation. These are *strain shadows*.

Quartz shows optical activity, that is the power of rotating the plane of polarized light. A plate 1 mm thick cut normal to the optic axis produces a rotation of about 19° for red light or 32° for blue, either to right or left (with reference to an observer looking towards the light source). This right or left rotatory power corresponds to right- or left-handed crystallographic type as defined above. In twinned crystals the rotation will be in opposite senses in the two parts of the twin. In convergent light the interference figure of thick plates has the centre of the black cross replaced by a colour which changes as the lower polar is rotated. Superposed crystals of right- and left-handed quartz, or interpenetrant twins, show a four-armed spiral curve in the centre of the figure, called Airy's spirals. The strength of these special effects increases with the thickness of the crystal plate, and none of them are seen in an ordinary microscope slide.

Tests: Combination of crystal form, lustre, hardness and fracture are usually distinctive. Infusible and dissolved only by hydrofluoric acid. Floats in bromoform, a distinction from diamond with which it has been confused. A more carefully controlled flotation test distinguishes it from beryl which is a little denser (see also under beryl for a chemical test).

Alteration: Very stable, a leading example of a resistate mineral.

Occurrence: The commonest of all minerals at the earth's surface. It is an essential constituent of acid igneous rocks and forms 25–40% by volume of granites. Quartz is one of the principal gangue minerals of veins radiating from intermediate and acid igneous rocks.

Because of its resistance to weathering and to attrition during transport it accumulates as a principal constituent of sediments, nearly all of which contain some proportion of quartz sand or silt. Sands and sandstones may contain over 95% of quartz.

Correspondingly, quartz is a major constituent of many metamorphic rocks derived from sediments. In metamorphism, however, part of the sedimentary quartz reacts with carbonates, oxides, etc. to produce silicate minerals.

Though regarded as a highly insoluble mineral, quartz shows considerable mobility in the earth's crust and recrystallizes in sandstones to form overgrowths on, and cement around detrital grains. In deep burial and/or deformation of sediments quartz segregates as veinlets as one of the earliest manifestations of metamorphic change.

Varieties: Crystalline material goes under various names. *Rock crystal* is colourless and water-clear. *Milky quartz* is milk white occurring in veins and pegmatites. The colour may be due to numerous minute fluid inclusions. *Blue* or *opalescent quartz* occurs in some metamorphic rocks, owing its appearance to minute included rods of rutile. *Amethyst* is transparent purple. *Rose quartz* is a delicate pink, usually massive. The colour fades on exposure. *Citrine* is yellow and transparent. It may be produced by heat treatment of amethystine quartz which is much commoner. *Smoky quartz* and *cairngorm* is transparent in shades of brown.

Cryptocrystalline forms of quartz are numerous. *Chalcedony* has a waxy lustre and is often in botryoidal or stalactitic forms. Microscopically it may be fibrous with the fibres elongated along z, though sometimes normal to z (*quartzine*). *Carnelian* or *sard* is clear red to orange brown chalcedony. *Prase*, *chrysoprase* and *plasma* are green chalcedony. When flecked with red jasper the material is called *bloodstone* or *heliotrope*. *Agate* is concentrically layered chalcedony filling cavities in lavas, the colours, usually white, grey and black or brown, being due to impurities. *Moss-agates* contain branching moss-like patches of dark impurity. *Onyx* is similar but the banding is straight. *Jasper* is red chalcedony stained by hematite. It is one of the commoner forms of vividly coloured silica, often found in veinlets in volcanic rocks. Its occurrence in extensive banded formations with pre-Cambrian sedimentary iron ores is noteworthy. *Flint* and *chert* are grey to black compact cryptocrystalline quartz, the first characteristically found as nodules

in chalk and the latter as irregular layers in limestone. *Radiolarian chert* differs in being a biogenic precipitate often associated with submarine pillow-lavas, but sometimes forming thick and extensive rock formations (e.g. the Monterey cherts of California).

Silicified wood is a replacement of wood by fine-grained chalcedony in which the fine structure of the cell walls is often preserved in amazing perfection.

Many of the cryptocrystalline varieties of silica are somewhat porous and can be coloured by artificial means.

Use: The uses of quartz sand are numerous. Enormous amounts are used in the building industry in concrete and mortar. It is used as a filler and diluent, as an abrasive and for making moulds for metal castings. Pure silica sand is the foundation of the glass industry, and is used in ceramics and refractories. Fused silica ware is highly resistant to heat and corrosion.

Coloured crystals and coloured and banded cryptocrystalline forms are used in jewellery.

The piezoelectric property leads to its use in oscillators and pressure gauges, for which highly perfect crystals are required that nowadays are grown artificially.

Cristobalite   8[SiO_2]

Sructure: High-cristobalite is cubic, F $m3m$, a 0·714, while the low-temperature form is tetragonal, P 42, a 0·503, c 0·822 with 4[SiO_2]. The former is shown in Fig. 20, p. 26.

Habit: Small octahedra often aggregated into tiny rounded granular groups. Other properties are given in Table 8, p. 23.

Optics: Colourless, transparent, n 1·484–1·487, very weakly birefringent. Relief moderate with $n <$ mounting cement. It possesses characteristic curving fracture lines.

Occurrence: In cavities of volcanic rocks such as obsidian and rhyolite.

Tridymite

Hexagonal (high-tridymite) P 6/*mmm*   4[SiO_2]
a 0·5046 c 0·8256

Orthorhombic (or hexagonal)   864[SiO_2]
(low-tridymite) a 3·008 c 4·908

Structure: The structure of high tridymite is shown in Fig. 19, p. 26, and some of its properties in

Table 8, p. 23. The low-temperature form has a unit cell six times as large. The open framework of tridymite may be stabilised by the presence of small amounts of interstitial Na^+ or Ca^{2+} with balancing Al^{3+} replacing Si^{4+} in the tetrahedra. Small amounts of these elements are a constant feature of its chemical analysis.

Habit: Thin hexagonal tablets flattened $||\{001\}$. Twins common on $\{110\}$ (orthorhombic). D 2·27.

Optics: Transparent, colourless. Optic axial plane $\perp$ plane of tablets and to one pair of prism faces. Taking the O.A.P. as 100, $\alpha = y$ 1·478, $\beta = x$ 1·479, $\gamma = z$ 1·481, $\gamma - \alpha$ 0·0025, $2V_\gamma$ 66–90° (mostly 80–89°). Aragonite-type twins give wedge-shaped sectors on $\{001\}$ between crossed polars, in twos and threes, and more complex twin-lamellae recalling leucite are also sometimes seen on this plane.

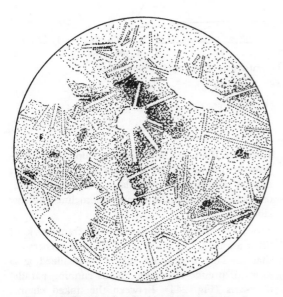

Fig. 242. Paramorphs of tridymite, now inverted to quartz, in thin section of contact-metamorphosed Torridonian Sandstone, Isle of Rhum.

Occurrence: In cavities and veins of trachyte and andesite lavas. Also as the product of partial fusion of sandstone near igneous intrusions. It may be represented by pseudomorphs (strictly *paramorphs*) of quartz preserving the form of tridymite (Fig. 242). It is believed that tridymite in lavas may grow metastably outside its true temperature/pressure stability field.

Opal $SiO_2.nH_2O$

Structure: One of the few amorphous minerals.

Habit: Amorphous fragments or botryoidal forms. Fracture conchoidal. H 5½–6½. D 2·0–2·2. Lustre waxy or resinous. Colour white, yellow, red, brown, etc. of pale shades. In precious opal a rich play of colours or 'fire' results from internal dispersive refraction of light from regular arrays of tiny spheres of amorphous silica (250–300 nm diam.) that make up the stone.

Optics: Transparent, isotropic or showing birefringence like that produced by strain in glasses. n 1·44–1·46 varying with water content.

Occurrence: Formed by weathering and alteration of siliceous rocks and deposited in fissures, or as coatings, and as replacements of fossils. Precious opal comes from altered trachyte in Czechoslovakia, and from sandstones at White Cliffs, New South Wales, Australia. Also found in Mexico. Siliceous sinter deposited around hot springs is opaline in nature. The skeletons of diatoms, radiolaria and sponges are of opaline silica.

Use: The varieties with a play of colours are gemstones. *Diatomaceous earth* (*diatomite, infusorial earth,* or *tripolite*) is used as a filtering agent, as an insulating material and a light-weight inert filler. Also as a mild abrasive.

Lechatelierite is silica glass produced by fusion of sand by lightning. It is found as tubes (*fulgurites*) penetrating downwards into loose sand deposits. Similar material (*Libyan glass*) has been found near meteorite craters in desert areas, and is thought to be quartz sand fused by the meteorite impact.

FELDSPAR GROUP

Feldspars fall into two main series: (*i*) K–Na feldspar or the Alkali Feldspar Series and (*ii*) Ca–Na feldspar or the Plagioclase Series. The composition fields of these series are shown in Fig. 243. Less important are Ba feldspar, celsian, and Ba–Na feldspar, hyalophane.

Alkali feldspars and plagioclase together make up between 50 and 60 wt% of all igneous rocks. Naturally, they have received intensive study, and the resulting body of knowledge allows a great deal of information about the history of a rock to be won by the thorough study of its feldspars.

The alkali feldspar series exhibits obvious features of polymorphism and inhomogeneity that have been revealed in the plagioclase series only by X-rays and the alkali feldspars will therefore be dealt with first.

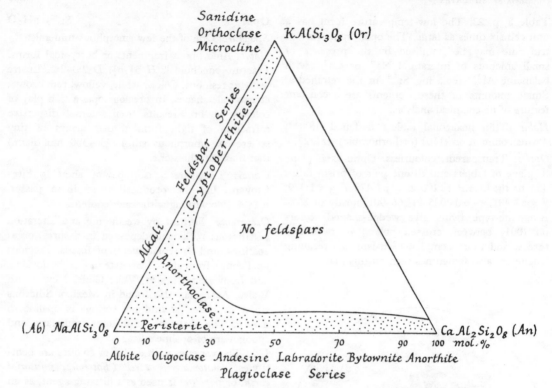

Fig. 243. Composition and nomenclature of common feldspars. Field of feldspar composition dotted.

ALKALI FELDSPAR SERIES

Sanidine (high) $4[(K, Na)\{AlSi_3O_8\}]$
Monoclinic C $2/m$ $a\,0\cdot8564$ $b\,1\cdot3030$ $c\,0\cdot7175$
                                              $\beta 115\cdot99°$

Orthoclase $4[(K, Na)\{AlSi_3O_8\}]$
Monoclinic C $2/m$ $a\,0\cdot8562$ $b\,1\cdot2996$ $c\,0\cdot7193$
                                              $\beta 116\cdot01°$

Microcline $4[K\{AlSi_3O_8\}]$
Triclinic C $\bar{1}$ $a\,0\cdot8561$ $b\,1\cdot2966$ $c\,0\cdot7216$
                          $\alpha 90\cdot65°$ $\beta 115\cdot83°$ $\gamma 87\cdot70°$

Adularia $K\{AlSi_3O_8\}$
Part monoclinic, part triclinic.

Anorthoclase $4[(Na, K)\{AlSi_3O_8\}]$
Triclinic $Na/Na + K > 0\cdot66$ Contains some Ca^{2+}

Albite $4[Na\{AlSi_3O_8\}]$
Triclinic C $\bar{1}$ (low)
                 $a\,0\cdot8138$ $b\,1\cdot2789$ $c\,0\cdot7156$
                 $\alpha 94\cdot33°$ $\beta 116\cdot57°$ $\gamma 87\cdot65°$
Triclinic C $\bar{1}$ (high)
                 $a\,0\cdot8149$ $b\,1\cdot2880$ $c\,7\cdot106$
                 $\alpha 93\cdot37°$ $\beta 116\cdot30°$ $\gamma 90\cdot28°$

In nature there are micro- or cryptoperthite intergrowths between high sanidine and high albite; normal sanidine, anorthoclase and high albite; orthoclase and low albite; and microcline and low albite.

Structure: Based on a continuous framework of SiO_4 tetrahedra in which all tetrahedral corners are shared. This framework may be visualized as a series of pseudo-tetragonal chains running parallel to *x*-axis (Fig. 244). Between the linked chains, tunnels traverse the structure parallel to *x*, in which lie the alkali-metal ions. The alkali ions are held in the structure by charges arising from the replacement of one in four of the Si^{4+} ions of the tetrahedra by Al^{3+}. Without this substitution all the O^{2-} of the framework would, of course, be valency-satisfied and unable to attach interstitial ions.

While in sanidine and orthoclase the structure has mirror planes parallel to {010}, in albite a slight collapse of the chains around the smaller Na^+ ion in the tunnel destroys this mirror plane and leads to triclinic symmetry.

In addition to the change of symmetry resulting

(a) *Idealised orthoclase framework: view ∥ x-axis.*

(b) *Perspective of idealised tetragonal chain — one of the four surrounding the "tunnel" in (a).*

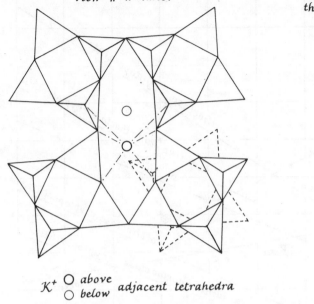

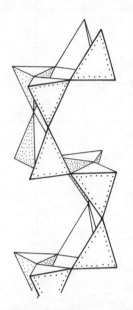

K⁺ ◯ above
   ◯ below *adjacent tetrahedra*

Fig. 244. Idealized structural pattern of orthoclase. In fact, the pseudotetragonal rings are somewhat twisted about their axis, producing the effect shown by the lower dotted tetrahedral ring in (*a*).

from replacement of K by Na, the change from monoclinic to triclinic symmetry between sanidine or orthoclase and microcline must be explained. In this case compositions may all be K-rich. The symmetry change from sanidine to microcline is progressive: some microcline is nearly monoclinic with angles α and γ close to 90°, and a series exists leading to *maximum microcline* in which α and γ angles reach 90·7° and 87·7°. This change is due to the ordering of the distribution of Al and Si over the tetrahedral sites. In sanidine they are distributed at random, while in microcline Al is concentrated in one out of the four different types of tetrahedral site. The Al tetrahedron is measurably bigger than the Si ones. Orthoclase represents an intermediate stage in the ordering process.

The difference between high and low albite, though not involving a change of symmetry, is due to the same process of ordering of Al, Si distribution. In high albite the two ions are distributed at random, and in low albite Al is concentrated in one particular site out of four kinds.

Exsolution phenomena: At high temperatures homogeneous mixed feldspars may form that contain Na and K in all proportions, randomly distributed over the sites in the tunnels of the structure. At lower temperatures, because the difference in size between Na and K imposes undue strains on the lattice, these crystals tend to unmix to form small volumes, or *domains*, of K⁺-rich or Na⁺-rich feldspar. This takes place essentially by migration and segregation of the alkali ions along the tunnels.

The *b* and *c* dimensions of potash feldspar and albite are very similar, hence the structures of the two types of domain fit neatly on the *bc* plane. The *a* dimension shortens or lengthens as the structure changes from K feldspar to Na feldspar and back again.

The resulting intergrowths, in lamellae parallel to {100}, are called *perthites* and they may be on various scales from visible lamellae down to those so fine that the presence of two phases is only revealed by the X-ray diffraction pattern. The scale depends upon the rate at which the crystal has been cooled—the *annealing time*. The longer the annealing time the coarser the lamellae will be. The structure of the members of the intergrowth will also differ, depending upon the temperature at the time of exsolution, which will control the degree of Al, Si ordering (see above) and hence the

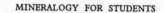

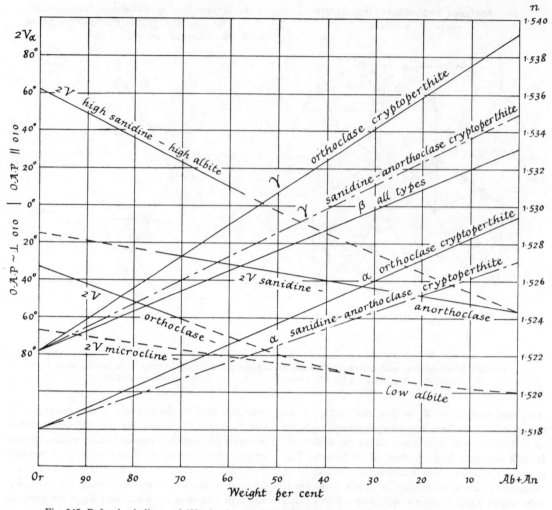

Fig. 245. Refractive indices and 2V related to composition in the alkali feldspar series. (Based on TUTTLE, O.F., 1952, pp. 557, 559.)

symmetry of the exsolved phases. The curves of 2V in Fig. 245 show the pairs of phases commonly intergrown; but many intergrowths probably contain more than two distinct phases.

Twinning: The feldspars are extremely susceptible to twinning, which takes place on a dozen different laws: the commoner ones are listed in Table 23.

The twins are classified in three groups:

(*i*) Normal twins in which the twin axis is normal to the composition plane, which is a possible crystal face.

(*ii*) Parallel twins in which the twin axis is a zone axis (in practice one of the crystallographic axes). It is parallel to the composition plane which need not always be a possible face.

(*iii*) Complex twins, in which part 1 of the crystal is related to part 1' by a normal twin law, and to part 2 by a parallel law with the same composition plane. Then part 2 is related to part 1 by a complex law in which the twin axis is normal to a zone axis and lies in the composition plane.

Twinning plays a large part in recognition of the different phases, and in estimation of their chemical composition. The following rules are generally a safe guide.

1. Monoclinic K-feldspars, sanidine and orthoclase, characteristically have simple twins, on the Carlsbad law.

2. Triclinic K-feldspars, microcline and anorthoclase, have fine-scale cross-hatched twinning, the

Table 23. The common feldspar twins.

Name	Twin axis	Composition plane		System
(i) Normal twins				
*Albite	⊥010	010	Repeated	Triclinic
Manebach	⊥001	001	Simple	Monoclinic or triclinic
Baveno	⊥021	021	Simple	Usually monoclinic
(ii) Parallel twins				
*Carlsbad	z-axis	(A)010, (B)100	Simple	Monoclinic or triclinic
*Pericline	y-axis	rhombic section ‖y	Repeated	Triclinic
Acline	y-axis	001, 100	Repeated	Triclinic
Estérel	x-axis	rhombic section ‖x	Repeated	Monoclinic or triclinic
Ala	x-axis	(A)001, (B)100	Repeated	Monoclinic or triclinic
(iii) Complex twins				
*Albite-Carlsbad	⊥z in 010	010	Repeated	Triclinic
Manebach-Acline (A)	⊥y in 001	001	Repeated	Triclinic

* These laws predominate by far over the others. One survey found $42\frac{1}{2}\%$ Albite, 24% Carlsbad and 16% Albite-Carlsbad twins in the sample.

combination of Albite and Pericline laws, giving the 'tartan pattern' (Fig. 246).

3. Plagioclase has repeated Albite twinning. This is often combined with Pericline twins, but the resulting sharply-defined 'cross-roads' pattern is quite distinct from the more shadowy tartan pattern (Fig. 246).

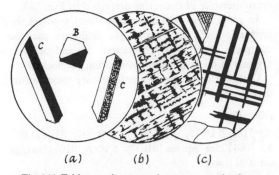

(a)   *(b)*   *(c)*

Fig. 246. Feldspar twins as seen between crossed polars.
(a) Carlsbad (C) and Baveno (B) laws in sanidine.
(b) Microcline showing twinning on Albite and Pericline laws with shadowy lamellae (tartan-pattern).
(c) Plagioclase, showing sharp Albite and Pericline lamellae.

The relations between the Albite and Pericline twinning are shown in Fig. 247. The *rhombic section*, which is the composition plane of the Pericline twin, is a plane containing y on which the traces of $\{110\}$ and $\{1\bar{1}0\}$ mark out a rhombus. One diagonal of this rhombus is y, the other must be normal to y and lie in the plane of the edges

between $\{110\}$ and $\{1\bar{1}0\}$, i.e. lie in 010. The orientation of the rhombic section is very sensitive to small changes in the interaxial angles. Its trace on $\{010\}$ measured against the $\{001\}$ cleavage, is characteristic of the different triclinic feldspar species (Fig. 247).

Cleavage: All feldspars have perfect $\{001\}$ and slightly less perfect $\{010\}$ cleavage. H $6-6\frac{1}{2}$. D 2·55–2·63. Lustre vitreous. *Colour*: colourless glassy (sanidine), white, pink, or green. In granitic rocks a pink colour often distinguishes K-feldspar from plagioclase.

Optical properties and diagnostic features follow.

Sanidine

α 1·520–1·523, β 1·525–1·529, γ 1·525–1·530, $\gamma - \alpha$ 0·005–0·007. In high-sanidine O.A.P. = 010, $\beta = y$, $\alpha \wedge x$ 5–9°, $2V_\alpha$ 11–63° (for artificially heated crystals). Normal sanidine has $\gamma = y$ $\alpha \wedge x$ 5°, $2V_\alpha$ 15–30°. O.A.P. ⊥010. Crystals are colourless, glassy, tabular ‖$\{010\}$; twinning on the Carlsbad law.

Occurrence: In trachytes and rhyolites rapidly cooled from initial high temperature of eruption. Usually natural crystals are members of the cryptoperthitic series towards anorthoclase (Fig. 245).

Anorthoclase

A member of a series extending from sanidine towards high albite, having $NaAlSi_3O_8$(Ab):$KAlSi_3O_8$(Or) > 67:33. Also feldspar in a series from

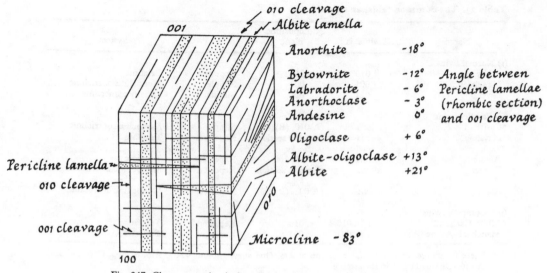

Fig. 247. Cleavages and twin-lamellae in plagioclase feldspar and microcline.

$Ab_{65}Or_{35}$ towards plagioclase $Ab_{35}An_{65}$ (Fig. 243). $\alpha\,1\cdot524$, $\beta\,1\cdot530$, $\gamma \simeq y\,1\cdot532$, $\gamma - \alpha\,0\cdot008$, $2V_\alpha 50-60°$. $\alpha \wedge \{001\}$ cleavage on $\{010\} = 9°$, $\alpha \wedge \{010\}$ cleavage on $\{001\} = 1-6°$. In specimens rich in Ca, indices and $2V$ move towards values for oligoclase (p. 290). The departure of γ from y is small and tartan-pattern twinning (indicative of triclinic symmetry) disappears on heating to temperatures of $100°-675°C$ depending upon composition; it reappears on cooling.

Tartan-pattern twinning on Albite and Pericline laws is diagnostic, the Pericline lamellae making an angle of $2-5°$ with the $\{001\}$ cleavage trace on $\{010\}$. This means that the tartan-pattern is best seen on $\{100\}$ which also shows both cleavages, and is a distinction from microcline where the tartan-pattern appears on $\{001\}$.

Occurrence: In alkali trachytes. Well-known examples come from Mt Kenya and from Mt Erebus in Antarctica. The crystals in these examples have canoe-shaped cross-sections.

Orthoclase

$\alpha \wedge x\,5-12°\,1\cdot519$, $\beta\,1\cdot523$, $\gamma = y\,1\cdot524$, $\gamma - \alpha\,0\cdot005-0\cdot007$. $2V_\alpha = 35°$.

Crystals tabular parallel to $\{010\}$ or elongated along x with a square cross-section (Baveno habit). Twinning on the Carlsbad law is common; on the Manebach, Baveno and other possible laws less so. Perthitic lamellae of albite parallel to $\{100\}$ are often visible under the microscope. The perthitic structure may produce a play of colours to the unaided eye in the variety *moonstone*.

Occurrence: Orthoclase is the characteristic K-feldspar of granites, granodiorites and syenites, where these have cooled at moderate depth. The orthoclase will generally be a microperthite or cryptoperthite of the series described in Fig. 245.

In more slowly-cooled granites and syenites the separation of the K and Na feldspars may be complete, with the formation of independent crystals of microcline and albite-rich plagioclase.

Orthoclase may grow in sedimentary rocks during diagenesis.

Microcline

$\alpha\,1\cdot518$, $\beta\,1\cdot522$, $\gamma\,1\cdot524$, $\gamma - \alpha\,0\cdot006$, $2V_\alpha 62-80°$. $\alpha \wedge \{001\}$ cleavage on $\{010\} = 5°$, $\alpha \wedge \{010\}$ cleavage on $\{001\} = 15°$.

In hand-specimen microcline sometimes has a distinctive bright green colour. Under the microscope the distinctive feature is the tartan-pattern twinning produced by the combination of Albite and Pericline twinning. The composition plane of the Pericline twin (the rhombic section) makes an angle of $80-83°$ with the $\{001\}$ cleavage on $\{010\}$. This means that the tartan-pattern is seen in sections near to $\{001\}$, which show only the $\{010\}$ cleavage, a distinction from anorthoclase.

Microcline may show perthitic intergrowth with albite.

Occurrence: In granites and syenites cooled slowly

at considerable depth. It often crystallizes with an albite plagioclase, there having been complete separation of the K and Na feldspars during long annealing. Also widely found in pegmatites.

Microcline is also characteristic of gneisses where equilibrium has been attained over a long period. Nevertheless, it is possible, and not uncommon, for microcline and orthoclase to occur in one rock.

Adularia

This is a habit modification of orthoclase, marked by the development of {110} to the virtual exclusion of {010}, giving crystals of rhombic cross-section $(110 \wedge 1\bar{1}0 = 61°)$. It is usually monoclinic with the same optic properties as orthoclase, but crystals have been found with some parts monoclinic and other regions slightly triclinic with microcline cross-hatched twinning.

Occurrence: In low-temperature veins in gneisses and schists (Alpine veins) and in altered wall rocks of epithermal gold-quartz veins. The monoclinic symmetry is thought to be anomalous in a low-temperature environment and the suggestion has been made that it is a metastable precipitate.

PLAGIOCLASE SERIES

Albite (low-temperature) $4[Na\{AlSi_3O_8\}]$

Triclinic C $\bar{1}$ a 0·8138 b 1·2789 c 0·7156
                         α94·33° β116·57° γ87·65°

Anorthite (low-temperature) $8[Ca\{Al_2Si_2O_8\}]$

Triclinic P $\bar{1}$ a 0·8177 b 1·2877 c 1·4169
                         α93·17° β115·85° γ91·22°

Composition: A continuous variation in composition takes place from anorthite to albite, by substitution of $Na^+ + Si^{4+}$ for $Ca^{2+} + Al^{3+}$. The series is divided into the sub-species shown in Fig. 243, but it is usual to designate composition by molecules per cent of anorthite written, *e.g.* An_{65}.

Structure: Fig. 248 summarizes the structural relationships.

At high temperatures a continuous series exists from high-albite to An_{90}, the members having the albite structure, which has a C face-centred cell, with $c \simeq 0·7$ nm, and is in all essentials like orthoclase, except for the lack of the monoclinic symmetry plane. Al and Si are randomly distributed over the tetrahedral sites. Pure anorthite has a P lattice with $c \simeq 1·4$ nm at *all* temperatures: in the pure compound AlO_4 and SiO_4 tetrahedra alternate regularly in the chains, each AlO_4 being surrounded by four SiO_4 groups.

In the intermediate plagioclases, as temperature falls (and earlier in the fall the more calcic the crystal) ordering of the Al, Si distribution takes place and the calcic component assumes a body-centred (I) lattice, with $c \simeq 1·4$ nm. At these lower temperatures, therefore, intermediate plagioclase between about An_{25} and An_{90} is made up of domains with a large unit cell (roughly a 0·81, b 9·41, c 5·15 nm) which are faulted one against another, to compensate this structural change. The *peristerites*, of the composition range An_{5-25}, have two exsolved phases, An_0 and An_{30} (the amounts of which are in the ratio 1:1 at An_{16}).

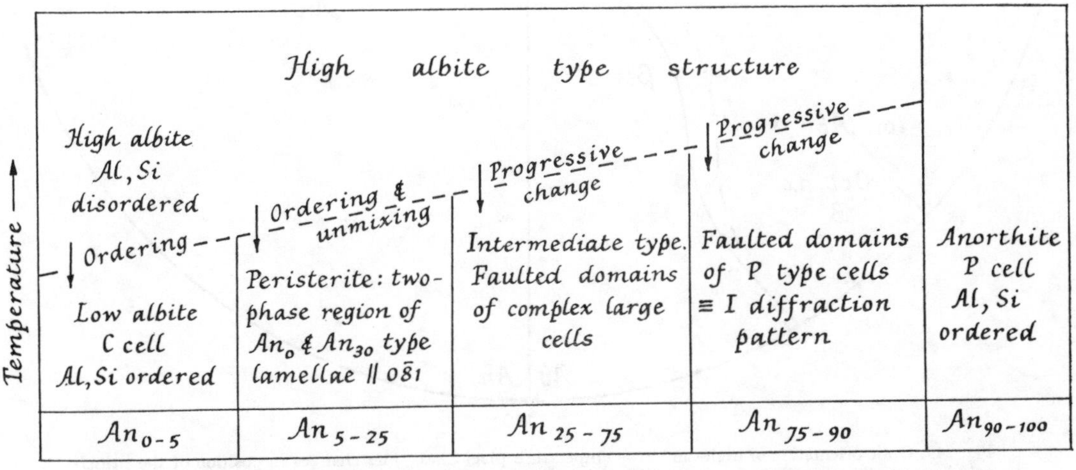

Fig. 248. Relationship between the various plagioclase structures. (Based on the studies of GAY, MEGAW and others.)

Pure low-albite has a homogeneous structure again with $c \simeq 0.7$ nm. In low-albite, Al is concentrated in one type of tetrahedral site instead of the Al and Si being randomly distributed over the tetrahedral sites as it is in the high form (cf. p. 283).

Habit: Prismatic crystals often flattened parallel to {010}. Rarely elongated along y (variety pericline). In anhedral grains and cleaved masses.

Twinned in a variety of ways as described under alkali feldspar (p. 285) Cleavage {001} perfect, {010} slightly less so. *H* 6–6½. *D* 2·62–2·76 increasing with Ca. Lustre vitreous, brilliant on cleavages. Colour white, greyish, pinkish or greenish, dark grey or bluish in some labradorite.

Optics: Properties vary continuously through the series and are summarized in Figs. 249–254.

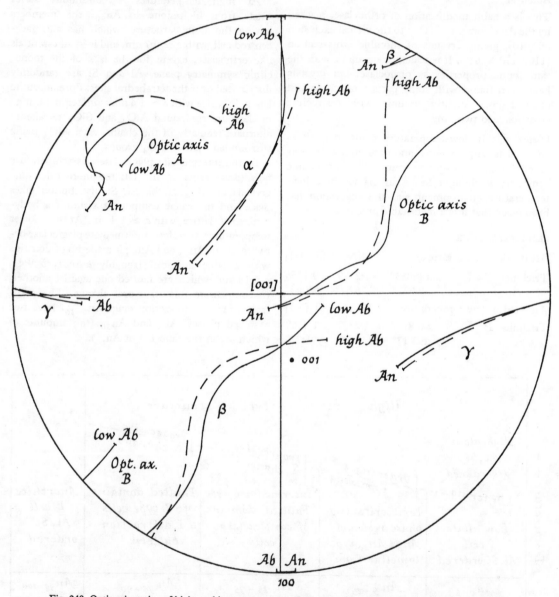

Fig. 249. Optic orientation of high- and low-temperature plagioclase. The changes in position of the principal vibration directions and the optic axes with composition are shown in full line for the low-temperature series and in broken line for the high-temperature series. (Mainly after MUIR, I. D., 1955, p. 549.)

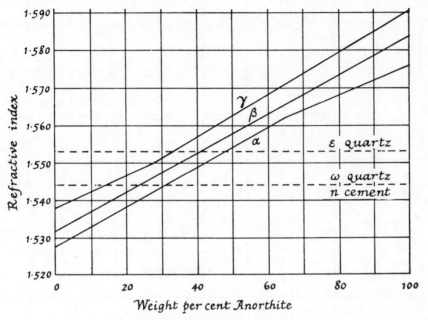

Fig. 250. Refractive indices in the plagioclase series.

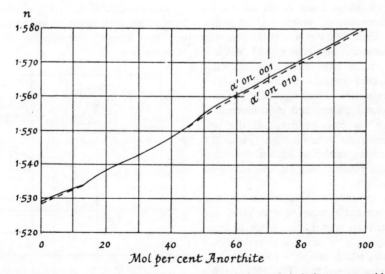

Fig. 251. Relation of the least refractive index on cleavage flakes of plagioclase to composition.

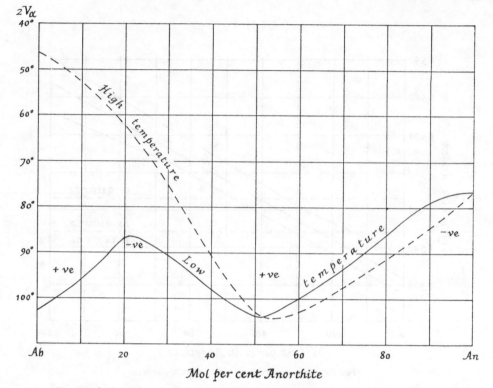

Fig. 252. Optic axial angle of plagioclase. (Based on data of SMITH, J. R., 1958, p. 1188.)

Because of the existence of high and low temperature modifications, orientation of the indicatrix determined by extinction angles is not reliable for composition determination, unless the structural state is known (Fig. 249). Plutonic rocks generally have low-temperature plagioclase, but volcanic and hypabyssal rocks have plagioclase of intermediate or high-temperature structural states.

Fortunately refractive index, especially α, varies little with structural change and measurement of this index, or α′ on cleavage flakes, is the best method of composition determination (Figs. 250, 251). When the composition is known, the other optical properties furnish data on the structural state (Figs. 249, 252).

Nevertheless, in routine examination of thin sections, extinction angles, coupled with relief and birefringence, give an approximate guide to composition, and especially to variation in composition in zoned crystals, which are very common. It is best to use sections normal to x (Fig. 253) or those showing Carlsbad-albite twinning (Fig. 254), which give a direct composition value for the grain under study, rather than the curve (Fig. 253) which depends on testing a sample of grains and taking a maximum value of extinction angle in the zone normal to 010.

Sections normal to x (Fig. 253) are recognized by showing both cleavages as sharp lines which do not move laterally on changing the focus of the microscope.

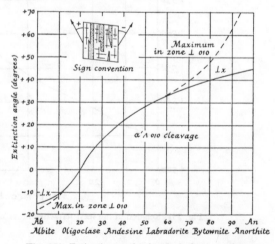

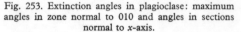

Fig. 253. Extinction angles in plagioclase: maximum angles in zone normal to 010 and angles in sections normal to x-axis.

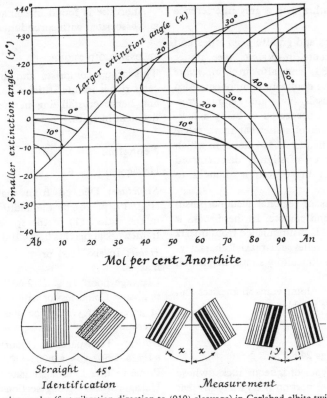

Fig. 254. Extinction angles (fast vibration direction to {010} cleavage) in Carlsbad-albite twins of plagioclase.

Sections with combined Carlsbad-albite twinning cut normal to 010 (Fig. 254) show equal illumination of the parts when the trace of the composition plane is parallel to a cross-hair, an appearance of simple Carlsbad twinning in 45° position, and equal and opposite values of the extinction angles of the two individuals of a twin.

The use of the curve of maximum extinction angle normal to 010 (Fig. 253) requires sections showing equal illumination when the composition plane is parallel with a cross-hair, and having equal and opposite extinction angles in the two individuals of the twin. A sample of about half-a-dozen or more such sections is examined and the *maximum* value of the means of the pairs of angles is taken to indicate the composition.

In the use of both Fig. 253 and Fig. 254 there is a region of indeterminancy when the extinction angle is less than 19°. This is resolved by regarding the extinction angle as +ve when α' is in the acute angle of intersection of the {010} and {001} cleavage traces and −ve when it is in the obtuse angle (Fig. 253, inset). Refractive index estimation also resolves the indeterminancy.

High albite

$\alpha 1 \cdot 527$, $\beta 1 \cdot 532$, $\gamma 1 \cdot 534$, $\gamma - \alpha 0 \cdot 007$. $2V_\alpha 42-55°$. $\alpha' \wedge x$ on {010} = 9°, on {001} = 0–2°. Maximum extinction angle $\perp 010$ ($\alpha' \wedge$ {010} cleavage trace) = 25–38°.

Distinguished from other Albite-twinned feldspars by low indices, moderate −ve $2V$, and relatively large maximum extinction angle $\perp 010$.

Occurrence: First described from material produced in the laboratory by heating low albite and by precipitation from melts, high albite has subsequently been reported from some rhyolites and trachytes.

Low albite

$\alpha 1 \cdot 529$, $\beta 1 \cdot 533$, $\gamma 1 \cdot 539$, $\gamma - \alpha 0 \cdot 010$. $2V_\alpha 97°$. $\alpha' \wedge x$ on {010} = 20°, on {001} = 3°. Maximum extinction angle $\perp 010$ ($\alpha' \wedge$ {010} cleavage trace = −19°).

Distinguished from other Albite-twinned feldspars by indices lower than 1·54 coupled with an extinction $\perp x$ (or max. $\perp 010$) > 11°, and slightly higher birefringence than intermediate plagioclase,

giving bright white rather than a grey interference colour.

Occurrence: In granites and granite pegmatites and some syenites. It occurs in *perthite* (see under alkali feldspar, p. 283) and also in *antiperthite* where plagioclase is the dominant component. Also in spilite (albitized basalt) lavas and keratophyres. In igneous rocks often suspected of being a low-temperature replacement of somewhat more calcic primary plagioclase. Chequer-board albite, in which the crystals are made up of slightly disorientated smaller units squarish in section, is thought to have this origin.

In low-grade schists of regional metamorphism with quartz, chlorite and epidote. In these rocks it is often untwinned and is distinguished from quartz by its cleavage, lower relief and biaxial character; but distinction may be difficult by simple inspection without an optical test.

Also a constituent of Alpine veins in gneisses and schists. Found in some marbles.

Intermediate plagioclase

Optics are given in Figs 249–254.

Occurrence: In most types of igneous rocks, whose classification depends in part on the variety present (p. 157). In igneous rocks it is often zoned from Ca-rich cores to Na-rich margins. Widespread in metamorphic rocks also, where the Ca content increases, in general, with increasing metamorphic grade.

The more calcic varieties are rather readily altered to zoisite, calcite and albite (an aggregate called *saussurite*). More sodic types occur as detrital grains in feldspathic sandstones and greywackes.

Anorthite

$\alpha 1{\cdot}577$, $\beta 1{\cdot}584$, $\gamma 1{\cdot}590$, $\gamma - \alpha\ 0{\cdot}013$. $2V_\alpha 77°$. $\alpha' \wedge x$ on $\{010\} = 39°$, on $\{001\} = 42°$. Maximum extinction angle $\perp 010$ $(\alpha' \wedge \{010\}) = 88°$.

Distinguished, amongst Albite-twinned feldspars, by high relief and relatively high birefringence (yellowish-white interference colours).

Occurrence: Anorthite is rare in igneous rocks where plagioclase is not usually more calcic than An_{80-90}. It occurs at Mt Somma, Vesuvius in ejected blocks amongst the old lavas, in some basic lavas and tuffs in Japan, and, rarely, in amygdaloidal cavities in basalt. In spite of the name, anorthosites are usually made of bytownite (An_{70-90}) or labradorite (An_{50-70}), and not of anorthite.

Anorthite is found in contact-altered limestones and associated with corundum deposits (Madras).

FELDSPATHOID GROUP

This group comprises anhydrous framework silicates richer in interstitial cations than the feldspars and hence crystallizing in their stead from melts deficient in silica.

Nepheline $\qquad\qquad\qquad 8[(Na, K)\{AlSiO_4\}]$

Hexagonal $P\ 6$ $\qquad\qquad\quad a\ 1{\cdot}001\quad c\ 0{\cdot}840$

Structure: Derived from that of high-tridymite (Fig. 19, p. 26) by the addition of Na, K (ideally in the ratio $3:1$) in the spaces of the tetrahedral framework at heights 0, 50, 100 along c (Fig. 255).

Habit: In six-sided prismatic crystals or anhedral masses.

Cleavage poor. $H\ 6$. $D\ 2{\cdot}60$–$2{\cdot}63$. Lustre vitreous to greasy. Colourless or white, sometimes dark green or greenish grey. May be orange-red as a result of alteration.

Optics: Transparent, colourless. $\omega\ 1{\cdot}530$–$1{\cdot}545$, $\varepsilon\ 1{\cdot}526$–$1{\cdot}542$, $\omega - \varepsilon\ 0{\cdot}003$–$0{\cdot}005$. Optically negative. Prone to alter to an aggregate of white mica flakes. Hexagonal and oblong sections, lack of cleavage or lamellar twinning, low birefringence and uniaxial character distinguish it from feldspars.

Tests: Gelatinizes with HCl. Treatment of thin sections with acid followed by methylene blue stains nepheline and picks it out from feldspar with which it may be intergrown.

Varieties: *Elaeolite* is coarse-grained or massive, greenish, brownish or reddish nepheline with a greasy lustre. The ratio of K to Na in nepheline varies and there is a series towards *kalsilite* $KAlSiO_4$ which is a rare mineral found in certain potassic lavas on the Uganda–Congo border.

Occurrence: In silica-deficient igneous rocks, nepheline-syenites, phonolites, and theralites, where it is an essential constituent. Urtite is an almost monomineralic nepheline rock. Well-known localities for nepheline-rich rocks are Kola Peninsula, U.S.S.R.; Oslo district, Norway; Julianehaab district, S. Greenland; and Magnet Cove, Arkansas.

Also found in gneissic rocks produced by metasomatic alteration of limestones and other sediments in the Haliburton–Bancroft district of Ontario, associated with corundum.

Use: In the manufacture of glass and ceramics. To a limited extent as an ore of aluminium.

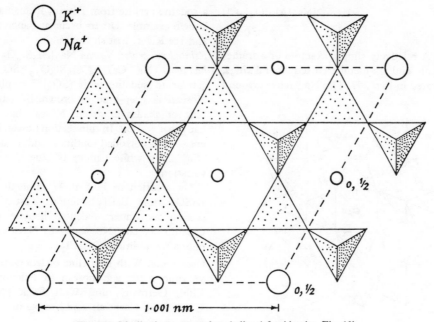

Fig. 255. Idealized structure of nepheline (cf. tridymite, Fig. 19).

Leucite   16[K{AlSi$_2$O$_6$}]

Cubic (above 625°C)   *I   m3m*   *a* 1·340

Tetragonal (below 625°C)   *I   4/m   a* 1·295   *c* 1·365

Structure: An arrangement of four-membered and six-membered rings of (Si, Al)O$_4$ tetrahedra linked to form a three-dimensional framework. The structure is similar to that of analcime.

Habit: Crystallizes in perfect trapezohedral {211} crystals. The cubic aspect is inherited from a high temperature of crystallization: at ordinary temperatures this aspect is maintained by repeated twinning of the tetragonal lattice in lamellae parallel to {110} of the cubic axes. These lamellae cause the faces to be striated.

Cleavage very poor. Fracture conchoidal. *H* 5½–6. *D* 2·47–2·50. Lustre vitreous. Colour white or ash-grey.

Optics: Transparent, colourless. *n* 1·508–1·511. Weak birefringence, 0·001, discloses twin lamellae in sets at right angles and at 45° (Fig. 256). Concentrically arranged inclusions are often present. The crystals are usually euhedral with a moderate relief *n* < mounting cement.

Occurrence: Found only in volcanic rocks, notably at Mt Vesuvius, localities near Rome, in the Bfumbira volcanic field Uganda and nearby Congo, in the Eifel district of Germany, and at Leucite Hills, Wyoming.

Leucite is unstable on slow cooling and is replaced by pseudomorphs made of K feldspar and

Fig. 256. Leucite crystal in thin section between crossed polars, showing birefringent twin lamellae, the result of inversion from cubic to tetragonal symmetry.

nepheline which are called *pseudoleucites*. The oval white or pinkish spots in the syenite of Loch Borolan in Sutherlandshire are supposed to be of this origin.

Sodalite $2[Na_4\{AlSiO_4\}_3Cl]$

Cubic P $\bar{4}3m$ a 0·887

Structure: A relatively simple example of a frame-work of four- and six-membered tetrahedral rings. It is illustrated in Fig. 257. The Na$^+$ ions surround

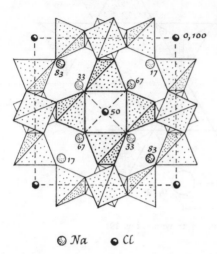

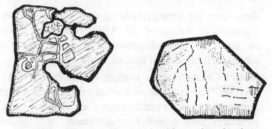

Fig. 257. The structure of sodalite.

the Cl$^-$ ions tetrahedrally as shown by chain-dotted lines around the central Cl$^-$ of the figure. The tetrahedra are at heights 0, 100 (heavy outline) 75, 50 and 25 (progressively lighter outlines). The four-membered rings are in the cube faces, the six-membered rings normal to the triad axes. Thermal expansion studies indicate untwisting of a frame-work partly collapsed at ordinary temperatures (cf. quartz, cristobalite).

Habit: Crystallizes in dodecahedra. Commonly massive anhedral.

Cleavage {110} only fair. *H* 5½–6. *D* 2·30. Lustre vitreous or greasy. Colour blue, white or pink.

Optics: Transparent, colourless or very pale pink or blue. $n = 1·483$. Marked relief, n much below that of the cement, isotropic. Difficult to distinguish from analcime unless by blue colour in hand-specimen.

Tests: After dissolving in dilute nitric acid, the addition of silver nitrate gives a precipitate of silver chloride indicating chlorine. If sodalite in an un-covered thin section is attacked by dilute nitric acid, when this evaporates sodium chloride crystals will form.

Occurrence: In alkali-syenites and trachytes. Ditroite,

a sodalite syenite from Ditrău, Rumania, is a well-known example. Others occur at Bancroft, Ontario, and Ice River, British Columbia.

Related minerals: *Nosean* $Na_8\{Al_6Si_6O_{24}\}SO_4$ and *hauyne* $(Na, Ca)_{4-8}\{Al_6Si_6O_{24}\}(SO_4)_{1-2}$ are similar to sodalite with $(SO_4)^{2-}$ replacing Cl$^-$. Nosean is grey-blue or brownish, hauyne bright blue or shades of green. Nosean has $n = 1·495$, hauyne $n = 1·50$. In thin section nosean is yellowish grey with corroded outlines and a dark border (Fig. 258), while hauyne is blue also with zonal variation.

On solution in HCl these minerals precipitate $BaSO_4$ from $BaCl_2$ solution. Tested on a thin section as under sodalite, hauyne precipitates gypsum crystals. Nosean gives NaCl and gypsum only after addition of $CaCl_2$.

Occurrence: With nepheline or leucite in alkali-rich volcanic rocks e.g. nosean in leucitophyre from Eifel, Germany, and Wolf Rock phonolite off Lands End, Cornwall, and hauyne in leucitic rocks from the Roman province.

Lazurite (Lapis lazuli) is like sodalite but with Na_2S in place of Cl. Its azure-blue colour makes it a valued ornamental stone. It is a contact meta-somatic mineral of limestone near granite, coming from Afghanistan, Bokhara in Turkestan, Persia and Chile.

Fig. 258. Crystals of nosean in thin section showing embayment, dark borders, canal-like markings and fine striations.

ZEOLITE GROUP

Zeolites are a family of framework alumino-silicates containing water molecules along with Na$^+$ and Ca^{2+} in the spaces of the framework. They are characteristically late-stage minerals deposited in the vesicles of lavas or shallow igneous intrusions from residual waters or vapours from the magma, or low-grade metamorphic minerals of tuffaceous sediments. They nearly all have refractive indices between 1·47 and 1·54 and densities between 1·9 and 2·4.

Structurally they are characterized by very open frameworks with large interconnecting spaces or

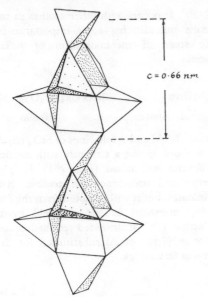

Fig. 259. The chain-linkage of $(Si,Al)O_4$ tetrahedra common to all the fibrous zeolites.

through-going channels. Those with interconnecting spaces, when dehydrated, readily absorb atoms or molecules of substances diffusing through the lattice from solution. They constitute *molecular sieves* and, depending upon the diameters of the tetrahedral rings in the structure, may permit the entry of groups below a certain size but prevent the diffusion of larger groups.

The zeolites may be divided into fibrous zeolites with interlinked chains of the type shown in Fig. 259 between which are channels through the structure, and an equant group—those with a framework of cages made up of rings of $(Si, Al)O_4$ tetrahedra, with 4, 6, 8 or more tetrahedra to a ring (cf. Fig. 260).

Analcime $16[Na\{AlSi_2O_6\}H_2O]$

Cubic *I m3m*   *a* 1·37

Structure: A linked series of 4- and 6-membered tetrahedral rings closely related to leucite, and comparable in a broad way with sodalite.

Habit: Trapezohedral {211} crystals. Also anhedral and replacing plagioclase in alkaline dolerites. Cleavage {100} imperfect. *H* 5–5½. *D* 2·29. Lustre vitreous. Colourless, occasionally tinted.

Optics: Transparent, colourless. *n* 1·48–1·49. May sometimes show very weak anomalous birefringence.

It has *n* lower than leucite, but may be difficult to distinguish from sodalite, except by chemical test.

Occurrence: In cavities in volcanic and shallow intrusive igneous rocks, along with other zeolites. As a primary groundmass mineral in teschenites and alkali-basalts. In these rocks it may also corrode and replace earlier plagioclase. Some pseudo-leucites (see under leucite) are partly of analcime.

Also found as an authigenic mineral in sandstones, but this is rare.

In the early stages of metamorphism of tuffaceous sandstones analcime may form, with laumontite and heulandite, from volcanic glass.

Related minerals: *Wairakite*, a calcium analogue of analcime, is known from the Taupo district, New Zealand.

Chabazite $2[(Ca, Na_2)\{Al_2Si_4O_{12}\}6H_2O]$

Trigonal *R* $\bar{3}m$ *a* 1·37 *c* 1·49
                                                    (hexagonal axes)

Structure: A cage-type zeolite (Fig. 260).

Habit: Simple rhombohedra nearly cubic in angle. Twins common with twin axis *z*, interpenetrant. Cleavage {10$\bar{1}$1} imperfect. *H* 4–5. *D* 2·75. Lustre vitreous. Colour white or flesh pink.

Optics: Transparent, colourless. *ω* 1·49, *ε* 1·48, *ω* − *ε* 0·002–0·010. Usually optically −ve but sometimes +ve. Interference figure may be anomalous.

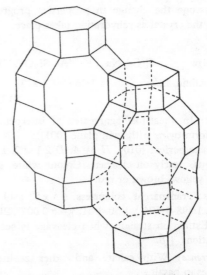

Fig. 260. Two cage-type units of the chabazite structure. The angles of the figure represent the positions of Si^{4+} ions; the O^{2-} ions lying about midway between them are not shown. The tetrahedra form 4-, 6- and 8-membered rings enclosing a large central cavity.

Occurrence: In cavities and fissures in basalts. It occurs in Carboniferous basalts S. and S.W. of Glasgow, and in Tertiary basalts in Antrim.

Related mineral: *Gmelinite* with Na dominating over Ca is very similar but occurs in crystals of more hexagonal aspect. It is found on the east coast of Antrim.

Laumontite $4[Ca\{Al_2Si_4O_{12}\}4H_2O]$

Monoclinic 2 or *m*   *a* 1·490 *b* 1·317 *c* 0·755
                                                              $\beta 111° 30'$

Habit: In prismatic crystals with oblique $\bar{2}01$ terminal face. Also radiating or anhedral.
Twins common on {100}. Cleavage {010}, {110} perfect. *H* 3½–4. *D* 2·25–2·35. Lustre vitreous to pearly on cleavage. Colour white. Pyroelectric.

Optics: Transparent, colourless. α 1·505–1·510, $\beta = y$ 1·514–1·518, $\gamma \wedge z$ 10–33° γ1·517–1·522, $\gamma - \alpha$ 0·012. $2V_\alpha$39–44°. The optical properties vary with the water content.

Occurrence: In fissures and cavities of igneous rocks. An important product of the early stages of metamorphism of tuffaceous sediments, forming from volcanic glass and plagioclase.

Related mineral: Laumontite dehydrates readily to a dull friable material *leonhardite* with a change in optical properties, and rehydrates again to laumontite. It is possible to observe under the microscope the change in properties progressing along the crystal as rehydration takes place.

Stilbite $4[(Ca, Na_2, K_2)\{Al_2Si_7O_{18}\}7H_2O]$

Monoclinic 2/*m*   *a* 1·363 *b* 1·817 *c* 1·131
                                                              $\beta 129° 10'$

Habit: In sheaf-like aggregates of interpenetrant cruciform twins with twin plane {001}.
Cleavage perfect {010}. *H* 3½–4. *D* 2·1–2·2. Lustre vitreous, pearly on cleavage. Colour white, sometimes yellow, brown or red.

Optics: Transparent, colourless. $\alpha \wedge x$ 5° 1·49–1·50. $\beta = y$1·498–1·504. γ 1·50–1·51. $\gamma - \alpha$ 0·007°. $2V_\alpha$30–50°. Extinction straight when cleavage is seen; 5° in sections $\perp \beta$.

Occurrence: With calcite and other zeolites in cavities in basalt.

Related minerals: *Heulandite*, lacking K^+, is very similar to stilbite but occurs in tabular crystals and has a better cleavage. It is optically positive with

$2V_\gamma$10–30°. It occurs with other zeolites in cavities in basalts but also has some importance in the zeolitic stage of metamorphism of tuffaceous sandstones.

Apophyllite $2[KFCa_4\{Si_8O_{20}\}.8H_2O]$

Tetragonal 4/*mmm*   *a* 0·896   *c* 1·578

Structure: 4-fold and 8-fold rings of SiO_4 tetrahedra lie in a plane, giving a structure with affinities to both zeolites and micas (Fig. 261). In physical properties and occurrence the zeolitic character predominates, but it will be noted from the formula that the interstitial cations are held by bonds from 'active' O^{2-} at unshared tetrahedral corners, and not in virtue of Al substitution for Si in a continuous framework.

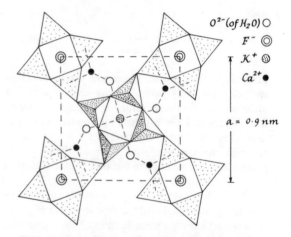

Fig. 261. Apophyllite: part of one of the sheets of linked tetrahedra. Free apices point up and down and are linked to interlayer cations.

Habit: In crystals of equant form with {110} {101} and {001}. {110} striated.
Cleavage {001} perfect in accord with the structure. {110} less perfect. *H* 4½–5. *D* 2·35. Lustre vitreous, pearly on {001}. Colourless, white, or tinted pink.

Optics: Transparent, colourless. ω 1·535, ε 1·537, $\omega - \varepsilon$ 0·002. Optically −ve. May show optical anomalies with sectors of varying behaviour not explicable by twinning.

Occurrence: As a secondary mineral in cavities of basalts and other igneous rocks, associated with calcite and other zeolites. Well shown at Traprain Law, Haddingtonshire with analcime, in the Isle of Skye (Talisker) and in Antrim.

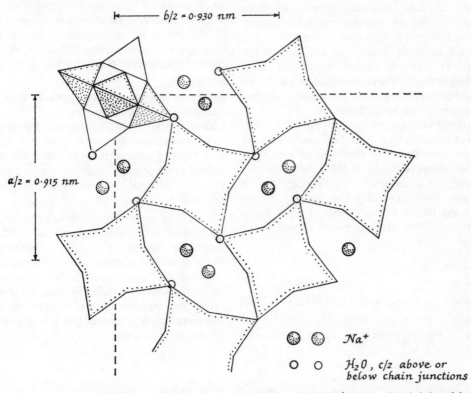

Fig. 262. The structural pattern of natrolite, showing channels containing Na⁺ between linked chains of the type shown in Fig. 259.

Table 24. Properties of some fibrous zeolites.

	Unit cell	Cleavage	n	Birefr.	Elongation	2V	Extinction
Natrolite $8[Na_2\{Al_2 Si_3 O_{10}\} 2H_2O]$	Orthorhombic $F\,mm$ $a\,1.83\ b\,1.86\ c\,0.657$	{110}	α 1.47 γ 1.49	0.012	slow	(+)60°	Straight
Thomsonite $4[Na\,Ca_2\{Al_5 Si_5 O_{20}\} 6H_2O]$	Orthorhombic $P\,mm$ $a\,1.30\ b\,1.30\ c\,1.32$	{010}	α 1.50 γ 1.54	0.012	fast or slow	(+)45° to 75°	Straight
Mesolite $8[Na_2 Ca_2\{Al_6 Si_9 O_{30}\} 8H_2O]$	Pseudo-Orthorhombic $F\,mm$ $a\,1.825\ b\,1.85\ c\,0.657$	{110}	α 1.50 γ 1.51	0.001	fast or slow	(+)80°	$\beta \wedge z$ 2°–5°
Scolecite $8[Ca\{Al_2 Si_3 O_{10}\} 3H_2O]$	Monoclinic $C\,m$ $a\,1.848\ b\,1.895\ c\,0.654$ $\beta\,90°\,39'$	{110}	α 1.51 γ 1.52	0.007	fast	(−)36° to 56°	$\alpha \wedge z$ ~16°

Fibrous Zeolites

Structure: Based on various arrangements of chains like that of Fig. 259 running parallel to *z*-axis. Natrolite, Fig. 262, furnishes an example.

Habit: All occur as bundles of radiating fibres. H 5–5½. D 2·15–2·4. Lustre vitreous or silky. Colourless or white. Transparent, colourless in thin section. The properties of the commoner members of the group are set out in Table 24.

Occurrence: In the amygdales and cavities of basalt lavas and related rocks. Natrolite and thomsonite may also form from the alteration of nepheline. The Carboniferous basalts of the Glasgow district (Renfrew and Dumbarton), the Tertiary basalts of Antrim, Skye (Talisker), the Faroe Isles and Iceland, the Bay of Fundy district of Nova Scotia and Paterson, New Jersey are well-known zeolite localities.

SCAPOLITE SERIES

Marialite $2[Na_4\{Al_3Si_9O_{24}\}Cl]$
Meionite $2[Ca_4\{Al_6Si_6O_{24}\}(SO_4, CO_3, Cl)]$
Tetragonal $4/m$ a 1·206–1·217 c 0·751–0·765

Structure: The framework of 4-membered tetra-hedral rings with interstitial Na, Ca has resemblances to that of feldspars (Fig. 244). Two types of rings occur and the superposition of two of type (b) in Fig. 263 at top and bottom of the cell leaves a large space to accommodate CO_3, SO_4 or Cl. The coupled substitution Na, Si for Ca, Al occurs as in feldspars.

Habit: In coarse, prismatic crystals with uneven faces, or granular and massive.

Cleavage {100}, {110} interrupted, giving a splintered appearance. H 5–6, D 2·50–2·78 increasing with Ca^{2+}. Lustre vitreous or somewhat resinous. Colour white or grey, sometimes tinted pale bluish, greenish, pink. May be fluorescent in orange or yellow in UV light.

Optics: Transparent, colourless. ω 1·55–1·60, ε 1·54–1·57, $\omega - \varepsilon$ 0·005–0·038, indices and birefringence increasing linearly with Ca^{2+}. Optically −ve. Relief about that of plagioclase, but the interference colour usually considerably higher, showing yellow or higher colours, while twinning is absent and extinction straight. More weakly birefringent members resemble quartz but are optically −ve. Cordierite is biaxial.

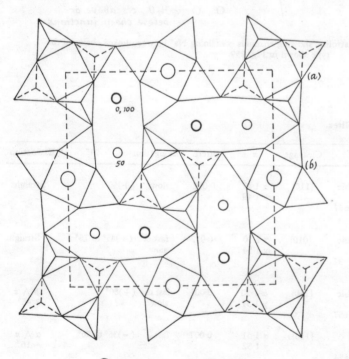

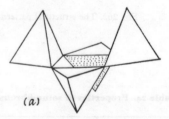

(a)

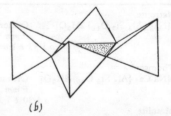

(b)

Types of $(Si, Al)O_4$ *tetrahedral ring*

○ CO_3, SO_4, Cl ○ Na, Ca

Fig. 263. The structural scheme of scapolite.

Occurrence: In contact-altered limestones near plutonic igneous intrusions. Also in regionally-metamorphosed gneisses, schists and amphibolites particularly those rich in Ca. Regional metasomatism may also cause widespread development of scapolite. In hydrothermally altered basic igneous rocks scapolite may replace calcic plagioclase. Sodic scapolite occurs as veins and replacements in the Sandsting granite-diorite rocks of the Shetland Isles.

Varieties: The scapolite series is subdivided by the amount of meionite into *marialite* to Me_{20}, *dipyre* to Me_{50}, *mizzonite* to Me_{80} and *meionite* to Me_{100}. The name *wernerite* is applied by some to common scapolite of an unspecified middle composition range.

Cancrinite $2[(Na, Ca)_4\{Al_3Si_3O_{12}\} (CO_3, SO_4, Cl).1-5H_2O]$

Hexagonal P $6/mmm$ a 1·26 c 0·518

Structure: A framework of $(Al, Si)O_4$ tetrahedra believed to be built of 4-, 6- and 12-membered rings with the other ions in the spaces. Not known in detail.

Habit: Usually massive.

Cleavage $\{10\bar{1}1\}$ perfect. H 5–6. D 2·4–2·5. Lustre vitreous. Colour white, light blue or greyish-blue.

Optics: Transparent, colourless. ω 1·50–1·53, ε 1·49–1·50, $\omega - \varepsilon$ 0·007–0·025. Optically $-$ve. Extinction parallel to cleavage; cleavage traces length-fast. The birefringence is generally fairly high, giving interference colours from yellow to second-order blue-green. This distinguishes cancrinite from most minerals with indices a little below the mounting cement.

Tests: Effervesces in warm HCl, but is fairly hard.

Occurrence: In nepheline syenites and their pegmatites. It may replace nepheline or form reaction rims between nepheline and calcite, as well as constituting veinlets and irregular masses.

Varieties: Sulphate-rich cancrinite is called *vishnevite*. Sulphatic cancrinite occurs at Loch Borolan, Sutherlandshire.

Further reading.

BRAGG, W. L. and CLARINGBULL, G. F., 1965. *The crystal structures of minerals*, London, G. Bell & Co.

DANA, E. S. and FORD, W. E., 1932. *A textbook of mineralogy*, New York, John Wiley & Sons.

DEER, W. A., HOWIE, R. A. and ZUSSMAN, J., 1962–3. *Rock-forming minerals*, vols. 1–5, London, Longmans Green & Co.

PALACHE, C., BERMAN, H., and FRONDEL, C., 1944, 1951, 1962. *Dana's system of mineralogy*, 7th ed. Vols. *I, II, III*, New York, John Wiley & Sons.

WINCHELL, A. N. 1951. *Elements of optical mineralogy*, Vols. I, II, III, New York, John Wiley & Sons

APPENDIX I

Names of the 32 symmetry classes
(see Fig. 68, p. 64)

The class names separated by semi-colons below are
alternative names used by different authors. In
nearly all cases the class name requires to be
preceded by the name of the system. Not all alter-
natives have been included.

Class	System	Class names
1	Triclinic	Asymmetric; pedial; hemihedral
$\bar{1}$		Holosymmetric; holohedral; pinacoidal; normal
2	Monoclinic	Hemimorphic; sphenoidal
m		Clinohedral, hemihedral; domatic
$2/m$		Holosymmetric; holohedral; prismatic; normal
mm	Orthorhombic	Hemimorphic; pyramidal
222		Holoaxial; sphenoidal; disphenoidal
mmm		Holosymmetric; holohedral; bipyramidal; normal
3	Trigonal	Tetartohedral hemimorphic; pyramidal
$\bar{3}$		Rhombohedral; tri-rhombohedral
$3m$		Rhombohedral-hemimorphic; ditrigonal hemimorphic; ditrigonal pyramidal
$\bar{3}m$		Holosymmetric; ditrigonal scalenohedral
32		Holoaxial; trapezohedral; hemihedral; ditrigonal dipyramidal
4	Tetragonal	Hemimorphic; pyramidal
$\bar{4}$		Sphenoidal; sphenoidal tetartohedral
$4/m$		Tripyramidal; bipyramidal
$4mm$		Hemimorphic; ditetragonal hemimorphic; ditetragonal pyramidal
$\bar{4}2m$		Ditetragonal scalenohedral; ditetragonal bisphenoidal
42		Holoaxial; trapezohedral
$4/mmm$		Holosymmetric; holohedral; ditetragonal dipyramidal; normal
6	Hexagonal	Hemimorphic; pyramidal
$\bar{6}$		Trigonal bipyramidal; trigonal tetartohedral
$6/m$		Hexagonal bipyramidal; tripyramidal
$6mm$		Hemimorphic; dihexagonal pyramidal
$\bar{6}m2$		Ditrigonal dipyramidal; trigonal
62		Holoaxial; trapezohedral
$6/mmm$		Holosymmetric; dihexagonal bipyramidal; normal
23	Cubic	Tetrahedral pentagonal dodecahedral; tetartohedral
$m3$		Diakisdodecahedral; pyritohedral
$\bar{4}3m$		Hexakistetrahedral; tetrahedral
43		Pentagonal icositetrahedral; plagiohedral; gyroidal
$m3m$		Holosymmetric; holohedral; hexakisoctahedral; normal

APPENDIX I

Names of the 32 symmetry classes
(see fig. 66, p. 61)

The class names separated by semi-colon below are
alternative names used by different authors. In
nearly all cases the class name requires to be
preceded by the name of the system; not all alter-
natives have been included.

Class	System	Class names
1	Triclinic	Asymmetric; pedial; hemihedral
1̄		Holosymmetric; holohedral; pinacoidal; normal
2	Monoclinic	Hemimorphic; sphenoidal
m		Clinohedral; hemihedral; domatic
2/m		Holosymmetric; holohedral; prismatic; normal
mm2	Orthorhombic	Hemimorphic; pyramidal
222		Holoaxial; sphenoidal; disphenoidal
mmm		Holosymmetric; holohedral; bipyramidal; normal
3	Trigonal	Tetartohedral hemimorphic; pyramidal
3̄		Rhombohedral; tri-rhombohedral
3m		Rhombohedral-hemimorphic; ditrigonal scalenohedral; ditrigonal pyramidal
3̄m		Holosymmetric; ditrigonal scalenohedral
32		Holoaxial; trapezohedral; trigonal trapezohedral
4	Tetragonal	Hemimorphic; pyramidal
4̄		Sphenoidal; sphenoidal; tetartohedral
4/m		Bipyramidal; bipyramidal
4mm		Hemimorphic; ditetragonal hemimorphic; ditetragonal pyramidal
4̄2m		Ditetragonal scalenohedral; ditetragonal bisphenoidal
422		Holoaxial; trapezohedral
4/mmm		Holosymmetric; holohedral; ditetragonal bipyramidal; normal
6	Hexagonal	Hemimorphic; pyramidal
6̄		Trigonal bipyramidal; trigonal tetartohedral
6/m		Hexagonal bipyramidal; bipyramidal
6mm		Hemimorphic; dihexagonal pyramidal
6̄m2		Ditrigonal bipyramidal; trigonal
622		Holoaxial; trapezohedral
6/mmm		Holosymmetric; dihexagonal bipyramidal; normal
23	Cubic	Tetartohedral pentagonal dodecahedral; tetartohedral
m3		Dyakisdodecahedral; pyritohedral
4̄3m		Hexakistetrahedral; tetrahedral
432		Pentagonal icositetrahedral; plagiohedral; gyroidal
m3m		Holosymmetric; holohedral; hexakisoctahedral; normal

APPENDIX II: METHODS OF CHEMICAL TESTING

Simple chemical tests are useful in identifying minerals, and if simple laboratory facilities are available a considerable number of tests can be applied, selected from manuals of qualitative chemical analysis. No general scheme is given here, but specific tests are suggested in the descriptions of minerals. It must be remembered that atomic substitution often leads to the presence of several types of cation in a mineral, so that qualitative tests may at times be hard to interpret, and schemes based on progressive elimination of possibilities may not be very effective. Notes on the commoner methods of testing mentioned in the mineral descriptions follow.

(*i*) Heating in the closed tube simply involves placing a little coarsely-ground mineral powder in a pyrex glass ignition tube, 4 to 5 cm long and 0·75 cm in diameter, closed at one end, and holding the closed end above the flame of a gas burner, with the mouth of the tube slanting up and to your right or left. Decrepitation can be observed and water, if liberated, condenses on the upper cool walls of the tube, and so may sublimates of different kinds.

(*ii*) The open tube is a length of pyrex glass tube 5 to 6 cm long and 0·75 cm in diameter, open at both ends. (A slight bend at one-third of its length helps to prevent the mineral powder falling out.) The powdered mineral is placed at about a third of the way up the tube, which is held slanting upwards from the horizontal at a small angle so that the powder does not drop out. The sample is nearer to the lower end of the tube, and the flame plays around the region of the sample, so that it is heated in a current of hot air passing over it through the tube. Oxide deposits and sublimates form on the cooler walls of the tube above the assay, and the smell of the gases is noted.

(*iii*) Acid solubility tests are carried out in a test tube on a small powdered sample of the mineral, with careful warming if necessary, holding the mouth of the tube slanting upwards to the right or left, and swirling gently from time to time.

(*iv*) *Blowpipe tests.* The blowpipe is a tapered metal tube about 15 cm long, angled about 3 cm from the narrow end which has a nozzle with a pin hole. It is used to direct the flame of a gas burner, spirit lamp or candle on to the sample. By blowing out the cheeks to keep the pressure up, and breathing through the nose, blowing can be sustained for several minutes. Two types of flame are used.

(*a*) The oxidizing flame, for which the nozzle is held within the flame, which is well mixed with air so that combustion is complete. The flame is quiet and finely pointed and the mineral is heated just beyond its tip, so that it is heated surrounded by air. This position is used for roasting (oxidizing) the sample. The hottest point of this flame, at the tip of the blue cone, is used for fusibility tests.

(*b*) The reducing flame is produced by holding the blowpipe nozzle a little outside the flame, so as to blow the flame and unburnt gases together over the sample as a ragged noisy envelope. This produces reducing conditions to convert oxides to metal.

For fusibility tests and flame colorations a chip of the mineral is held in the flame with a pair of forceps. For roasting and reduction it is powdered and placed in a shallow pit scraped in a block of charcoal. For reduction, some charcoal powder is mixed with the mineral powder, and sodium carbonate ($Na_2CO_3.10H_2O$) may be used as a flux. Be careful not to blow hard at first and scatter the powder. The colours of some deposits forming on the charcoal near the assay are mentioned under individual minerals.

(*v*) *Bead tests.* The colours given to a bead of flux by dissolving the mineral in it are diagnostic of some elements. The usual flux is borax (sodium tetraborate, $Na_2B_4O_7.10H_2O$); microcosmic salt ($NaNH_4HPO_4.4H_2O$) is also used.

An inch or so of platinum wire is fused into a glass tube as a handle, and the free end is bent into a loop 2 to 3 mm in diameter. The loop is heated and plunged hot into the flux, and then put back into the flame so that the flux fuses to a clear bead hanging in the loop. This is touched, hot, on *a very small amount* of the powdered mineral—a speck or two only—and heated to dissolve it. The colour will be seen on withdrawing the bead from the flame and allowing it to cool. Beads sometimes show fluorescence under ultra-violet light.

APPENDIX III

TABLE OF HARDNESS

Minerals of H ≤ 6 arranged in order of increasing hardness. M denotes metallic lustre, (M) sub-metallic lustre. Others have non-metallic lustre.

	Talc	1		Cryolite	2½		Kieserite	3½	M	Niccolite	5-5½
	Pyrophyllite	1-2		Carnallite	2½		Polyhalite	3½		Pyrochlore—	
(M)	Graphite	1-2		Brucite	2½	(M)	Sphalerite	3½-4		microlite	5-5½
M	Molybdenite	1-1½	M	Copper	2½-3	M	Chalcopyrite	3½-4	(M)	Goethite	5-5½
	Vivianite	1-1½	M	Silver	2½-3	M	Pyrrhotite	3½-4		Monazite	5-5½
	Erythrite	1-1½	M	Gold	2½-3	(M)	Cuprite	3½-4		Sphene	5-5½
M	Sylvanite	1-1½	M	Chalcocite	2½-3		Boehmite	3½-4		Analcime	5-5½
(M)	Covellite	1-1½	M	Calaverite	2½-3		Rhodo-			Fibrous	
	Orpiment	1-1½	M	Boulangerite	2½-3		chrosite	3½-4		zeolites	5-5½
	Realgar	1-1½	M	Bournonite	2½-3		Dolomite	3½-4	(M)	Hematite	5-6
	Soda nitre	1-1½		Anglesite	2½-3		Aragonite	3½-4	M	Ilmenite	5-6
	Sulphur	1½-2½		Trona	2½-3		Azurite	3½-4		Uraninite	5-6
M	Stibnite	2		Biotite	2½-3		Malachite	3½-4	(M)	Psilomelane	5-6
	Gypsum	2		Gibbsite	2½-3½		Pyromorphite	3½-4		Turquoise	5-6
	Sylvine	2		Jarosite	2½-3½		Alunite	3½-4		Pyroxene	5-6
	Carnotite	2		Serpentine	2½-3½		Laumontite	3½-4		Amphibole	5-6
	Alum	2	M	Bornite	3		Stilbite	3½-4		Scapolite	5-6
(M)	Pyrolusite	2-6	M	Enargite	3		Fluorite	4		Cancrinite	5-6
	Proustite	2-2½		Calcite	3	(M)	Manganite	4	M	Arsenopyrite	5½
M	Argentite	2-2½		Vanadinite	3		Magnesite	4		Periclase	5½
	Cinnabar	2-2½	M	Antimony	3-3½		Siderite	4		Perovskite	5½
	Autunite—		M	Millerite	3-3½		Smithsonite	4-4½	M	Chromite	5½
	torbernite	2-2½		Atacamite	3-3½	M	Wolframite	4-4½	M	Skutterudite—	
	Epsomite	2-2½		Cerussite	3-3½		Chabazite	4-5		smaltite—	
	Borax	2-2½		Witherite	3-3½		Scheelite	4½-5		chloanthite	5½-6
	Kaolinite	2-2½		Barytes	3-3½		Wollastonite	4½-5		Anatase	5½-6
	Muscovite	2-2¼		Celestine	3-3½		Apophyllite	4½-5		Brookite	5½-6
	Chlorite	2-2½	M	Tetrahedrite—		(M)	Lepido-			Amblygonite	5½-6
M	Krennerite	2-3		tennantite	3-4½		crocite	5		Allanite	5½-6
M	Bismuth	2½	M	Arsenic	3½		Apatite	5		Sodalite	5½-6
M	Galena	2½	M	Tenorite	3½		Hemi-			Leucite	5½-6
	Pyrargyrite	2½		Strontianite	3½		morphite	5	M	Magnetite	5½-6½
	Halite	2½		Wavellite	3½		Åkermanite—			Rhodonite	5½-6½
	Cerargyrite	2½		Anhydrite	3½		gehlenite	5			

TABLE OF REFLECTANCE

Some minerals of metallic and sub-metallic lustre arranged in order of increasing reflectance percentage.

Graphite	6–17	Proustite	25–28	Pyrrhotite	38–45	Millerite	54–60
Covellite	7–22	Cuprite	27	Galena	43	Skutterudite	
Chromite	14	Hematite	28	Bornite	44	group	56
Manganite	14–20	Pyrargyrite	28–31	Chalcopyrite	44	Krennerite	61
Molybdenite	15–37	Argentite	29	Sylvanite	48–60	Calaverite	63
Wolframite	16–18	Tetrahedrite—		Marcasite	49–55	Bismuth	68
Lepidocrocite	16–25	tennantite	30	Arsenic	49½	Platinum	70
Sphalerite	17	Stibnite	30–40	Pentlandite	52	Gold	74
Ilmenite	18–21	Pyrolusite	30–42	Arsenopyrite	52–56	Antimony	74½
Tenorite	20–27	Chalcocite	32	Niccolite	52–58	Copper	81
Magnetite	21	Bournonite	36–48	Pyrite	54	Silver	95
Psilomelane	23–24	Boulangerite	37–44				
Enargite	25–28						

TABLE OF REFRACTIVE INDEX

Some translucent minerals, grouped as isotropic, uniaxial or biaxial, arranged in order of increasing refractive index.

(a) Isotropic minerals

(B) indicates weak birefringence

	n		n		n		n
Cryolite (B)	1·338	Cristobalite (B)		Spinel	1·719	Andradite	1·887
Fluorite	1·434		1·484–1·487	Grossular (B)	1·734	Pyrochlore—	
Opal	1·440–1·460	Sylvine	1·490	Periclase	1·735	microlite	1·930–2·180
Alum	1·450	Leucite (B)	1·508–1·511	Spessartine	1·800	Chromite	2·160
Halloysite	1·470–1·550	Halite	1·544	Almandine	1·830	Perovskite	2·340
Analcime	1·480–1·490	Chlorite (B)	1·600	Hercynite	1·830	Diamond	2·417
Sodalite	1·483	Pyrope	1·714	Uvarovite	1·870		

(b) Uniaxial minerals

An entry in *italic* denotes the minimum value and an entry in roman type the maximum value for each mineral. The optic sign is shown before each mineral name.

± *Chabazite*	1·480	+ Quartz	1·553	− Apatite	1·667	− Jarosite	1·817
− *Calcite*	1·486	+ *Brucite*	1·560	± Melilite	1·669	+ Xenotime	1·827
± Chabazite	1·490	− *Beryl*	1·564	− Tourmaline	1·675	− Siderite	1·873
− *Cancrinite*	1·490	+ *Alunite*	1·572	− Dolomite	1·679	+ *Zircon*	1·923
− *Dolomite*	1·500	+ Brucite	1·580	− Magnesite	1·700	+ *Cassiterite*	2·006
− *Magnesite*	1·509	+ Alunite	1·592	− *Vesuvianite*	1·703	+ Zircon	2·015
− *Nepheline*	1·526	− *Rhodochrosite*	1·597	− *Jarosite*	1·715	+ Cassiterite	2·097
− *Ankerite*	1·530	− Scapolite	1·600	− Ankerite	1·720	− *Anatase*	2·488
− Cancrinite	1·530	− Beryl	1·601	+ *Xenotime*	1·720	− Anatase	2·561
− *Apophyllite*	1·535	− *Tourmaline*	1·610	− Vesuvianite	1·746	+ *Rutile*	2·612
− Apophyllite	1·537	− *Apatite*	1·629	− *Corundum*	1·759	+ Rutile	2·899
− *Scapolite*	1·540	± *Melilite*	1·632	− Corundum	1·772	− *Hematite*	2·930
+ *Quartz*	1·544	− *Siderite*	1·633	− Rhodochrosite	1·816	− Hematite	3·220
− Nepheline	1·545	− Calcite	1·658				

(c) Biaxial minerals

The order is that of increasing refractive index β. Where β has a range of values the entry giving the minimum value is in *italics*.

	β	$\gamma-\alpha$		β	$\gamma-\alpha$
Epsomite	1·455	0·028	Thomsonite	1·520	0·012
Borax	1·469	0·030	*K feldspar*	1·522	0·005–0·007
Tridymite	1·479	0·0025	Gypsum	1·523	0·009
Natrolite	1·480	0·012	*Wavellite*	1·526	0·025
Trona	1·492	0·128	K feldspar	1·529	0·005–0·007
Stilbite	1·500	0·007	Serpentine	1·530	0·004–0·017
Mesolite	1·500	0·001	*Plagioclase*	1·532	0·007–0·013
Scolecite	1·510	0·007	Kieserite	1·533	0·064
Laumontite	1·514–1·518	0·012	*Cordierite*	1·535	0·008–0·011

	β	$\gamma-\alpha$		β	$\gamma-\alpha$
Wavellite	1·543	0·025	Glaucophane	1·670	0·008–0·022
Kaolinite	1·559	0·005	Lawsonite	1·675	0·020
Polyhalite	1·560	0·020	Witherite	1·676	0·148
Biotite	1·560	0·030–0·080	*Oxyhornblende*	1·680	0·020–0·070
Gibbsite	1·568	0·019	*Pigeonite*	1·680	0·020–0·030
Kaolinite	1·569	0·005	Tremolite—actinolite	1·680	0·020–0·027
Chlorite	1·570	0·000–0·020	Aragonite	1·681	0·155
Serpentine	1·570	0·004–0·017	*Zoisite*	1·688	0·004–0·008
Cordierite	1·574	0·008–0·011	Orthoamphibole	1·690	0·015–0·025
Anhydrite	1·575	0·044	*Barkevikite*	1·690	0·014–0·018
Muscovite	1·580	0·036–0·050	*Kaersutite*	1·690	0·030–0·032
Plagioclase	1·583	0·007–0·013	Chlorite	1·690	0·000–0·020
Pyrophyllite	1·586	0·050	*Allanite*	1·700	0·013–0·036
Talc	1·589–1·594	0·050	Biotite	1·700	0·030–0·080
Orthoamphibole	1·600	0·015–0·025	Zoisite	1·710	0·004–0·008
Tremolite—actinolite	1·610	0·020–0·027	Riebeckite	1·710	0·003–0·016
Topaz	1·610	0·010	Pigeonite	1·720	0·020–0·030
Chamosite	1·610	0·000–0·008	Hornblende	1·720	0·020–0·027
Muscovite	1·610	0·036–0·050	*Chloritoid*	1·720	0·006–0·020
Hemimorphite	1·617	0·022	*Rhodonite*	1·720	0·011–0·014
Hornblende	1·620	0·014–0·030	Kyanite	1·722	0·015
Glaucophane	1·620	0·008–0·022	Diaspore	1·722	0·048
Celestine	1·624	0·009	Oxyhornblende	1·730	0·020–0·070
Wollastonite	1·630	0·014	*Aegirine*	1·730	0·030–0·060
Topaz	1·631	0·010	Chloritoid	1·730	0·006–0·020
Andalusite	1·633	0·011	Rhodonite	1·740	0·011–0·014
Pectolite	1·635	0·035	Clinopyroxene	1·740	0·018–0·031
Barytes	1·637	0·012	Barkevikite	1·740	0·014–0·018
Boehmite	1·650	0·020	Kaersutite	1·740	0·030–0·032
Forsterite (Olivine group)	1·650	0·035	Staurolite	1·745–1·753	0·014
Orthopyroxene	1·653	0·009–0·020	Orthopyroxene	1·770	0·009–0·020.
Andalusite	1·653	0·011	Epidote	1·780	0·005–0·050
Sillimanite	1·658	0·020	Aegirine	1·820	0·030–0·060
Spodumene	1·660	0·015–0·025	Allanite	1·820	0·013–0·036
Riebeckite	1·660	0·003–0·016	*Sphene*	1·870	0·100–0·160
Chamosite	1·660	0·000–0·008	Fayalite (Olivine group)	1·870	0·050
Jadeite	1·663	0·015	Anglesite	1·883	0·017
Strontianite	1·667	0·148	Sphene	2·030	0·100–0·160
Sillimanite	1·670	0·020	Sulphur	2·038	0·287
Epidote	1·670	0·005–0·050	Lepidocrocite	2·200	0·570
Clinopyroxene	1·670	0·018–0·031	Goethite	2·390	0·140
Spodumene	1·670	0·015–0·025	Brookite	2·600	0·130

APPENDIX IV

Charts for microscopic identification of anisotropic rock-forming minerals

The charts on the following six pages display the mean refractive index and birefringence of the commoner translucent, anisotropic minerals found in thin sections. A range of values is indicated for some minerals or mineral groups.

The procedure for mineral identification is given below. Remember to use *a sample of several grains*, thought from general appearance to be the same mineral, in making your determination. A single grain may not provide all the information needed, or may be misleading.

1 *Mineral is opaque.* Use reflected light and the Table of Reflectance on p. 306.

2 *Mineral is translucent but isotropic.* Estimate its refractive index from its relief (see p. 119) and by Becke White Line tests (p. 106) against the cement or contiguous known grains, and use the Table of Refractive Index, p. 306. Colour and form will be a help.

3 *Mineral is translucent and anisotropic.*

(*a*) Decide its colour. If it is markedly pleochroic this helps. Many pleochroic changes, however, are slight. As colour is a variable property many minerals appear in more than one colour category, and this allows also for pleochroic changes.

(*b*) Determine whether the extinction is straight or inclined to crystal length or cleavage, using a sample of several grains.

(*c*) If the grains are at all elongated, determine with the quartz wedge (p. 124) or sensitive tint (p. 126) whether they are length-fast, length-slow or both length-fast and length-slow in different sections. If no elongation is shown refer later to charts for all the possibilities.

(*d*) Estimate its refractive index (as under 2, above). Estimate its birefringence from the maximum interference colour shown by a sample of its grains, using the quartz wedge (p. 124) to decide the order of the colour.

(*e*) With this data, enter the appropriate chart(s) which will suggest minerals that it may be. The sign of the interference figure is given after each mineral name. Clearly, if an interference figure can be obtained it is most helpful. It will disclose whether the mineral is uniaxial or biaxial and the sign is often a good discriminator. But generally it should not be necessary to resort to interference figures. Remember that the position of the symbol for each mineral is at the mid-point of a range of values. The higher the birefringence the greater the range of the refractive index.

(*f*) Refer to the description of the mineral in Part II of the main text, and to the optic orientation diagrams in Appendix V, to decide whether the mineral indicated corresponds in other particulars with your sample and whether its paragenesis is appropriate.

310

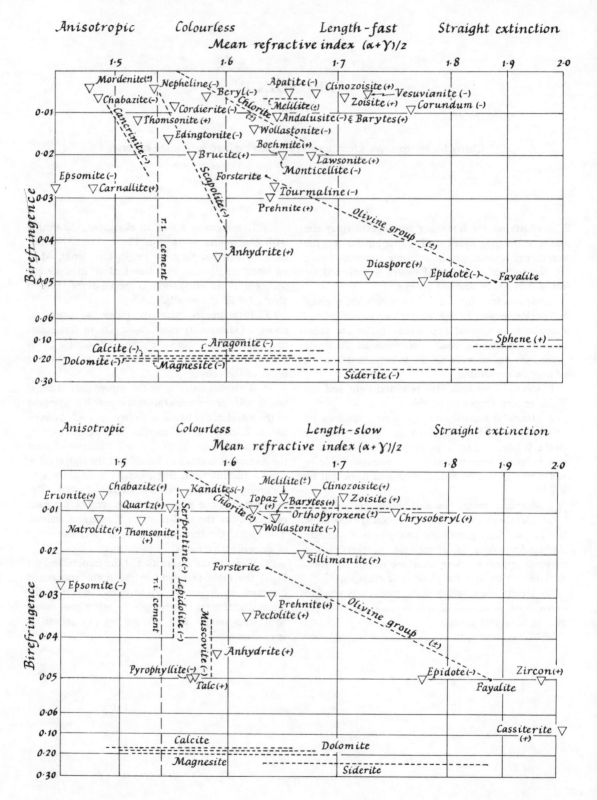

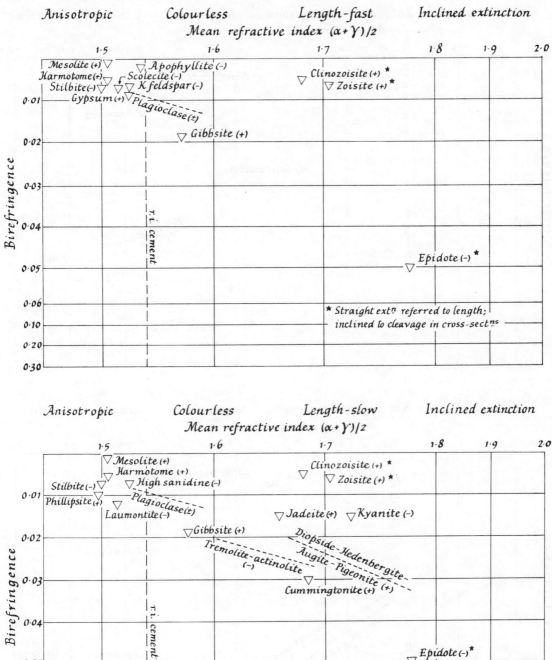

Anisotropic   Colourless   Length-fast   Inclined extinction

Mean refractive index $(\alpha + \gamma)/2$

Mesolite (+) ▽   ▽ Apophyllite (−)   Clinozoisite (+) *
Harmotome(+)   ▽ Scolecite (−)   ▽ Zoisite (+) *
Stilbite(−) ▽   ▽ K.feldspar (−)
Gypsum (+) ▽   Plagioclase (±)

r.i. cement

▽ Gibbsite (+)

Epidote (−) *

★ Straight extⁿ referred to length;
inclined to cleavage in cross-sectⁿˢ

Birefringence

Anisotropic   Colourless   Length-slow   Inclined extinction

Mean refractive index $(\alpha + \gamma)/2$

▽ Mesolite (+)   Clinozoisite (+) *
▽ Harmotome (+)   ▽ Zoisite (+) *
Stilbite (−) ▽   ▽ High sanidine (−)
Phillipsite (+) ▽   Plagioclase (±)
Laumontite (−) ▽

▽ Gibbsite (+)   Jadeite (+) ▽   ▽ Kyanite (−)

Diopside - Hedenbergite
Tremolite - actinolite (−)   Augite - Pigeonite
Cummingtonite (+)   (+)

r.i. cement

Epidote (−)*

▽ Monazite (+)

★ Straight extⁿ referred to length;
inclined to cleavage in cross-sectⁿˢ

Birefringence

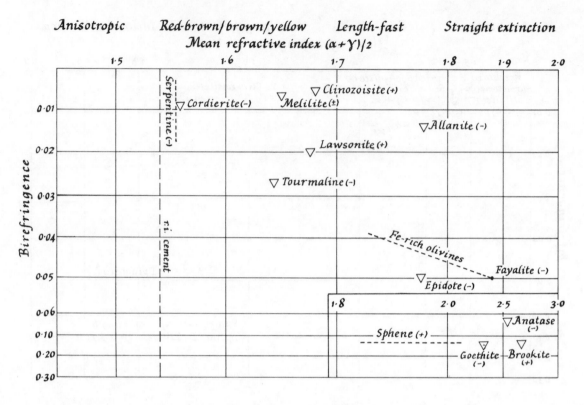

Anisotropic   Red-brown/brown/yellow   Length-fast   Straight extinction
Mean refractive index $(\alpha + \gamma)/2$

Anisotropic   Red-brown/brown/yellow   Length-slow   Straight extinction
Mean refractive index $(\alpha + \gamma)/2$

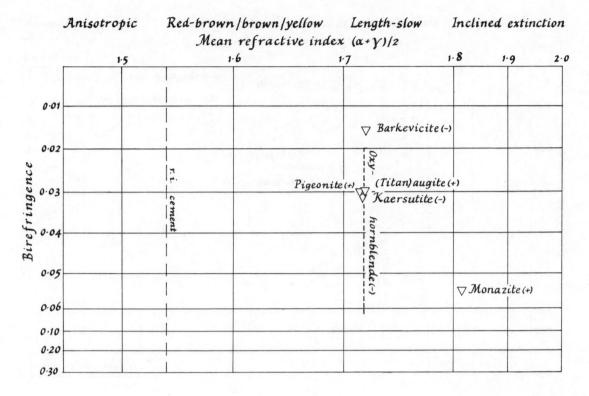

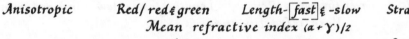

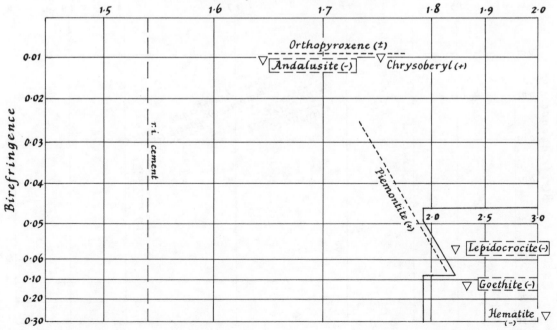

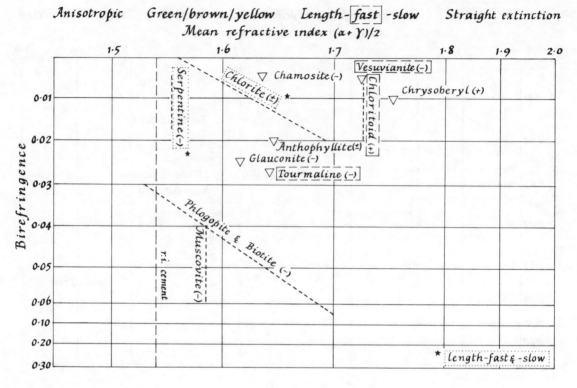

Anisotropic   Green/brown/yellow   Length-`fast`*-slow   Straight extinction*

Mean refractive index $(\alpha + \gamma)/2$

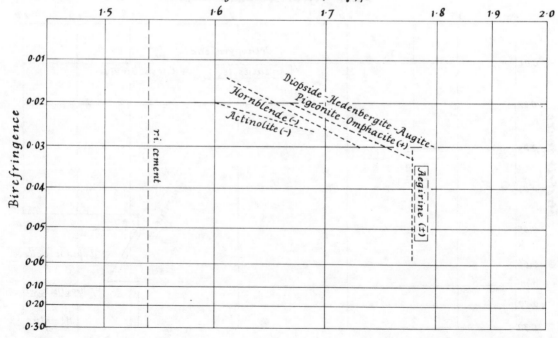

Anisotropic   Green/brown/yellow   Length-`fast`*-slow   Inclined extinction*

Mean refractive index $(\alpha + \gamma)/2$

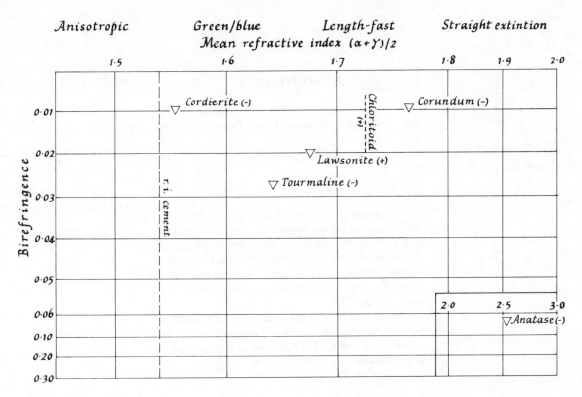

APPENDIX V

Optic orientation diagrams of rock-forming minerals

The diagrams that follow (except for the first one) show the optic orientations of the commoner *biaxial* rock-forming minerals, that is, those belonging to the orthorhombic, monoclinic and triclinic crystallographic systems. The refractive indices and optic signs of uniaxial minerals will be found on p. 307 (in Appendix III).

The diagrams are arranged in the same systematic order as the descriptions of the minerals in Part II of the main text. They are listed alphabetically below for ease of reference.

Calcite   CaCO₃   Trigonal   Dolomite   CaMg(CO₃)₂

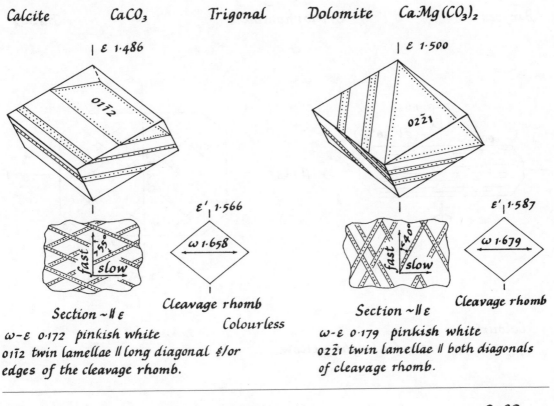

ε 1·486			ε 1·500

0̄11̄2

02̄2̄1

ε' 1·566

ω 1·658

Cleavage rhomb
Colourless

Section ~∥ε

ε' 1·587

ω 1·679

Cleavage rhomb

Section ~∥ε

ω–ε 0·172 pinkish white
01̄12 twin lamellae ∥ long diagonal &/or
edges of the cleavage rhomb.

ω–ε 0·179 pinkish white
02̄21 twin lamellae ∥ both diagonals
of cleavage rhomb.

Aragonite   Orthorhombic   CaCO₃

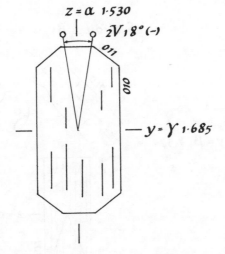

z = α 1·530
2V 18° (–)
011
010

— y = γ 1·685

Colourless
γ–α 0·155 high order pinkish white
Acicular or fibrous crystals

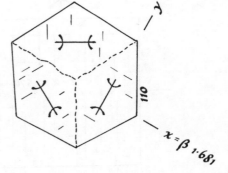

γ

110

x = β 1·681

Section ⊥ z
Sector twinning on 110
γ ≃ β ≫ α

Barytes   **Orthorhombic**   **BaSO₄**

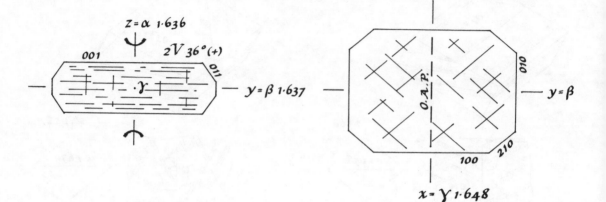

Colourless
$\gamma - \alpha$ 0.012   1st order bright yellow

Anhydrite   **Orthorhombic**   **CaSO₄**

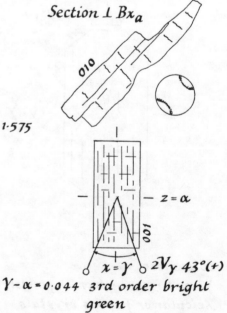

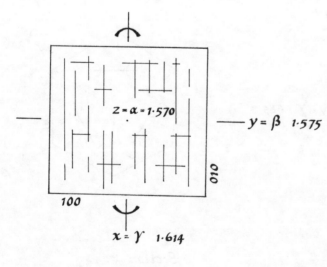

$\gamma - \alpha = 0.044$ 3rd order bright green

Gypsum   Monoclinic $CaSO_4 \cdot 2H_2O$

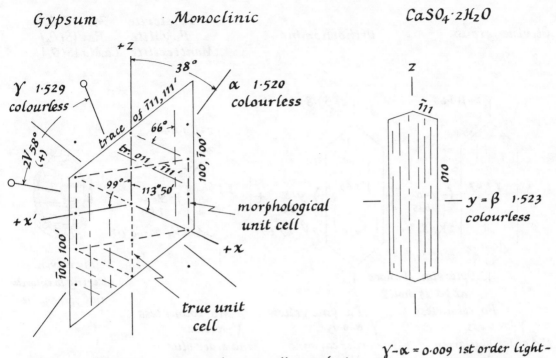

Dots are the lattice array with unit cells marked.
Primed symbols (x', $100'$ etc.) refer to morphological cell.

$\gamma - \alpha = 0.009$ 1st order light-yellow

Epsomite   Orthorhombic $MgSO_4 \cdot 7H_2O$

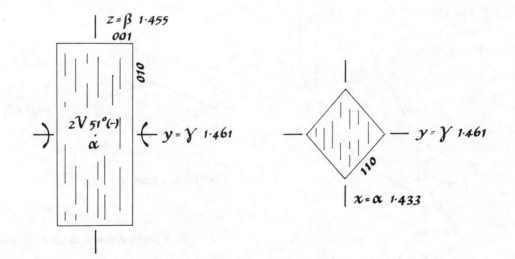

Colourless. $\gamma - \alpha$ 0.028 2nd order yellow
Usually in needles which may be length-fast or length-slow.

Olivine group   Orthorhombic

Forsterite $Mg_2\{SiO_4\}$
Fayalite $Fe_2\{SiO_4\}$
Monticellite $CaMg\{SiO_4\}$

$z = \beta\ 1.65$

021
010

$\gamma\ 1.67$

$y = \alpha$
1.635

$2V_\gamma\ 82°(+)$

$z = \beta\ 1.87$

$\gamma\ 1.88$

$y = \alpha$
1.83

$2V_\alpha\ 47°(-)$

$z = \beta\ 1.65$

011
021 010

$\gamma\ 1.66$

$y = \alpha$
1.64

$2V_\alpha\ 70-80°(-)$

Mg olivine
altering to chlorite
and serpentine

Optic sign changes
at Fa 25 mol.%

Fo colourless
$\gamma - \alpha\ 0.035$
3rd order blue/green

Fa pale yellow
$\gamma - \alpha\ 0.05$
3rd order carmine

Mo colourless
$\gamma - \alpha\ 0.02$
2nd order blue

continuous change   no continuous change

Sphene   Monoclinic $CaTi\{SiO_4\}(O,OH,Cl,F)$

$\approx 51°$ z

$\gamma\ 1.94 - 2.11$

001

$2V\ 20-50°(+)$

102

111

x

$\approx 21°$

100

$\alpha\ 1.84 - 1.95$

γ'

111

$y = \beta\ 1.87 - 2.03$

44°

Section $\perp$ edge $111 : 1\bar{1}1$

May show simple twins on $\{100\}$

Neutral tint to yellowish- or purplish-brown : may be slightly pleochroic.
$\gamma - \alpha\ 0.10 - 0.16$ high order pale pinkish colours. Dispersion strong, $r \gg v$.

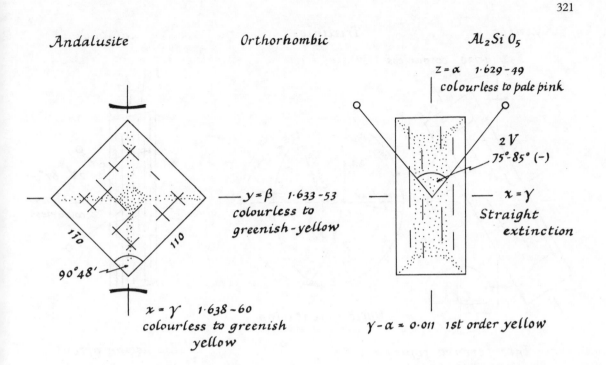

Andalusite   **Orthorhombic** Al_2SiO_5

$z = \alpha$ $1.629 - 49$
colourless to pale pink

$2V$
$75° - 85°$ (−)

$— y = \beta$ $1.633 - 53$
colourless to
greenish-yellow

$— x = \gamma$
Straight
extinction

110 110

$90°48'$

$x = \gamma$ $1.638 - 60$
colourless to greenish
yellow

$\gamma - \alpha = 0.011$ 1st order yellow

Special feature: inclusions in diagonal cross or central oblong, in chiastolite

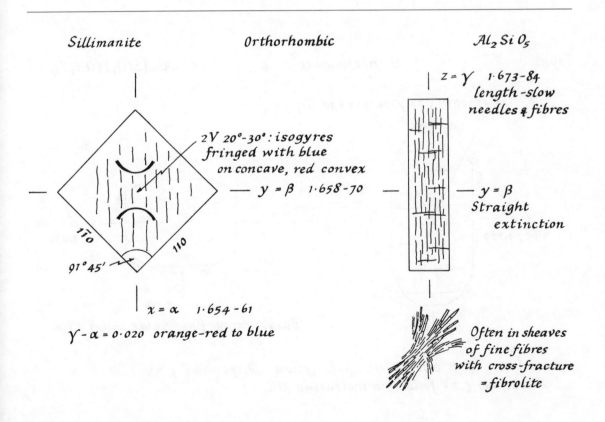

Sillimanite   **Orthorhombic** Al_2SiO_5

$z = \gamma$ $1.673 - 84$
length-slow
needles & fibres

$2V$ $20° - 30°$: isogyres
fringed with blue
on concave, red convex

$— y = \beta$ $1.658 - 70$

$— y = \beta$
Straight
extinction

110 110

$91°45'$

$x = \alpha$ $1.654 - 61$

$\gamma - \alpha = 0.020$ orange-red to blue

Often in sheaves
of fine fibres
with cross-fracture
= fibrolite

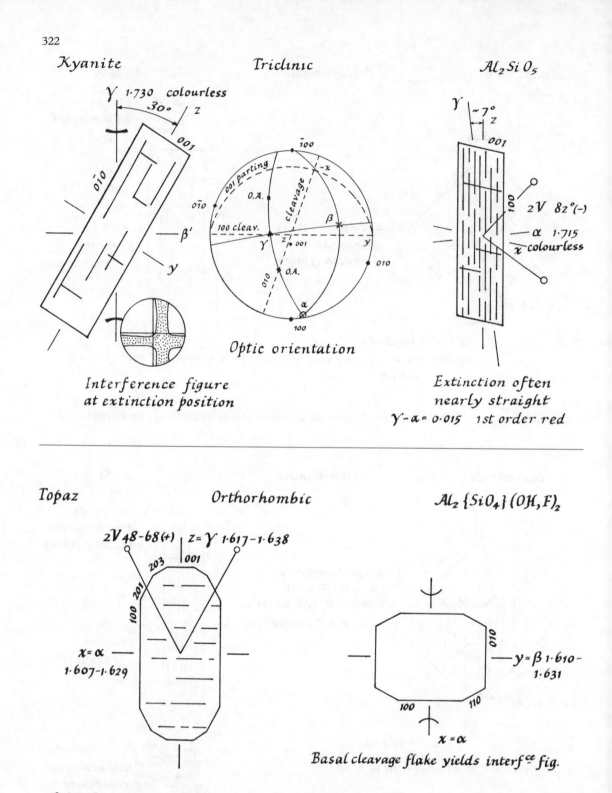

Kyanite Triclinic $Al_2\,Si\,O_5$

γ 1·730 colourless

30°

Interference figure
at extinction position

$\overline{1}00$

001 parting

O.A.

cleavage

100 cleav.

β

$0\overline{1}0$

z 001

010

γ

O.A.

α

100

Optic orientation

γ ~7° z

001

100

$2V$ 82°(−)

α 1·715
x colourless

Extinction often
nearly straight
$\gamma-\alpha=$ 0·015 1st order red

Topaz Orthorhombic $Al_2\,\{SiO_4\}\,(OH,F)_2$

$2V$ 48−68(+) $z=\gamma$ 1·617−1·638

203 001
201
100

$x=\alpha$
1·607−1·629

$y=\beta$ 1·610−
1·631

010

100 110

$x=\alpha$

Basal cleavage flake yields interf^{ce} fig.

Colourless $\gamma-\alpha$ 0·009−0·010 pale yellow. Dispersion $r>v$.
Indices rise & $2V$ falls with increasing OH.

Staurolite   *pseudo-Orthorhombic* $2Al_2SiO_5 \cdot Fe(OH)_2$

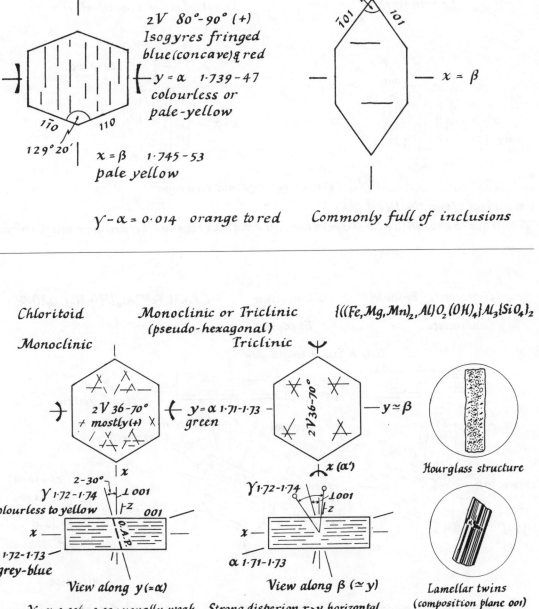

$z = \gamma$ 1·752-61
golden yellow

69°28'

$\bar{1}01$ 101

$x = \beta$

$2V$ 80°-90° (+)
Isogyres fringed
blue (concave) & red

$y = \alpha$ 1·739-47
colourless or
pale-yellow

$\bar{1}10$ 110

129°20'

$x = \beta$ 1·745-53
pale yellow

$\gamma - \alpha = 0.014$ orange to red   Commonly full of inclusions

---

Chloritoid   Monoclinic or Triclinic $\{((Fe,Mg,Mn)_2,Al)O_2(OH)_4\}Al_3\{SiO_4\}_2$
(pseudo-hexagonal)

Monoclinic   Triclinic

$2V$ 36-70°
+ mostly (+)

$y = \alpha$ 1·71-1·73
green

$2V$ 36-70° $y \simeq \beta$

x $x (\alpha')$

2-30°

γ 1·72-1·74
colourless to yellow

$\perp 001$

001

x

β 1·72-1·73
grey-blue

View along $y (= \alpha)$

γ 1·72-1·74

$\perp 001$

x

α 1·71-1·73

View along $\beta (\simeq y)$

Hourglass structure

Lamellar twins
(composition plane 001)

$\gamma - \alpha$ 0·006-0·02 : usually weak. Strong disperion r>v, horizontal.
Anomalous interf. colours. Numerous inclusions. Ubiquitous twinning & zoning.

Zoisite — Orthorhombic — $Ca_2 Al Al_2 \{SiO_4\}\{Si_2 O_7\} O (OH)$

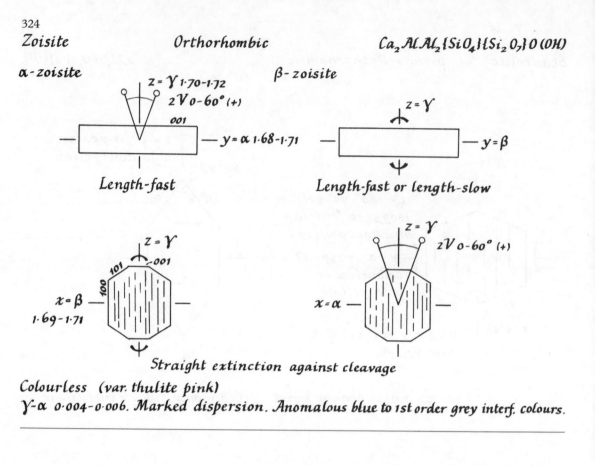

α-zoisite

$z = \gamma$ 1·70–1·72
$2V$ 0–60° (+)
001
$y = \alpha$ 1·68–1·71

Length-fast

β-zoisite

$z = \gamma$
$y = \beta$

Length-fast or length-slow

$z = \gamma$
-001
101
100
$x = \beta$ 1·69–1·71

$z = \gamma$
$2V$ 0–60° (+)
$x = \alpha$

Straight extinction against cleavage

Colourless (var. thulite pink)
$\gamma-\alpha$ 0·004–0·006. Marked dispersion. Anomalous blue to 1st order grey interf. colours.

Clinozoisite – Epidote — Monoclinic — $Ca_2 (Al, Fe^{3+}) Al_2 \{SiO_4\}\{Si_2 O_7\} O (OH)$

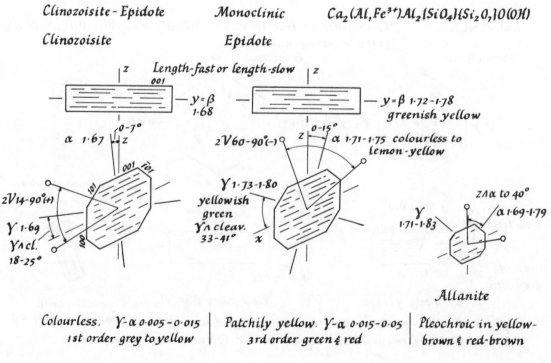

Clinozoisite

Length-fast or length-slow

z
001
$y = \beta$ 1·68

α 1·67 $\wedge z$ 0–7°

$2V$ 14–90° (+)
101
001 101
100
γ 1·69
$\gamma \wedge$ cl. 18–25°

Epidote

z
$y = \beta$ 1·72–1·78 greenish yellow

$2V$ 60–90° (−)
$z \wedge \alpha$ 0–15°
α 1·71–1·75 colourless to lemon-yellow

γ 1·73–1·80 yellowish green
$\gamma \wedge$ cleav. 33–41°
x

Allanite

$z \wedge \alpha$ to 40°
α 1·69–1·79
γ 1·71–1·83

Colourless. $\gamma-\alpha$ 0·005–0·015
1st order grey to yellow

Patchily yellow. $\gamma-\alpha$ 0·015–0·05
3rd order green & red

Pleochroic in yellow-brown & red-brown

Pumpellyite   *Monoclinic* $Ca_4(Mg,Fe^{2+})(Al,Fe^{3+})_5(OH)_3\{Si_2O_7\}_2\{SiO_4\}_2.2H_2O$

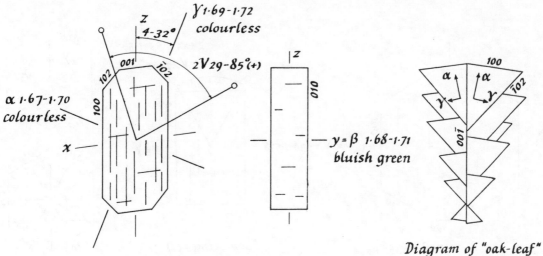

Diagram of "oak-leaf"
twins on {001}

$\gamma - \alpha$ 0·012-0·022   1st order yellow to 2nd blue.   Dispersion $r < v$.

Lawsonite   *Orthorhombic* $CaAl_2(OH)_2\{Si_2O_7\}H_2O$

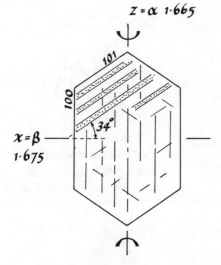

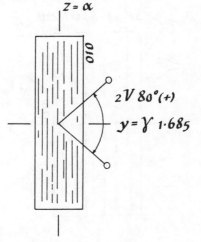

Colourless
$\gamma - \alpha$ 0·021   2nd order blue

Lamellar twins //{101}: $\beta \wedge$ lamellae 34° (in 010)
May be intergrown with pumpellyite

Cordierite   *Orthorhombic* $Al_3(Mg, Fe^{2+})_2 \{Si_5 Al O_{18}\}$

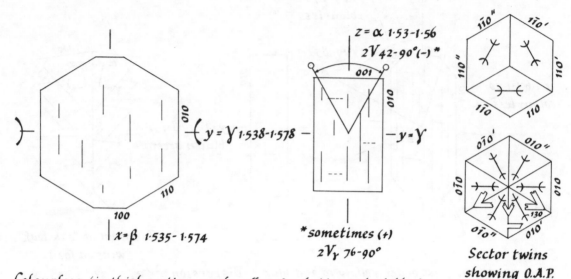

$z = \alpha$ 1·53-1·56
$2V$ 42-90°(-) *

$y = \gamma$ 1·538-1·578 $-y = \gamma$

$x = \beta$ 1·535-1·574

sometimes (+)
$2V_\gamma$ 76-90°

Sector twins
showing O.A.P.

Colourless (in thick section α pale yellow, β pale blue, γ dark blue).
$\gamma - \alpha$ 0·008 - 0·011   white to yellow

Enstatite - Hypersthene   *Orthorhombic* $(Mg, Fe) SiO_3$

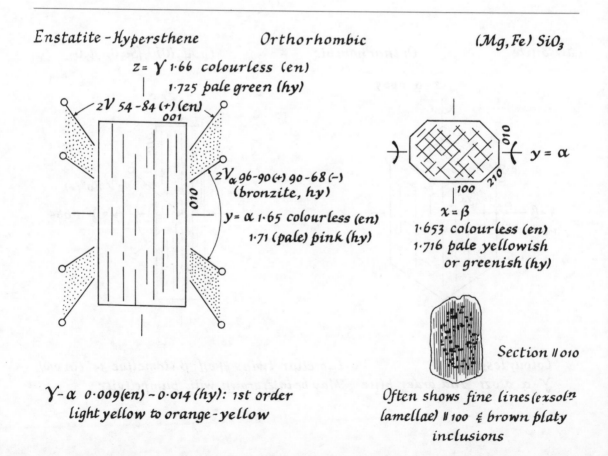

$z = \gamma$ 1·66 colourless (en)
1·725 pale green (hy)

$2V$ 54-84 (+) (en)

$2V_\alpha$ 96-90 (+) 90-68 (-)
(bronzite, hy)

$y = \alpha$ 1·65 colourless (en)
1·71 (pale) pink (hy)

$y = \alpha$

$x = \beta$
1·653 colourless (en)
1·716 pale yellowish
or greenish (hy)

Section ∥ 010

$\gamma - \alpha$ 0·009(en) - 0·014 (hy): 1st order
light yellow to orange-yellow

Often shows fine lines (exsol^n
lamellae) ∥ 100 & brown platy
inclusions

Diopside – Hedenbergite   Monoclinic $CaMg\{Si_2O_6\} - CaFe\{Si_2O_6\}$

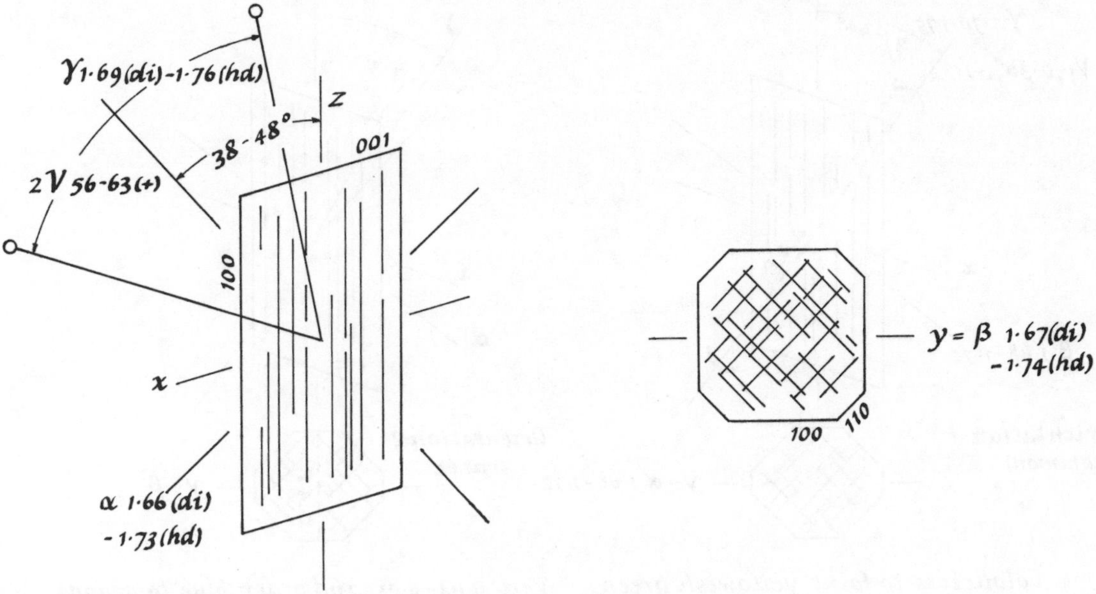

γ 1.69(di) – 1.76(hd)

38 – 48°

z

001

$2V$ 56-63(+)

100

x

α 1.66 (di)
– 1.73 (hd)

$y = \beta$ 1.67(di)
– 1.74(hd)

100 110

Colourless (di), brownish-green (hd)
$\gamma - \alpha$ 0.030 – 0.025   2nd order green to yellow

Augite   Titanaugite   Monoclinic $Ca(Mg,Fe^{2+})(Al,Fe^{3+}Ti)\{(Al,Si)_2O_6\}$

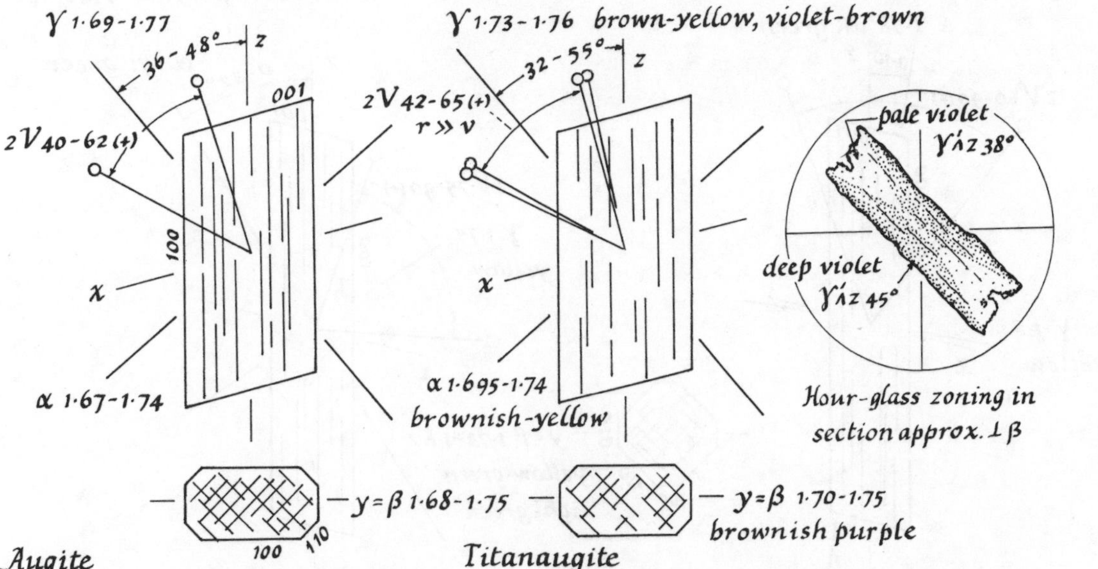

γ 1.69 – 1.77

36 – 48°   z

001

$2V$ 40-62(+)

100

x

α 1.67 – 1.74

γ 1.73 – 1.76 brown-yellow, violet-brown

32 – 55°   z

$2V$ 42-65(+)
$r \gg v$

x

α 1.695 – 1.74
brownish-yellow

pale violet
$\gamma \wedge z$ 38°

deep violet
$\gamma \wedge z$ 45°

Hour-glass zoning in
section approx. $\perp \beta$

$y = \beta$ 1.68 – 1.75

100 110

$y = \beta$ 1.70 – 1.75
brownish purple

Augite
Colourless to pale green or brown
$\gamma - \alpha$ 0.02 – 0.03 2nd ord. blue to red

Titanaugite
$\gamma - \alpha$ 0.021 – 0.033: strong dispersion + absorption
gives incomplete extinction & anomalous colours
in some sections

Pigeonite   Monoclinic $Ca(Mg,Fe)\{Si_2O_6\}$ (Ca-poor)

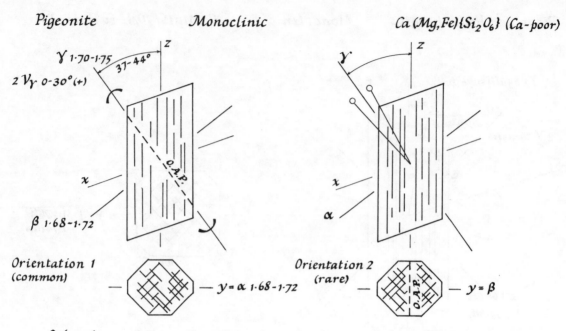

γ 1·70–1·75

37°–44°

$2V_\gamma$ 0–30° (+)

β 1·68–1·72

O.A.P.

γ

z

z

α

x

x

Orientation 1 (common)   — $y = \alpha$ 1·68–1·72

Orientation 2 (rare)   O.A.P.   — $y = \beta$

Colourless to faint yellowish green. $\gamma - \alpha$ 0·02–0·03: 2nd order blue to orange

---

Aegirine (Acmite)   Monoclinic $NaFe^{3+}\{Si_2O_6\}$
Aegirine–augite $(Na,Ca)(Fe^{3+},Mg,Fe^{2+})\{Si_2O_6\}$

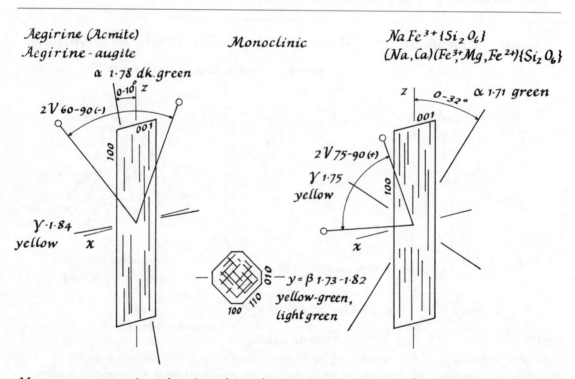

α 1·78 dk. green

0–10° z

$2V$ 60–90 (–)

100   001

γ 1·84 yellow   x

— $y = \beta$ 1·73–1·82 yellow-green, light green

100 110 010

z   0–32° α 1·71 green

001

$2V$ 75–90 (+)

γ 1·75 yellow

100

x

$\gamma - \alpha$ 0·03–0·06 2nd to 4th order colours, but often anomalous yellow-orange to brown

Jadeite   *Monoclinic* $NaAlSi_2O_6$

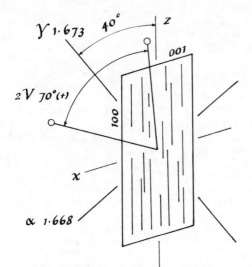

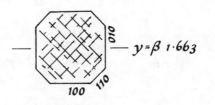

Colourless
$\gamma-\alpha$ 0.015   *1st order orange*

Usually anhedral, as fibrous masses

Wollastonite   *Triclinic* $Ca_3\{Si_3O_9\}$

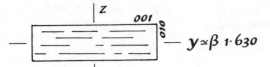

*Columnar or fibrous $\parallel y$: crystals may
be both length-fast & length-slow*

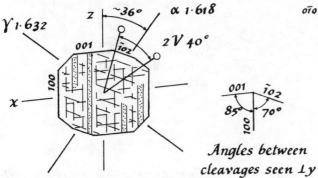

*Angles between
cleavages seen $\perp y$*

Optic orientation

Colourless. $\gamma-\alpha$ 0.014   *1st order orange-yellow. Twins $\parallel$ 100 (lamellar)*

Pectolite   Triclinic $NaHCa_2\{Si_3O_9\}$

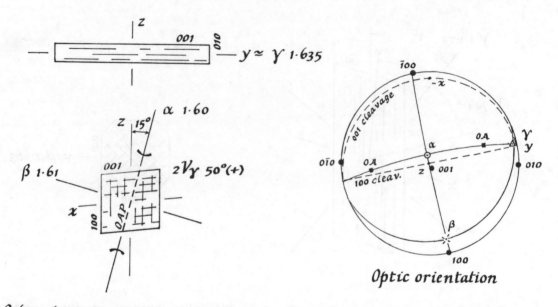

Optic orientation

Colourless. $\gamma-\alpha$ 0.035 2nd order red. Length-slow needles elongated $\parallel y$.

Anthophyllite   Orthorhombic $(Mg,Fe^{2+})_7\{Si_8O_{22}\}(OH,F)_2$

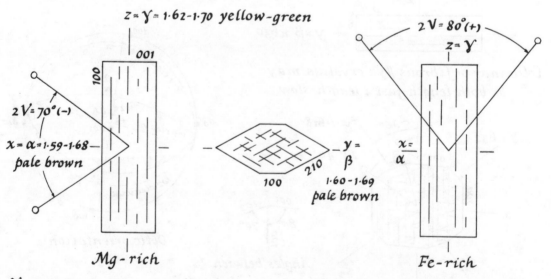

Mg-rich

Fe-rich

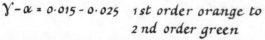

$\gamma-\alpha = 0.015 - 0.025$ 1st order orange to 2nd order green

Gedrite, Al & Fe-rich, is optically similar, but may have higher r.i.'s

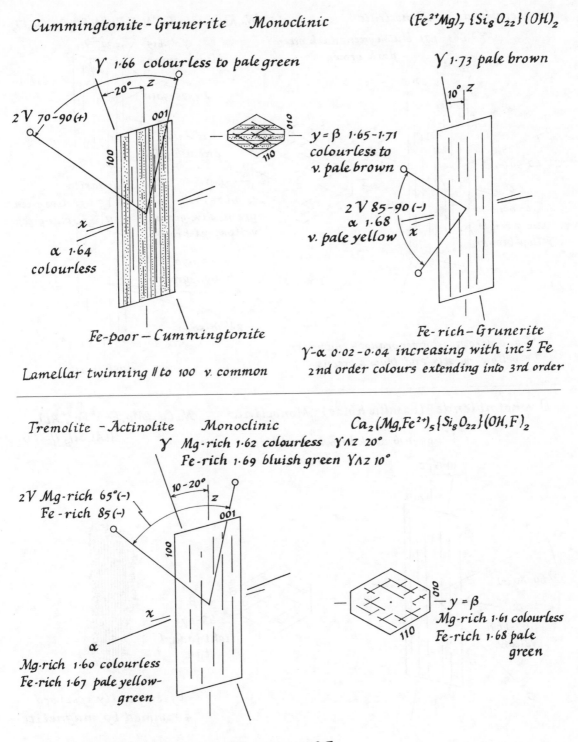

Cummingtonite - Grunerite   Monoclinic $(Fe^{2+}Mg)_7\{Si_8O_{22}\}(OH)_2$

γ 1.66 colourless to pale green

20° z

2V 70-90 (+)

001

100

$y = \beta$ 1.65-1.71 colourless to v. pale brown

x

α 1.64 colourless

γ 1.73 pale brown

10° z

2V 85-90 (-)
α 1.68
v. pale yellow

x

Fe-poor – Cummingtonite

Fe-rich – Grunerite

Lamellar twinning ∥ to 100 v. common

γ-α 0.02-0.04 increasing with incᵍ Fe
2nd order colours extending into 3rd order

Tremolite – Actinolite   Monoclinic $Ca_2(Mg,Fe^{2+})_5\{Si_8O_{22}\}(OH,F)_2$

γ Mg-rich 1.62 colourless $\gamma\wedge z$ 20°
Fe-rich 1.69 bluish green $\gamma\wedge z$ 10°

10-20° z

2V Mg-rich 65°(-)
Fe-rich 85 (-)

001

100

x

α

Mg-rich 1.60 colourless
Fe-rich 1.67 pale yellow-green

$y = \beta$
Mg-rich 1.61 colourless
Fe-rich 1.68 pale green

γ-α 0.020-0.027 decreasing with increase of Fe
2nd order blue, green, yellow-green

332

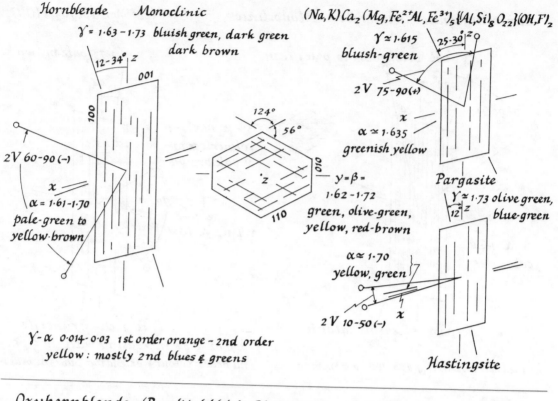

Hornblende   Monoclinic $(Na,K)Ca_2(Mg,Fe^{2+},Al,Fe^{3+})_5\{(Al,Si)_8O_{22}\}(OH,F)_2$

$\gamma = 1.63 - 1.73$ bluish green, dark green
dark brown

12–34° z

001

100

2V 60–90 (–)

x

$\alpha = 1.61 - 1.70$
pale-green to
yellow-brown

124°

56°

010

110

z

$\gamma \simeq 1.615$
bluish-green

25–30° z

2V 75–90 (+)

x

$\alpha \simeq 1.635$
greenish yellow

Pargasite

$y = \beta = 1.62 - 1.72$
green, olive-green,
yellow, red-brown

$\gamma \simeq 1.73$ olive green,
blue-green

12 z

$\alpha \simeq 1.70$
yellow, green

2V 10–50 (–)   x

Hastingsite

$\gamma - \alpha$ 0.014–0.03 1st order orange – 2nd order
yellow : mostly 2nd blues & greens

Oxyhornblende (Basaltic hblde.) Monoclinic $NaCa_2(Mg,Fe^{2+},Fe^{3+},Al)_5$
$\{(Al,Si)_8O_{22}\}O_2$

γ 1.69–1.76 dark brown, reddish
brown

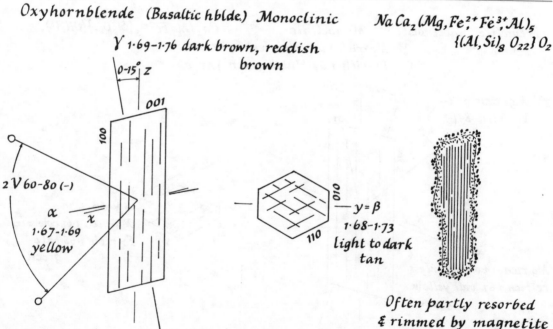

0–15° z

001

100

2V 60–80 (–)

α
1.67–1.69
yellow

x

010

110

$y = \beta$
1.68–1.73
light to dark
tan

Often partly resorbed
& rimmed by magnetite

$\gamma - \alpha$ 0.02–0.07 3rd & 4th order interf. colours, but masked by absorption colour

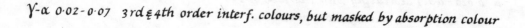

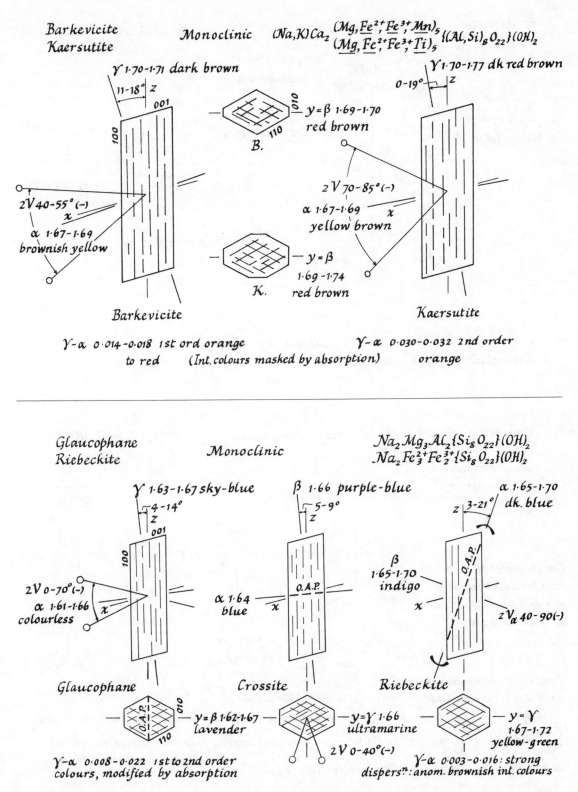

Barkevicite
Kaersutite

Monoclinic $(Na,K)Ca_2 \dfrac{(Mg,\underline{Fe}^{2+},\underline{Fe}^{3+},\underline{Mn})_5}{(\underline{Mg},\underline{Fe}^{2+},\underline{Fe}^{3+},\underline{Ti})_5} \{(Al,Si)_8 O_{22}\}(OH)_2$

γ 1·70-1·71 dark brown

11-18° z

001

100

B.

$y=\beta$ 1·69-1·70
red brown

010

110

$2V$ 40-55° (−)

x

α 1·67-1·69
brownish yellow

K.

$y=\beta$
1·69-1·74
red brown

γ 1·70-1·77 dk red brown

0-19° z

$2V$ 70-85° (−)

α 1·67-1·69
yellow brown

x

Barkevicite

$\gamma-\alpha$ 0·014-0·018 1st ord orange
to red   (Int. colours masked by absorption)

Kaersutite

$\gamma-\alpha$ 0·030-0·032 2nd order
orange

Glaucophane
Riebeckite

Monoclinic

$Na_2 Mg_3 Al_2 \{Si_8 O_{22}\}(OH)_2$
$Na_2 Fe_3^{2+} Fe_2^{3+} \{Si_8 O_{22}\}(OH)_2$

γ 1·63-1·67 sky-blue

4-14°
z
001

100

$2V$ 0-70° (−)
α 1·61-1·66
colourless

x

β 1·66 purple-blue

5-9°
z

α 1·64
blue

x

O.A.P.

α 1·65-1·70
dk. blue

z 3-21°

O.A.P.

β
1·65-1·70
indigo

x

zV_α 40-90(−)

Glaucophane

$y=\beta$ 1·62-1·67
lavender

O.A.P.

010

110

Crossite

$y=\gamma$ 1·66
ultramarine

$2V$ 0-40° (−)

Riebeckite

$y=\gamma$
1·67-1·72
yellow-green

$\gamma-\alpha$ 0·008-0·022 1st to 2nd order
colours, modified by absorption

$\gamma-\alpha$ 0·003-0·016: strong
dispers^n: anom. brownish int. colours

Arfvedsonite   Monoclinic $Na_3 Fe_4^{2+} Fe^{3+} \{Si_8 O_{22}\}(OH)_2$

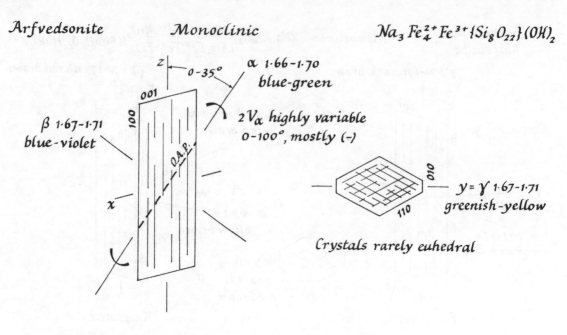

α 1·66–1·70 blue-green

β 1·67–1·71 blue-violet

$2V_\alpha$ highly variable 0–100°, mostly (–)

$y = \gamma$ 1·67–1·71 greenish-yellow

Crystals rarely euhedral

$\gamma - \alpha$ 0·005–0·018 : int. colours masked by absorption.
Extinction wavy & anomalous.

Muscovite & related layer silicates   Monoclinic $K Al_2 \{Al Si_3 O_{10}\}(OH)_2$

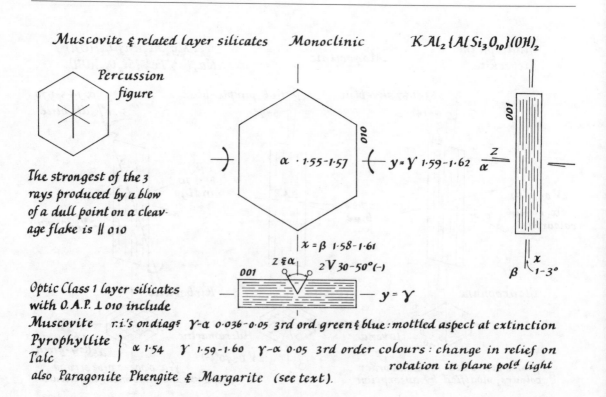

Percussion figure

The strongest of the 3 rays produced by a blow of a dull point on a cleavage flake is ∥ 010

α · 1·55–1·57

$y = \gamma$ 1·59–1·62

$x = \beta$ 1·58–1·61

$2V$ 30–50°(–)

$y = \gamma$

β 1–3°

Optic Class 1 layer silicates with O.A.P. ⊥ 010 include

Muscovite   r.i.'s on diags $\gamma - \alpha$ 0·036–0·05 3rd ord. green & blue : mottled aspect at extinction

Pyrophyllite } α 1·54 γ 1·59–1·60 $\gamma - \alpha$ 0·05 3rd order colours : change in relief on rotation in plane pold light
Talc

also Paragonite Phengite & Margarite (see text).

Biotite
Phlogopite   *Monoclinic*
Lepidolite

$K(Mg, Fe^{2+})_3 \{AlSi_3 O_{10}\}(OH)_2$
$K \; Mg_3 \{AlSi_3 O_{10}\} (OH)_2$
$K(Li, Al)_3 \{AlSi_3 O_{10}\} (OH)_2$

Optic Class 2 layer silicates: O.A.P. ∥ 010

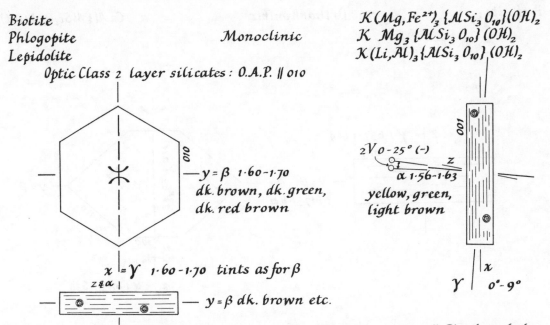

$y = \beta$ 1·60-1·70
dk. brown, dk. green,
dk. red brown

$2V$ 0-25° (-)
α 1·56-1·63

yellow, green,
light brown

$x = \gamma$ 1·60-1·70 tints as for β

$y = \beta$ dk. brown etc.

γ 0°-9°

Biotite $\gamma-\alpha$ = 0.04 - 0.08 3nd order blue to 4th order: Mottled at extinct^n: Pleochroic haloes.
Phlogopite α 1.53-1.59 pale yellow, $\beta = \gamma$ 1.56-1.64 brownish red or yellow: $2V$ = 0-15° (-)
Lepidolite α 1.52-1.55 colourless, $\beta = \gamma$ 1.55-1.58 pale pink or violet: $2V$ = 0-58° (-)

Stilpnomelane   *Monoclinic* $(K, Na, Ca)_{0-0.7} (Fe^{3+}, Fe^{2+}, Mg, Al(Mn)_{3-4} \{Si_4 O_{10}\}(OH)_2$
$(O, OH, H_2O)_{2-4}$

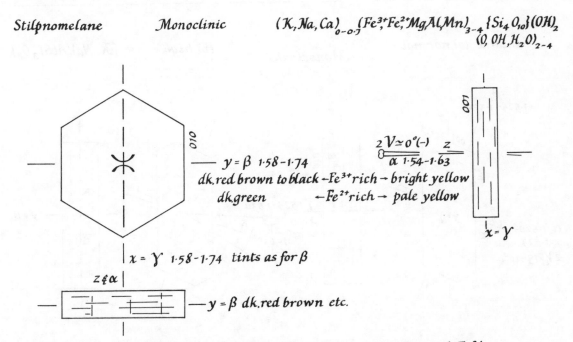

$y = \beta$ 1·58-1·74
dk. red brown to black ←Fe^{3+} rich → bright yellow
dk. green   ←Fe^{2+} rich → pale yellow

$2V \simeq 0°(-)$
α 1·54-1·63

$x = \gamma$ 1·58-1·74 tints as for β

$y = \beta$ dk. red brown etc.

$x = \gamma$

$\gamma-\alpha$ 0·03-0·11   2nd order orange to high order colours, increasing with Fe^{3+}.
Differences from biotite: no mottling at extinct^n, less perf. cleav., brittle flakes.

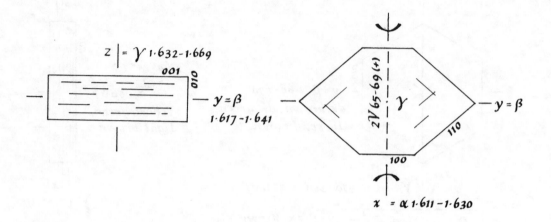

Colourless $\gamma-\alpha$ 0·021-0·039   2nd order green to 3rd blue : but anomalous interf. colours & small $2V$ may result from submicr. twinning.

Sanidine (a) normal   Monoclinic   (b) high $(K,Na)\{AlSi_3O_8\}$

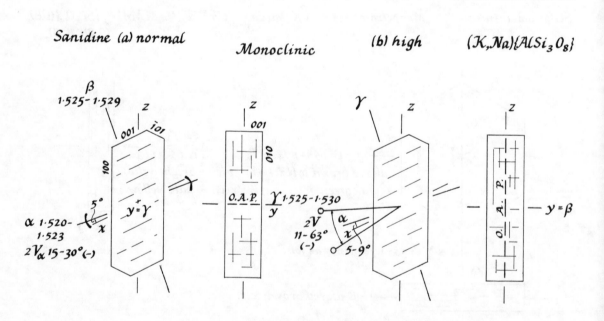

Colourless $\gamma-\alpha$ 0·005-0·007   1st order grey to white. Clear glassy crystals. Single twins (Carlsbad Law).

Orthoclase   Monoclinic $(K,Na)\{AlSi_3O_8\}$

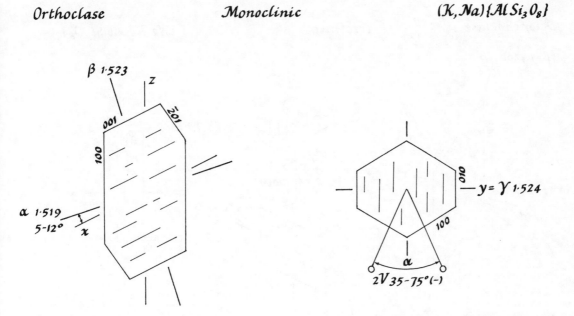

Colourless $\gamma-\alpha$ 0·007 1st order grey-white. Single Carlsbad twins common.
Forms a partial series towards low-albite & exsolves lamellae of albite ‖ 100 (perthite).

Microcline   Triclinic $K\{AlSi_3O_8\}$

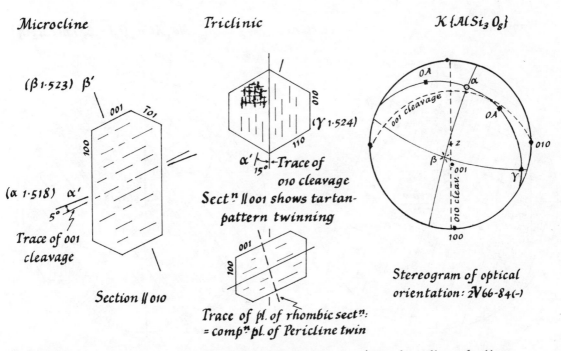

Section ‖ 010

Sect^n ‖ 001 shows tartan-pattern twinning

Trace of pl. of rhombic sect^n:
= comp^n pl. of Pericline twin

Stereogram of optical
orientation: $2V$ 66-84(-)

Colourless $\gamma-\alpha$ 0·006 1st order grey. May show perthitic lamellae of albite.

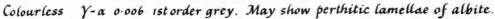

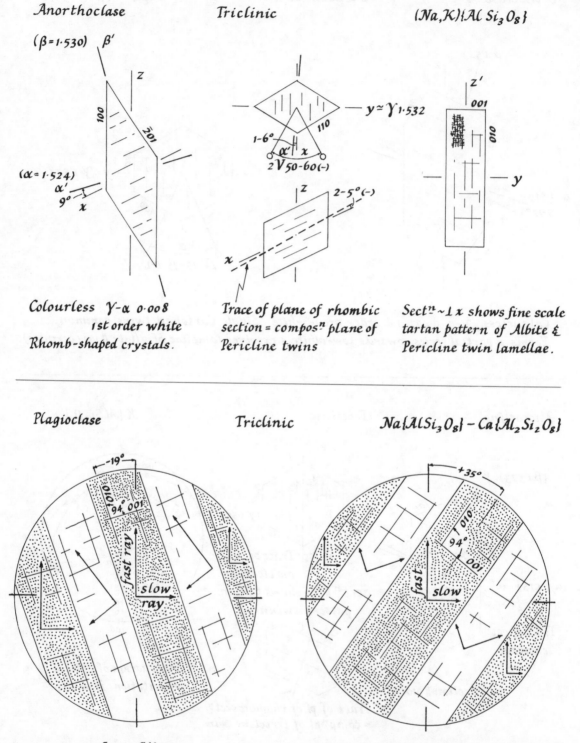

Anorthoclase   Triclinic $(Na,K)\{Al\,Si_3\,O_8\}$

$(\beta = 1.530)$ β'

$(\alpha = 1.524)$

$y \simeq \gamma\ 1.532$

$2V\ 50\text{-}60\,(-)$

$2\text{-}5°\,(-)$

Colourless $\gamma - \alpha\ 0.008$
    1st order white
Rhomb-shaped crystals.

Trace of plane of rhombic
section = compos^n plane of
Pericline twins

Sect^n $\sim \perp x$ shows fine scale
tartan pattern of Albite &
Pericline twin lamellae.

Plagioclase   Triclinic $Na\{AlSi_3O_8\} - Ca\{Al_2Si_2O_8\}$

Low Albite   Labradorite

Vibration directions in relation to cleavage angle & albite twin plane in section $\perp x$.
For further diagrams see pp. 288-291.

Laumontite $Ca\{Al_2Si_4O_{12}\}\cdot 4H_2O$ Heulandite $(Ca,Na_2)\{Al_2Si_7O_{18}\}\cdot 6H_2O$

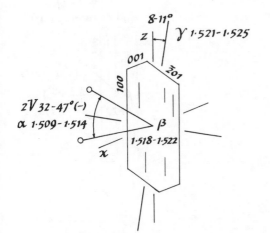

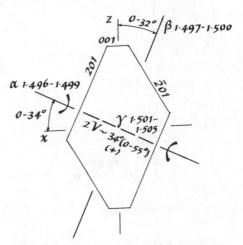

Laumontite labels:
8-11° γ 1.521-1.525
z
001 $\bar{2}01$
100
$2V$ 32-47°(-)
α 1.509-1.514
β 1.518-1.522
x

Heulandite labels:
z 0-32° β 1.497-1.500
001
201 $\bar{2}01$
α 1.496-1.499
0-34°
γ 1.501-1.505
$2V \sim 34°(0-55°)$ (+)
x

Colourless $\gamma-\alpha$ 0.012 Stout prisms
Dehydrates readily to leonhardite, with
lower indices (1.502-1.518) & γ:z reaching 35°

Colourless $\gamma-\alpha$ 0.006 Cleavage $\perp$ γ
Clinoptilolite is Si-rich heulandite
with β mostly ⩽ 1.485.

Stilbite (Desmine)
$(Ca,Na_2,K_2)\{Al_2Si_7O_{18}\}\cdot 7H_2O$

Phillipsite
$(K,Na,Ca)_5\{Al_5Si_{11}O_{32}\}10H_2O$

Harmotome
$Ba\{Al_2Si_6O_{16}\}\cdot 6H_2O$

all Monoclinic
cruciform interpenetrant twins characteristic,
giving pseudo-tetrag./o'rhomb. forms.

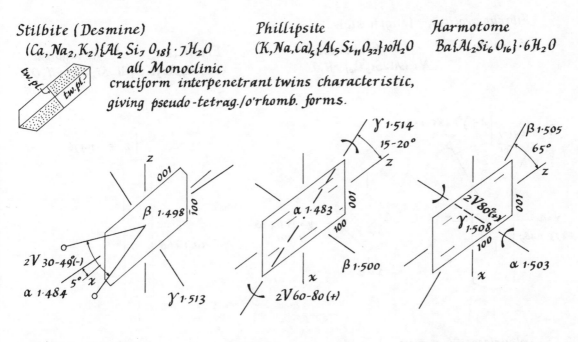

Stilbite labels:
z
001
β 1.498
100
tw.pl. tw.pl.
$2V$ 30-49(-)
5° x
α 1.484
γ 1.513

Phillipsite labels:
γ 1.514
15-20°
z
001
α 1.483
100
x
β 1.500
$2V$ 60-80 (+)

Harmotome labels:
β 1.505
65°
z
001
$2V$ 80(+)
γ 1.508
100
x
α 1.503

Colourless $\gamma-\alpha$ 0.010
In sheaf-like forms

Colourless $\gamma-\alpha$ 0.010
Columnar, tabular // 010

Colourless $\gamma-\alpha$ 0.005
Columnar, tabular // 010
length-fast/-slow

Fibrous zeolites – length-fast

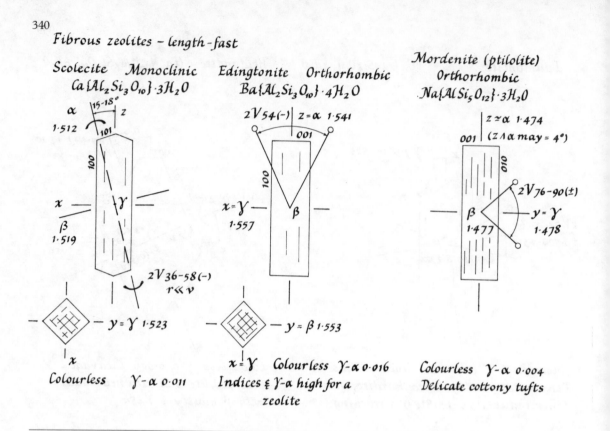

Scolecite Monoclinic
$Ca\{Al_2Si_3O_{10}\}\cdot 3H_2O$

α 1·512
15-18° z
101
100
x
β 1·519
γ
2V 36-58(-)
r ≪ v
y = γ 1·523
x
Colourless   γ-α 0·011

Edingtonite Orthorhombic
$Ba\{Al_2Si_3O_{10}\}\cdot 4H_2O$

2V 54(-)   z = α 1·541
001
100
x = γ 1·557
β
y = β 1·553
x = γ   Colourless γ-α 0·016
Indices & γ-α high for a
zeolite

**Mordenite (ptilolite)
Orthorhombic**
$Na\{AlSi_5O_{12}\}\cdot 3H_2O$

z ≈ α 1·474
001   (z∧α may = 4°)
010
β 1·477   2V 76-90(±)
y = γ 1·478
Colourless γ-α 0·004
Delicate cottony tufts

Fibrous zeolites – length-slow

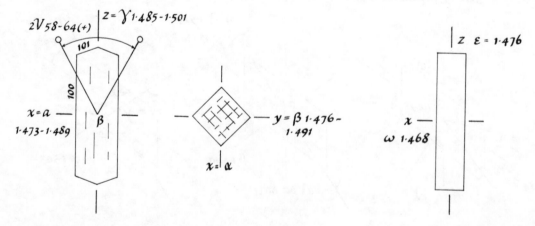

Natrolite   Orthorhombic
$Na_2\{Al_2Si_3O_{10}\}\cdot 2H_2O$

2V 58-64(+)   z = γ 1·485-1·501
101
100
x = a 1·473-1·489
β
y = β 1·476-1·491
x = α
Colourless γ-α 0·012
Length-slow needles or fibres

Erionite   Hexagonal
$(Ca,Mg,Na_2K_2)\{Al_2Si_4O_{12}\}\cdot 6H_2O$

z ε = 1·476
x
ω 1·468
ε-ω 0·008 Cottony fibres
(structurally related to chabazite)

Fibrous zeolites – length-fast/slow

Thomsonite   *Orthorhombic*
$$NaCa_2\{Al_5 Si_5 O_{20}\} \cdot 6H_2O$$

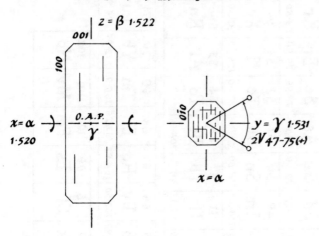

Colourless $\gamma-\alpha$ 0·011
Needles, fibres, often radiating.

Mesolite   *Monocl. (pseudo-o'rhomb.)*
$$Na_2 Ca_2 \{Al_6 Si_9 O_{30}\} \cdot 8 H_2O$$

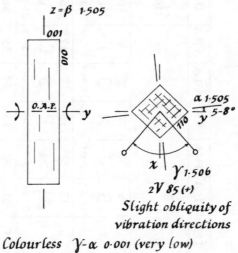

Colourless $\gamma-\alpha$ 0·001 (very low)
Fibres, delicate silky tufts,
also massive.

APPENDIX VI: THE PERIODIC TABLE

Periodic Table of the Elements

Periods \ Groups	IA	IIA	IIIB	IVB	VB	VIB	VIIB	VIII			IB	IIB	IIIA	IVA	VA	VIA	VIIA	0
1	1 H 1·008																	2 He 4·003
2	3 Li 6·94	4 Be 9·01											5 B 10·81	6 C 12·01	7 N 14·007	8 O 15·999	9 F 18·998	10 Ne 20·18
3	11 Na 22·99	12 Mg 24·31											13 Al 26·98	14 Si 28·09	15 P 30·97	16 S 32·06	17 Cl 35·45	18 A 39·95
4	19 K 39·10	20 Ca 40·08	21 Sc 44·96	22 Ti 47·90	23 V 50·94	24 Cr 51·99	25 Mn 54·94	26 Fe 55·85	27 Co 58·93	28 Ni 58·71	29 Cu 63·54	30 Zn 65·37	31 Ga 69·72	32 Ge 72·59	33 As 74·92	34 Se 78·96	35 Br 79·90	36 Kr 83·80
5	37 Rb 85·47	38 Sr 87·62	39 Y 88·90	40 Zr 91·22	41 Nb 92·91	42 Mo 95·94	43 Tc 98·91	44 Ru 101·07	45 Rh 102·90	46 Pd 106·4	47 Ag 107·87	48 Cd 112·40	49 In 114·82	50 Sn 118·69	51 Sb 121·75	52 Te 127·60	53 I 126·90	54 Xe 131·30
6	55 Cs 132·90	56 Ba 137·34	57 La 138·91 (L)	72 Hf 178·5	73 Ta 180·95	74 W 183·85	75 Re 186·2	76 Os 190·2	77 Ir 192·2	78 Pt 195·09	79 Au 196·97	80 Hg 200·59	81 Tl 204·37	82 Pb 207·2	83 Bi 208·98	84 Po	85 At	86 Rn
7	87 Fr	88 Ra 226·03	89 Ac (A)															

L	58 Ce 140·12	59 Pr 140·91	60 Nd 144·24	61 Pm	62 Sm 150·4	63 Eu 151·96	64 Gd 157·25	65 Tb 158·92	66 Dy 162·50	67 Ho 164·93	68 Er 167·26	69 Tm 168·93	70 Yb 173·04	71 Lu 174·97
A	90 Th 232·04	91 Pa 231·04	92 U 238·03	93 Np 237·05	94 Pu	95 Am	96 Cm	97 Bk	98 Cf	99 Es	100 Fm	101 Mv	102 No	

INDEX

References in **bold** type are to the principal descriptions
of mineral species. The frequent references to pages between
152 and 168 are to the mineral associations of species.